Environmental Geography

Environmental Geography

People and the Environment

Leslie A. Duram

Understanding Our World

An Imprint of ABC-CLIO, LLC
Santa Barbara, California • Denver, Colorado

Copyright © 2018 by ABC-CLIO, LLC

All rights reserved. No part of this publication may be reproduced, stored in a retrieval system, or transmitted, in any form or by any means, electronic, mechanical, photocopying, recording, or otherwise, except for the inclusion of brief quotations in a review, without prior permission in writing from the publisher.

Library of Congress Cataloging-in-Publication Data

Names: Duram, Leslie A., author.
Title: Environmental geography : people and the environment / Leslie A. Duram.
Description: First edition. | Santa Barbara, California : ABC-CLIO, an
   imprint of ABC-CLIO, LLC, [2018] | "Understanding our world." | Includes
   bibliographical references and index.
Identifiers: LCCN 2018014154 (print) | LCCN 2018019902 (ebook) |
   ISBN 9781440856112 (ebook) | ISBN 9781440856105 (Hardcopy : acid-free paper)
Subjects: LCSH: Environmental geography.
Classification: LCC G143 (ebook) | LCC G143 .D87 2018 (print) |
   DDC 304.2—dc23
LC record available at https://lccn.loc.gov/2018014154

ISBN: 978-1-4408-5610-5 (print)
       978-1-4408-5611-2 (ebook)

22 21 20 19 18   1 2 3 4 5

This book is also available as an eBook.

ABC-CLIO
An Imprint of ABC-CLIO, LLC

ABC-CLIO, LLC
130 Cremona Drive, P.O. Box 1911
Santa Barbara, California 93116-1911
www.abc-clio.com

This book is printed on acid-free paper ∞

Manufactured in the United States of America

To Jon for years of laughter and wonderful maps; and to Mom for all the giggles and careful editing. Kyle and Maggie, I am sure that your generation is smart enough to address these environmental problems.

Contents

Introduction xi

SECTION I

Introduction: How Humans Affect the Environment 1

Case Studies
    1. Cars Rule—American Dependence on the Automobile 15
    2. Great Barrier Reef—Human Activities Endanger Coral Reefs 22
    3. Dead Zones—The Gulf of Mexico 27
    4. Great Pacific Garbage Patch 32
    5. Nigeria's Oil Causes Human Rights Abuses and Environmental Degradation 39

Key Concepts
    Agriculture 47
        Food Miles 48
    Air Pollution 50
    Animal Agriculture 52
    Biodiversity Loss 54
        Agrobiodiversity Loss 55
    Climate Change 57
    Climate Change Policies 59
    Deforestation 61
    Endangered Species 63
    Energy 65
        Corn Ethanol from Mining the Soil 66
    E-waste 67
    Fracking 70
    Genetically Modified Crops 72
    Hazardous Waste 74
        Brownfield Sites 76
    Mining 78
    Overfishing 80

Solid Waste (Garbage!)   82
Superfund   84
Technology: Innovation and Consequences   86
  Assembly Line Manufacturing   87
Urbanization   89
Water Pollution   92
Water Scarcity   94

SECTION 2

Introduction: How the Environment Affects Humans   97

Case Studies
  6. Climate Change Is Occurring—Regardless of Politics   109
  7. Climate Refugees—Island Nations Disappear Because of Rising Seas   115
  8. Endangered Snow Leopard in Afghanistan—Local Efforts to Promote Conservation   121
  9. The Power of Hurricane Katrina (2005)—Evacuation and the Aftermath   126
  10. The 2004 South Asian Tsunami Natural Disaster   131

Key Concepts
  Adaptation   137
  Anthropocene   139
    Small-Scale Solar Power in Africa   140
  Drought   143
  Earthquake   145
  Ecosystem Services   147
  Ecotourism   150
  Flood   153
  Global North and Global South   155
    Environmental Crime   157
  Habitat and Wildlife   158
  Human Modification of Ecosystems   161
    Global Environmental Agreements   163
  Human Population   165
  Hurricane   167
  Mitigation   170
    U.S. Global Change Research Program   171
  Natural Hazards   173
  Parks and Urban Green Space   176
    Happiness and Sustainability   178
  Protected Areas and National Parks   180
  Tornado   183
  Volcano   185
  Wildfire   188

SECTION 3

Introduction: Sustainable Management of Natural Resources   191

Case Studies
    11. China's Bold Steps toward Renewable Energy—Better Late than Never   209
    12. Citizen Science—Helping Scientists Understand Migratory Birds   215
    13. Costa Rica's Peace with Nature: Conservation, Biodiversity, and Sustainability   220
    14. Denmark's Achievements in Organic Agriculture   226
    15. Great Green Wall of Africa   232

Key Concepts
    Alternative Agriculture   239
    Assessments and "Footprints"   241
    Composting   243
        Zero-Waste Communities   244
    Earth Day   246
    Electric Cars   249
    Environmental Justice   251
        Professor Maathai and Kenya's Green Belt Movement   253
    Environmental Nongovernmental Organizations (ENGOs)   255
    Environmental Policy   256
    Green Buildings   259
    Green Consumerism   262
        Fleece Jackets from Recycled Bottles!   263
    Green Political Party   265
    Green Technology   268
    Recycling   270
        Zero-Waste Home   272
    Renewable Energy   273
    Sustainable Cities   276
    Sustainable Development   279
        Auroville, City of Peace   280
    Sustainable Diet   282
    Water Conservation   284

Glossary   289

Bibliography   297

Selected Books Related to Environmental Geography   301

Index   305

Introduction

Environmental geography is the core of geography. It describes the complex interactions between people and the environment. Place and location are key factors, which is why geography is so valuable. It makes us think about the entire planet, from our own local neighborhood all the way up to the massive, complex whole Earth.

Environmental geography: Where do these words come from and what is their historical meaning? GEO is "Earth," and GRAPHY is "to describe"; ENVIRON is "surroundings or to enclose" and MENT is "the result of." So environmental geography describes the Earth, specifically focused on the result or outcome of our surroundings. It is interesting to note that the words ENVIRONMENTALIST and ENVIRONMENTALISM are very recent additions to our language, both coming into our vocabulary in the 1970s (Merriam-Webster, 2018; Oxford Dictionary, 2018).

Of course, other disciplines also cover environmental topics, but only geography fully integrates the two realms of *people* and *nature*. Each of these are gigantic topics alone—think about people's cultures, beliefs, politics, education, and economies and then think about natural systems from the tiniest living insect to the Earth's huge climate system. This indicates how tricky it is to put all of this together and understand the greater whole! People and the environment are both extremely complex; plus they do indeed have a complicated relationship with each other.

COMPLEXITY

It is not simple to explain these complicated interactions between Earth and its human inhabitants. But, in fact, that makes it more important for people to learn about and take action to promote environmental sustainability! Understanding the intersection between humans and nature is extremely important. The future of both the planet and people depends on it.

The Butterfly Effect is a useful way to think about Earth's complexities. Edward Lorenz, a math whiz and meteorology professor, developed a theory and wrote a paper called "Predictability: Does the Flap of a Butterfly's Wings in Brazil Set Off a Tornado in Texas?" In this paper, he described how changing a very small variable at the beginning of a system could have huge changes at the end of that system. He showed that when many variables are used to predict weather patterns, one small change can have large consequences (Dizikes, 2011; Wolchover, 2011). This basic idea was expanded into what is known as Chaos Theory, which is now one of the most important concepts in modern science. Rather than assume that everything in the world is predictable and linear, Chaos Theory states that small errors and changes early on make long-term future prediction impossible (Palmer, 2008). Humans cannot always predict or control nature because it is simply too complex.

In the past, we have often separated our studies into distinct disciplines: history, biology, economics, chemistry, and so forth. But to understand the relationships between people and the environment, we need a bigger, holistic view. Environmental geography provides this overarching perspective, especially because geography works so well at the intersection where people and the environment interact. Location matters for both the environment (for example: Where is the drought located?) and people (for example: What cultural and economic values are associated with water?) and thus geography is inherently necessary. In addition, environmental issues demand collaboration between natural science (for example: understanding the water cycle) and social science (for example: developing policies to promote water conservation).

TAKE ACTION

The complex environmental problems facing society will not be solved with simple solutions that easily please everybody. Rather, there will be trade-offs, because pros and cons are inherent in complex environmental decision making.

This book takes a very honest, straightforward approach in describing current environmental problems and the decisions and activities that led to these crises. People need to understand the entire situation: past, present, and future. It is not enough to see current ecological dilemmas in isolation without acknowledging the many background forces at play. Understanding these multifaceted factors that led to current problems is important because only then can options be fully developed. This book will help readers identify possible actions to promote a more sustainable future.

Overall, this book identifies problems and concerns and outlines ways to take action. This is not a gloom-and-doom text! In this book, readers will see that our environmental crisis is very real and cannot be ignored. But this book emphasizes that humans are smart and can use science, technology, and community—together—to fix it!

GEOGRAPHY IDEAS

Place matters. That is a simple statement, but it encompasses integrated notions of local communities, regional variations, national policies, and even global interactions. Geographers understand that place matters, so their studies and research are based on this very concept. Spatial thinking is special thinking! Spatial is the term that means space, place, and location. It is the idea that location matters. Geographers use place-based reasoning to understand the current world, its problems, and how people might take action to fix them.

Introduction

xiii

Every place has both a site (and absolute location), which is the actual location on the globe, and a situation, which is its location in relation to other places, also known as relative location. So the site of London, England, is 51°N, 0.12°W. But its situation is more descriptive. It is along the Thames River, on a large island called Great Britain, and in the Atlantic Ocean. This situation protected London from invasion from mainland Europe through much of its history. Now, however, modern technology has changed the situation, as London is quite accessible to Europe through speedy travel by plane or even train due to the 31-mile-long "Chunnel" or *channel tunnel* beneath the English Channel from Europe to the UK.

Geographers use spatial thinking to understand site and situation and other factors that influence human interactions with the environment. Diffusion is one such key concept, and it means the movement of ideas, traits, innovations, or natural properties. Rarely do things or ideas move once and remain static. Instead, things move around, influence other things to move, increase changes on the landscape, and spread out on the landscape. This is diffusion, or the spread of something more widely. Diffusion can be either good, like an innovative idea, or bad, like a negative ecological problem.

A great example of the diffusion of an idea is refillable water bottles. Two decades ago, it was very common to use single-use plastic water bottles and toss them in the trash without a care. But then some people realized this needed to change and started to get the word out that producing plastic water bottles requires 17 million barrels of oil each year, and bottles never really go away, since it takes up to 1,000 years to decompose. So a few people started using refillable water bottles. They told their friends, who told more friends, until the demand for metal and hard-plastic refillable water bottles increased significantly and now it is common to see water "Refill Stations" attached to drinking fountains at airports, schools, and businesses. Today, the United States alone consumes over 50 billion plastic water bottles each year and only 23% of them are recycled. So, this diffusion of an environmentally friendly action still has a long ways to go (Ban the Bottle, 2017).

Just as ideas and actions can diffuse, so can weeds or unwanted new pests. For example, the emerald ash borer is an Asian beetle that has killed hundreds of millions of ash trees across the United States. The United States Forest Service describes the geographic diffusion of this pest:

> Emerald ash borer (EAB), *Agrilus planipennis Fairmaire*, is an exotic beetle that was discovered in southeastern Michigan near Detroit in the summer of 2002. The adult beetles nibble on ash foliage but cause little damage. The larvae (the immature stage) feed on the inner bark of ash trees, disrupting the tree's ability to transport water and nutrients. Emerald ash borer probably arrived in the United States on solid wood packing material carried in cargo ships or airplanes originating in its native Asia. Emerald ash borer is also established in Windsor, Ontario, was found in Ohio in 2003, northern Indiana in 2004, northern Illinois and Maryland in 2006, western Pennsylvania and West Virginia in 2007, Wisconsin, Missouri and Virginia in the summer of 2008, Minnesota, New York, Kentucky in the spring of 2009, Iowa in the spring of 2010, Tennessee in the summer of 2010, Connecticut, Kansas, and Massachusetts in the summer of 2012, New Hampshire in the spring of 2013, North Carolina and Georgia in the summer of 2013, Colorado in the fall of 2013, New Jersey in the spring of 2014, Arkansas in the summer of 2014, Louisiana in the winter

of 2015, Texas in the spring of 2016, Nebraska and Delaware in the summer of 2016, and Oklahoma and Alabama in Fall 2016. (USFS, 2017)

Both the water bottle and the beetle provide examples of diffusion that are relevant to environmental geography and our study of places and patterns. But keep in mind that because of our digital world, nowadays diffusion can occur and not be related to proximity. In other words, a teenager might be thrilled with her new pair of running shoes made from recycled rubber and plastics and post a picture on social media. Then, if each of her friends share that post, these recycled shoes could spread in popularity to other places that are far away. Given these interconnections among places, both in location and in personal connections, geographers understand the importance of both maps and modeling.

MAPS AND MODELS

Geographers love maps. Maps allow geographers to see the world at various scales (up close, far away, or in between) and pick out key factors that are relevant to a specific topic. Maps are a key tool for geographers, who enjoy seeing the world both in reality and "zoomed in" to the neighborhood, region, or specific factor that can be the theme of our map. But a cell phone allows people to do so much more than a paper map—they can now get directions, real-time traffic information, street view images, satellite imagery, and much more on a mobile device.

The "Internet of Things" is the term describing a system of smart devices, electronics, sensors, buildings, vehicles, and software that are networked together to exchange data. In the future, "anything that can be connected, will be connected" (Morgan, 2014).

Geographic technology helps people understand the relationships between their daily lives and the environment. Geographers can also take data—numbers, statistics, facts, and information—that describe a place and create a model to build a better understanding of an environmental problem. A model is a "simplification of reality that is constructed to gain insights into select attributes of a physical, biological, economic, or social system" (EPA, 2016).

A model can promote understanding and help build solutions to complicated problems that involve both human and environmental systems. Geographers often use information gathered from remote sensing, such as data from satellites, and work with a GIS (geographic information system) to build environmental models. For example, we can use remote sensing data to build a GIS model of an agricultural watershed. Data about the landscape, elevation, crops, pesticide applications, and soils would be used to build a model that investigates how storm water runoff might carry pesticide pollution from a farm field to a stream and into a river, bay, or ocean. Then we can study different variables within the model to determine the best environmental management options to reduce or even stop this type of pollution.

Other environmental models might use data and information gained from surveying people about their opinions and perceptions. Some models are quantitative, like the watershed pollution example, and depend on numbers or quantities to build their simulations. Other models are qualitative and use words, ideas, or even policies as components. Think about how such qualitative models would be very useful in the complex world of environmental modeling and decision making. To put it simply, it would be fairly meaningless to develop a solution to a pollution problem based only on numbers and physical environmental data when, in fact, the required policy could never be implemented due to political or social barriers to change.

Oftentimes, environmental geographers realize that it is best to combine both quantitative and qualitative methods to study a problem and develop solutions. That means they use numbers and words combined with place, location, and mapping to understand how an environmental problem developed and what possible choices exist for fixing it.

INFORMATION, KNOWLEDGE, AND CONCERN

A well-known geography professor named Gilbert White (1911–2006) was one of the first people to explain that a "good" scientific solution to an environmental problem might not be accepted by people in the local region, and thus cannot be considered a solution at all. He called this the "range of choice," and it is a valuable concept to consider. There are typically many solutions for an environmental problem, but any one solution will only be implemented if people see it as a viable option. This early idea in environmental management is relevant today, as we must communicate science to influence policy. Professor White had a long career, beginning in the 1930s as member of the administration of President Franklin Delano Roosevelt (FDR). Professor White was a very down-to-earth person who lived well into his 90s. Much later, when he was a geography professor at the University of Colorado, he still recalled stories about how he and his colleagues would joke around with FDR and even played some pranks with silly maps they had made. That is a funny image! But overall, Professor White provides a worthy example of how geographers can study, understand, and solve environmental problems that truly affect our daily lives. His work in natural hazards will be discussed in the second section of this book.

"Stakeholders" are all the people affected by a certain environmental issue. Dr. White and many geographers understand that the purely logical, scientific, or engineering "fixes" are not effective if stakeholders do not support them. That means an environmental policy may not have legitimacy if the affected population does not "buy in" to the rule changes, especially if the environmental problem is complex, as many are.

It is worth thinking about the fact that there are different types of environmental problems: wicked versus tame. Wicked problems are those that cannot be solved by scientific approaches alone because either 1) the problem cannot be clearly defined or 2) people's opinions on the problem vary too much (Balint et al., 2011). In this sense, wicked does not mean evil, but rather wild or untamed. So tame environmental problems are those with less scientific uncertainty and low levels of conflicting values among the stakeholders (Rittel and Webber, 1973). An example may be a tornado: there is clear evidence that it occurred, and people all agree that it needs to be cleaned up. But wicked environmental problems like climate change

Environmental Issues: Wicked versus Tame Problems. (Duram, 2017)

are defined by scientific complexity and variations in people's values (Hulme, 2009). This, combined with misinformation used to gain profit or exploit science for political gains, makes it very difficult to tackle a wicked problem.

Certainly people must be informed about possible solutions to environmental problems, but first they must care enough to get informed. Many environmental problems vary by time and geographic place. People tend to care more about problems that are closer to home and in the present time. As an environmental problem gets further away geographically or occurs in the future, people have less concern, and thus less motivation to address it. For example, people are more likely to get involved in trying to protect a forest park in their neighborhood that is scheduled to be cut down next month. People are less likely to get involved in an issue like climate change because it seems far away, affecting the whole planet into the future (O'Neill and Hulme, 2009). These can be faulty perceptions, however, as local effects of climate change are certainly being felt in the present time, in everything from increased wildfires to larger tropical storms and droughts.

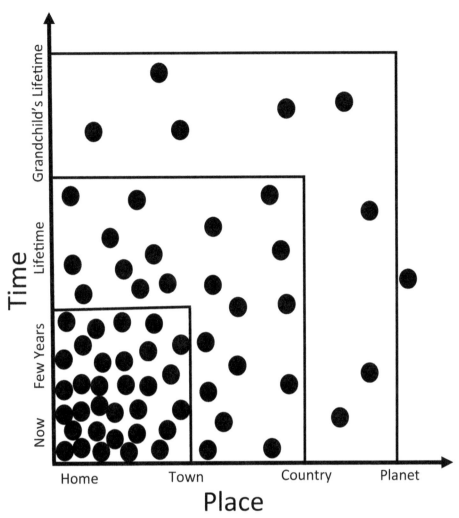

Environmental Concern Is Affected by Geography and Time Frame. (Duram, 2017)

The time-place concept is joined by several other factors, such as perception, memory, and experience, that affect people's concern about environmental problems and their motivation to take action for environmental protection. For example, how many people remember that the 2010 BP oil spill disaster in the Gulf of Mexico caused 4.9 million barrels (210 million U.S. gallons) of oil and huge amounts of methane gas to flow for 87 days from an uncapped well that was 1 mile deep in the ocean? There was significant environmental damage to marine wildlife and the fishing industry along the entire Gulf Coast, and it was a big story in the news. But people who were directly affected certainly remember this event more than someone with no personal experience there. Or here is another example: Do people still remember the four-month-long methane gas leak in 2015 in southern California (called the Alison Canyon or Porter Ranch gas leak), which was the worst natural gas leak in U.S. history, in terms of carbon released into the atmosphere?

Again, location, experience, and media coverage affect whether a person remembers these events or others like them. This is called a person's environmental memory. Certainly, it is affected by the news sources they are exposed to, but also their concern about various issues and even if they are emotionally inclined to relate to a given topic.

In news reporting, there is competition among issues, so even a really important story might not get reported on TV or in print because it is crowded out by other topics. Specifically, economic issues and news on war often crowd out environmental issues (Djerf-Pierre, 2012).

Sometimes a company or an organization tries to keep certain topics out of the news because it could harm their profits. A sneaky way to do this is to create the idea that scientists do not really know all the facts about a topic. Chronic uncertainty can be created to make people unsure about an issue and thus not feel motivated to take action. The best current example of this is petroleum corporations paying to discredit climate science (Jennings et al., 2015).

LOCAL TO GLOBAL

The key point here is: get informed! Do not assume that environmental issues will be addressed by other people. You should learn as much as possible about environmental problems and take action to find solutions.

People who are knowledgeable about their local environment are more likely to become part of a broader environmental movement. The first step may be to learn about your own daily actions and how they affect the environment. Each individual action adds up to create environmental impact on your community, region, and country. Together each person affects the planet, just as the butterfly wings in Brazil affect a tornado in Texas.

This is a great example of geographic scale. Each individual is part of a greater whole that is all interconnected. Geographers know that people must understand environmental issues at the local level in order to build an understanding of national and global problems. Likewise, geographers know that global issues affect national problems that can influence local people. Geographic scale is integrated so it goes both ways from a local community to the global world and from the whole Earth back down to regions and communities. That is why geographers are so readily able to address environmental problems—because these issues are linked to and balanced by this dimension of interacting places. Causes and solutions are found locally, globally, and in between.

Let's think about some specific examples of these geographic linkages. A guy wakes up in the morning with his cell phone alarm, takes a 15-minute hot shower, microwaves a breakfast burrito, drives his usual hour commute in his large SUV to the office, uses a computer and other digital technology all day at work, drives to a restaurant for a large burger at lunchtime, drives about an hour to the gym after work, takes another hot shower, grabs a bite to eat (steak and cheesy fries), and then meets a friend for a movie before heading home to bed. The next day is a similar routine. Think about the energy this lifestyle demands. Every hot shower, digital device, meat meal, and vehicle mile takes fuel. Right now this fuel is mostly

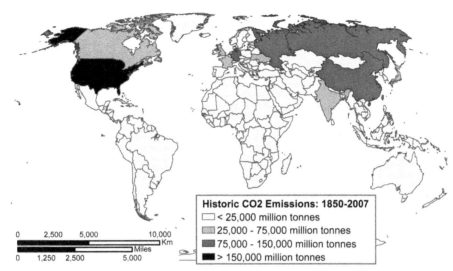

Estimated Carbon Dioxide Emissions by Country, 1850–2007. (Bathgate, 2017)

from fossil fuels: coal-burning power plants create the electricity to turn on his lights, warm his shower, and boot up his computer. Petroleum (gas and oil) makes his SUV start and operate for hours each day and also grow his food.

On the one hand, this life seems pretty normal. But on the other hand, every time people turn on the light switch, eat a beefsteak, or drive their car a mile—they are creating greenhouse gas (GHG) emissions. Burning fossil fuels creates carbon dioxide (CO_2), which has increased significantly in the atmosphere since humans began using fossil fuels in the 1800s (Clark, 2011).

If this guy, and every other person with a similar lifestyle, made some different choices, the use of fossil fuels, and thus the GHG emissions, could be reduced. First, people can demand that renewable energy be used to produce the electricity that powers their lives. Press electric companies to develop wind farms and solar arrays rather than rely on coal. Second, people can make more sustainable choices regarding transportation. They can demand better public transit options, which reduces the use of fossil fuels. They can push auto makers to build more energy-efficient vehicles that use less gasoline, and eventually push society to fully embrace the shift to electric vehicles, which would recharge by renewable energy from the new power plants. Third, dietary choices have a huge impact on GHG emissions, so eating more local, plant-based foods can drastically reduce individuals' impact on the climate.

If this guy, and billions of people like him, make individual choices that pull society away from fossil fuel dependence, the global climate crises could be addressed. Different countries, for example, have very different levels of GHG emissions. Environmental geographers emphasize that individual decisions and local actions have a huge impact on the global environment. They study these complex local-to-global interrelationships, and they realize that an underlying driving factor is climate change.

CLIMATE CHANGE

The main global initiative to study and understand climate change is the Intergovernmental Panel on Climate Change (IPCC). Since 1988, this organization of thousands of expert scientists has come together to work on sharing data and studying climate changes across our planet. The IPCC was formed from two United Nations organizations: the World Meteorological Organization and the United Nations Environment Programme. The mission of the IPCC is to assess "the scientific, technical and socio-economic information relevant to understanding of the scientific basis of risk of human-induced climate change" (IPCC, 2018). All documents from the IPCC are reviewed and approved by experts and governments, so its large and significant reports show the current facts that we know about climate change.

The first IPCC report came out in 1990 and was followed by reports in 1995, 2001, 2007, and 2014, and will continue with a sixth report in 2022. The IPCC reports are based on available scientific information and provide "assessments on all aspects of climate change and its impacts with a view of formulating realistic response strategies" (IPCC, 2018). Thus the reports help inform all global conventions or meetings on climate change. Acknowledging this group's global importance, the IPCC was awarded the Nobel Peace Prize in 2007.

The first assessment report of the IPCC served as the basis for negotiating the United Nations Framework Convention on Climate Change (UNFCCC), which went into force in 1994. Now 197 countries, nearly every nation, have ratified the agreement and are thus "parties" to the convention. The big climate meetings—for example, in Paris in 2015, Copenhagen in 2009, and so on—are called COP which stands for Convention of Parties, meaning that all the countries come together to discuss global agreements to combat climate change.

The IPCC reports are the most important source of scientific, technical, and socio-economic information for the COP meetings. This is really a key relationship because it represents the importance of the interaction between the policy makers and scientists. "Preventing 'dangerous' human interference with the climate system is the ultimate aim of the UNFCCC," and this requires international cooperation and agreement (UNFCCC, 2014)

Thus, from the 1990s on, scientists and global decision makers have known about the impact that humans have on the Earth's climate system. For example, the 2007 IPCC Report stated facts with frightening clarity: 1) global warming is occurring; 2) human activities are causing the increase in global temperatures; 3) if current trends continue, there will be increases in temperature extremes, heat waves, and heavy precipitation; and 4) the earth's temperature and seas will continue to rise into the next century (IPCC Synthesis, 2007; Rosenthal and Revkin, 2007).

With all this scientific information and international awareness for nearly 30 years, why do some people still question the overwhelming evidence that climate change is occurring, is caused by humans, and that we must do something to slow it down? In a few words: money talks.

There are powerful industries that earn a lot of money from our current use of natural resources. Our production system is based on fossil fuels, our transportation systems depend on fossil fuels, and many consumer products contain fossil fuels (from synthetic cloth to cosmetics to cell phones). The corporations that control these fossil fuels want to continue selling oil, coal, and natural gas. They do not profit from changing to alternative forms of energy.

Climate denial is when people or groups spread "disinformation" (lies or faulty data) or "manufacture uncertainty" (saying that climate scientists are not sure) about climate change (Dunlap, 2013). This influences our policies, because it encourages people to think that we do not need to take action. As noted in the well-respected journal *Scientific American* and elsewhere, there is "dark money" that cannot be tracked, something like $550 million, that has gone from the fossil fuel industry to foundations whose money cannot be traced to fund campaigns that question climate science (Fischer, 2013).

To be clear, 97% of climate scientists concur that global warming is occurring and is due to human actions (Cook et al., 2013). But climate denialists continue to promote the notion that we do not have strong science about climate change.

It goes even further than providing misinformation, as new evidence points to the fact that oil industries knew of climate change 45 years ago, but have continued to deny or lie about the fact. Researchers warned American Petroleum Institute (the main industry group) in 1968 that the release of carbon dioxide from fossil fuels could lead to "worldwide environmental changes." ExxonMobil, the world's largest oil company, knew that their industry had an impact on climate change in 1981. But both denied human-caused climate change and spent millions of dollars to spread climate denial information (Milman, 2016).

For these companies and others, it seems that profit comes above environmental concern. They think that their best chance of staying profitable is to increase our society's use of fossil fuels. And they point to the fact that many people depend on jobs in these industries. But could there be jobs instead in the renewable energy sector, working on wind turbines and solar power plants? And why is monetary profit so often the most important factor driving resource use? People can decide, for example, that happiness or health or sustainability is actually the measurement that determines environmental decisions. This will take a massive shift in society's goals.

Yes, as discussed throughout this book, individual people determine societal norms, which drive consumer demand and industrial growth. This determines natural resource use and environmental sustainability. Modern society has chosen a path that places unsustainable demands on the environment. Some would also argue that the current system is not sustainable in terms of economic or social equity. This manifests in various forms of environmental degradation and social inequalities at the local level and global climate change. In the future, people as individuals and society as a whole could, and must, opt to take action to reduce their negative impacts on the Earth and create a more sustainable system of natural resource use.

OVERVIEW OF THIS BOOK

The goal of this book is to inform and motivate each reader who flips through the pages to become engaged in environmental geography. It offers useful information, relevant sources to gain further knowledge, and ideas for how individuals can be informed and take action. Most current environmental problems are related to (either as a factor of or caused by) climate change, so this topic is a recurring theme throughout this book.

Following this introduction, the book is organized in three sections: 1) How Humans Affect the Environment, 2) How the Environment Affects Humans, and 3) Sustainable Management of Natural Resources.

Each section is set up in the same way: Introduction, Case Studies, and Key Concepts. The three sections each start with an introduction that outlines the big picture of what is encompassed by that particular area of environmental geography. Next, five Case Studies are used to illustrate the key themes that are important in the section. Then 25 Key Concepts describe relevant terms within the section. Throughout the text, maps, figures, and sidebars provide additional information to help readers visualize the various environmental geography topics.

Case Study Format

Case Studies allow a deeper understanding of complex global environmental issues by providing detailed local "real-world" examples. Learning about tangible examples of people who are currently addressing an environmental problem in their specific region can motivate each of us to act locally. Often our local actions will have a significant impact on global environmental problems.

Topics in environmental geography are complex, with numerous social and ecological components, differing viewpoints, and diverse geographic variables. Case Studies help us compare and contrast various topics and settings.

HOW TO USE THIS BOOK

With the ultimate goal of motivating readers to get informed and take action, this book can be approached from several angles. Option 1: Because it provides detailed examples of how environmental geography is relevant for addressing current environmental problems, it could be read from cover to cover. Option 2: The organization of the book lends itself to "target" reading, in which one picks out a hierarchy of specific topics of interest and then reads the relevant Case Studies and Key Concepts within those fields. Finally, with Option 3, a reader flips through the book and finds topics that spark their interests, especially sidebars of brief examples, which leads to further reading on related Case Studies and Key Concepts. Whichever type of reader you are, I hope you enjoy this book and gain both factual knowledge and personal motivation. Learn geographically, act locally, and be a global citizen!

FURTHER READING

Balint, Peter J., Ronald E. Stewart, Anand Desai, and Lawrence C. Walters. 2011. *Wicked Environmental Problems: Managing Uncertainty and Conflict.* Washington D.C.: Island Press.

Ban the Bottle. 2017. "Bottled Water Facts." www.BanTheBottle.net.

Clark, Duncan. 2011. "Which Nations Are Most Responsible for Climate Change?" *The Guardian*. April 21. www.theguardian.com/environment/2011/apr/21/countries-responsible-climate-change.

Cook, John, Dana Nuccitelli, Sarah A. Green, Mark Richardson, Barbel Winkler, Rob Painting, . . . Andrew Skuce, 2013. "Consensus on Consensus: A Synthesis of Consensus Estimates on Human-Caused Global Warming." *Environmental Research Letters* 8(2): 7 pp.

Dizikes, Peter. 2011. "When the Butterfly Effect Took Flight." MIT Technology Review. February 22. https://www.technologyreview.com/s/422809/when-the-butterfly-effect-took-flight/.

Djerf-Pierre, Monika. 2012. "The Crowding-Out Effect: Issue Dynamics and Attention to Environmental Issues in Television News Reporting over 30 Years." *Journal of Journalism Studies* 13(4): 499–516.

Dunlap, Riley E. 2013. "Climate Change Skepticism and Denial: An Introduction." *American Behavioral Scientist* 57(6): 691–698.

EPA. 2016. "Environmental Modeling 101: Training Module." U.S. Environmental Protection Agency. https://www.epa.gov/modeling/environmental-modeling-101-training-module.

Fischer, Douglas. 2013. "'Dark Money' Funds Climate Change Denial Effort." *Scientific American*. December 23. https://www.scientificamerican.com/article/dark-money-funds-climate-change-denial-effort/.

Hulme, Mike. 2009. *Why We Disagree About Climate Change: Understanding Controversy, Inaction and Opportunity*. Cambridge, UK: Cambridge University Press.

IPCC. 2018. "History." Intergovernmental Panel on Climate Change. https://www.ipcc.ch/organization/organization_history.shtml.

IPCC Synthesis. 2007. "Climate Change 2007: Synthesis Report." Intergovernmental Panel on Climate Change. http://www.ipcc.ch/report/ar5/syr/.

Jennings, Katie, Dino Grandoni, and Susanne Rust. 2015. "How Exxon Went from Leader to Skeptic on Climate Change Research." *Los Angeles Times*. October 23. http://graphics.latimes.com/exxon-research/.

Merriam-Webster. 2018. "Geography." https://www.merriam-webster.com/dictionary/geography.

Milman, Oliver. 2016. "Oil Industry Knew of 'Serious' Climate Concerns More than 45 Years Ago." *The Guardian*. April 13. https://www.theguardian.com/business/2016/apr/13/climate-change-oil-industry-environment-warning-1968.

Morgan, Jacob. 2014. "A Simple Explanation of 'The Internet of Things.'" *Forbes*. May 13. https://www.forbes.com/sites/jacobmorgan/2014/05/13/simple-explanation-internet-things-that-anyone-can-understand.

O'Neill, Saffron J. and Mike Hulme. 2009. "An Iconic Approach for Representing Climate Change." *Global Environmental Change* 19(4): 402–410.

Oxford Dictionary. 2018. "Environment." https://en.oxforddictionaries.com/definition/environment.

Palmer, Tim. 2008. "Edward Norton Lorenz." *Physics Today* 9(81): 61.

Rittel, Horst W. J. and Melvin M. Webber. 1973. "Dilemmas in a General Theory of Planning." *Policy Sciences* 4(2): 155–169.

Rosenthal, Elisabeth and Andrew C. Revkin. 2007. "Science Panel Calls Global Warming 'Unequivocal.'" *New York Times*. February 3. http://www.nytimes.com/2007/02/03/science/earth/03climate.html.

UNFCCC. 2014. "Introducing the Framework Convention on Climate Change." United Nations Framework Convention on Climate Change. http://unfccc.int/essential_background/convention/items/6036.php).

USFS. 2017. "Emerald Ash Borer Information Network." United States Forest Service and Michigan State University. http://www.emeraldashborer.info/.

Wolchover, Natalie. 2011. "Can a Butterfly in Brazil Really Cause a Tornado in Texas?" Live Science. December 13. http://www.livescience.com/17455-butterfly-effect-weather-prediction.html.

Section I
Introduction: How Humans Affect the Environment

Throughout most of human history, people believed that all human activities were positive and beneficial for the environment. Everything done to tame the wilds of nature was considered good. People didn't even consider that they could have a negative impact on the Earth. Of course, this is linked to their survival instincts: wild beasts and severe weather were truly a threat to human survival throughout much of history. So people were driven to civilize the wilderness and control nature for their own use.

Then in the late 1800s, environmentalism began. Although the glories and benefits of industrialization were still emphasized, writers such as Henry David Thoreau became popular for their ideas about the benefits of nature, and this movement was called Transcendentalism. Thoreau presented new anti-urban themes and stated, for example, that New York City was fit for "no real and living person" (Kline, 2011). In contrast, he eloquently depicted the beauty and emotion of nature in works such as *Walden: or, Life in the Woods* (Thoreau, 1854).

Scientists were investigating these new themes and also began to publish books. For example, George Perkins Marsh wrote *Man and Nature: Physical Geography* in 1864, in which he stated that "Man is everywhere a disturbing agent." Keep in mind that in that time frame "man" meant "human." Marsh's book was important because it was one of the first works that questioned whether human action was beneficial to the Earth (Marsh, 1864). Also during this era, John Wesley Powell, a decorated Civil War veteran, was tasked with exploring and documenting the Colorado River in 1875. Amazingly, the United States was composed of vast tracts of land in the West, but the government knew very little about it. So Powell's exploration provided a detailed, scientific view of the Western landscapes for the first time.

Humans have had an increasing impact on the environment since the 1800s because of industrialization. Think about how much the actions of people changed: North America and Europe went from agrarian societies with simple tools to an industrial society with powerful machines. This shift demanded raw materials (natural resources) as inputs for the emerging factories. And although agricultural practices had affected the landscape for thousands of years—land-use change can be seen on maps and through photos—the changes accelerated once society built machines and factories. The geography of this social and ecological change occurred as the use of natural resources expanded greatly. Natural resources are materials that occur in nature, such as forests, minerals, water, and arable land, that can be used for economic gain.

CONSERVATION VERSUS PRESERVATION

There is an important debate about natural resource use that began in the late 1800s and is still relevant today. This is the division between conservation and preservation. In general, people are not motivated by sensible, wise use of natural resources because most people take a very short-term view of economic gain. This was based on a frontier mentality, or belief that humans had the right to control nature and use the "endless" natural resources they believed were always at the edge of the frontier. Initially, this was European colonists going to Africa and the Americas to forcefully obtain raw materials; then in the United States this became conquering the western frontier and seeking to exploit every natural resource along the way.

Luckily, a few key people stood up and began to change this view. President Theodore Roosevelt was an avid hunter and nature lover, so he personally saw the benefit in managing wildlife for the long term. Drawing from the tragedy of the American West, in which white, European settlers killed about 60 million buffalo in 40 years, President Roosevelt said: "The extermination of the buffalo has been a veritable tragedy of the animal world." (NPS, 2018). Rather than seeing only the economic value of the buffalo hides, Roosevelt and others saw the need for careful planning and use of natural resources in the United States so that our nation could be prosperous in the future.

President Roosevelt appointed Gifford Pinchot to be the first Chief Forester of the United States, a new position specifically set up to begin careful management of the nation's natural resources. Together Roosevelt and Pinchot developed a policy of scientifically based, efficient, planned use of natural resources. The government workers were taught to be objective and base their management on science to promote the long-term use of natural resources for the economic health of the United States. They used the term "conservation," which in 1901, Pinchot defined as using natural resources for "the greatest good to the greatest number of people for the longest time" (USFS, 2007). So conservation is based on a utilitarian approach—using the resources—not protecting or saving them. Pinchot said: "The object of our forest policy is not to preserve forests because they are beautiful. . . . Forests are to be used by man" (Kline, 2011: 64).

Preservation is a stark contrast to conservation. John Muir was a wilderness expert and experienced naturalist who rightly earned the nickname "John of the Mountains." Muir often spent months alone in the wilderness and hiked thousands of miles over his lifetime. He believed that nature provided both emotional and physical health for humans. To him, when hiking in the mountains, "nature's peace will flow into you as sunshine flows into trees . . . while cares will drop off like autumn leaves" (Muir, 1901: 56). The environment was a religious experience for Muir, and he believed that God created nature's beauty and it was humans' duty to protect it. He stood in absolute opposition to the utilitarian approach to using natural resources. Muir wanted protection—preservation—of nature for its own sake. He knew that most people wanted to use our natural resources, cutting down the trees for timber, for example. And he knew that the government needed to act to preserve wildlands in the United States. Nature and forests were created by almighty God, "but he cannot save them from fools—only Uncle Sam can do that." (Muir, 1901: 365).

This demonstrates the political and management conflict: conservation pursues the planned use of natural resources, whereas preservation seeks protection of entire natural landscapes for current and future generations to enjoy. This conflict came to a head over the Hetch Hetchy Dam in Yosemite Park, which had initially been a California state park and then later a U.S. National Park. Yosemite is a stunningly beautiful valley with massive breathtaking mountains, sparkling waterfalls, and bubbling rivers. It strikes awe in those who visit.

The Hetch Hetchy River flowed through the northern portion of the park. In the late 1800s, the city of San Francisco wanted to dam the river to create a reservoir and then pipe this fresh mountain water 145 miles to the city to supply its water. From a conservation viewpoint, this was a good use of the water, as it would allow

Beautiful Yosemite Valley is one of the areas John Muir sought to protect. (Bathgate, 2017)

the city to grow and prosper. From a preservation viewpoint, this was completely unjustifiable and absolutely unacceptable. Muir fought for decades to keep Hetch Hetchy naturally flowing, saying that damming the river was like damming and flooding "the people's cathedrals and churches." He blamed politicians and managers, who "instead of lifting their eyes to the God of the mountains, lift them to the Almighty Dollar" (Muir, 1912: 262).

In the end, however, the U.S. Congress voted in 1913 to allow the huge dam to be built, so the beautiful Hetch Hetchy valley was flooded, and the water was transported to San Francisco by the 1930s. Thus, the preservationists lost this fight, and conservation—meaning the use of natural resources—became the central U.S. policy, and thus America's impact on the environment.

LAND-USE CHANGE

When we think of how humans affect the environment, the most significant impacts over the longest time frame come from agriculture. Humans have cut down over half of the Earth's forests to make room for cropland and grazing lands (Billington et al., 1996; UNEP, 2001). Certainly humans started altering our landscape thousands of years ago, but the rate of deforestation today is shocking. It is estimated that 13 million hectares (50,000 square miles) of forest are destroyed each year, according to the United Nations' Food and Agriculture Organization (FAO, 2010). For geographic reference, this means that a land area the size of the state of North Carolina is lost each year, which is equivalent to about 48 football fields every minute (WWF, 2018). However, these shocking numbers are actually lower than in the 1990s due to local and global efforts to combat deforestation. Further, about 15% of the total land area of the Earth is in parks and preserves, and this protection gives us hope for the future (World Bank, 2017).

Deforestation is particularly concentrated in the tropical forests of Latin America, Sub-Saharan Africa, and parts of Asia (FAO, 2015) which are biologically rich regions. The importance of forests is their biodiversity, ability to combat climate change, and cultural significance.

Biodiversity is "the variety of life," which can be understood from many geographical scales. From the global level, scientists can study different species on the entire Earth, and down at the local level, people can study biodiversity within a neighborhood park or their own backyard. The key thing is the realization that everything is interconnected, so changing one thing will cause changes to all other things. Scientists have identified 1.5 million species on Earth, which sounds like a lot, until you hear that they estimate there are 8 million or more species on our planet (Giller, 2014; Pappas, 2016). So there is a lot more to learn! And there is urgency because of the current rates of deforestation. Species are becoming extinct before people have time to discover them! Earth's current rate of extinction is the highest since the loss of the dinosaurs 65 million years ago. Dozens of species are lost to extinction every day, and some scientists estimate that 30 to 50% of all species may become extinct by 2050 (Chivian and Bernstein, 2008; Center for Biological Diversity, 2018). Biological hotspots are regions with particularly high levels of biodiversity.

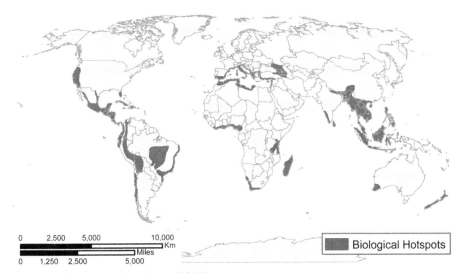
Biodiversity Hotspots. (Bathgate, 2017)

E. O. Wilson is a well-respected biologist known for studying, explaining, and protecting biodiversity. His research helped scientists understand how habitat destruction has a direct impact on numbers and types of species. He has worked hard to protect habitats so whole communities of plants and animals can continue to thrive and live in unison (French, 2011).

On the other hand, forests are very important to humans, both socially and economically. "The forest sector contributes about $600 billion annually to global GDP and provides employment to over 50 million people" (FAO, 2015). Further, 2 billion people rely on forests for wood fuel as their only source of fuel for cooking and heating. In addition, nearly all people use forest products for their daily lives, from their buildings to their papers. Thus planting forests for human use, forest management, and long-term planning are crucial. Silviculture is the science of growing, tending, and managing forests for tree crops. Currently 7% of all forests on Earth are these planted forests, specifically managed for human use (FAO, 2012).

Just as trees are planted for human use, of course, people also depend on plant food and fiber crops. For example, cotton crops provide us with clothing and household items; imagine life without cotton jeans or coffee filters!

By volume (measured in metric tons, or 1,000 kg), the main crops and livestock products produced in the world today are sugar cane, maize (corn), rice, wheat, milk, potatoes, vegetables, sugar beets, cassava, and soybeans. By value, the top 10 list looks different: milk, rice, cattle, pig, chicken, wheat, soybeans, tomatoes, sugar cane, and eggs (FAOSTAT, 2017). Clearly animal products have high monetary value in current societies, but they also have a high environmental and social cost.

ANIMAL AGRICULTURE

Looking at human impact on the environment, it is impossible to ignore animal agriculture. To produce a cow, it takes a lot of "inputs," in the form of natural resources and energy, to grow the huge amounts of grains the livestock needs to eat from birth to slaughter. This resource use is much higher than the resources it takes to grow a grain crop that people can eat directly.

"It takes about 28 calories of fossil fuel energy to produce 1 calorie of meat protein for human consumption, [whereas] it takes only 3.3 calories of fossil fuel energy to produce 1 calorie of protein from grain for human consumption," according to Cornell University expert, David Pimentel (WWI, 2017). Further, it takes 100 calories of grain to produce just 1 calorie of beef (Foley, 2014). This is a very inefficient use of natural resources, and humans eat animal products more often and in greater portions than at any time in our history. So it now takes many more hectares (acres) of land to grow the ever-increasing and vast amounts of grains needed to feed animals to be slaughtered.

For example, in the United States, very little of the corn or soybeans that farmers grow actually goes to feed humans. Some of it goes to make fuel (see sidebar on corn ethanol). Most of it goes to feed livestock, especially cows, pigs, and chickens. Globally, it is estimated that between 30 and 45% of the total land area on Earth is devoted to livestock and livestock feed, making it the largest land-use sector on Earth (FAO, 2006a; Herrero and Thornton, 2013). There is even a thought-provoking joke about this: if aliens came from another planet and saw the geography of land-use, they would assume the cattle were in control of the Earth based on sheer numbers and overuse of land on our planet.

Stated a different way, growing crops that were "food exclusively for direct human consumption could, in principle, increase available food calories by as much as 70%, which could feed an additional 4 billion people" (Cassidy et al., 2013).

If humans ate lower on the food chain, we could provide enough food for more people on Earth, given the amount of arable land available. So if we ate grains and legumes instead of beef, we would free up land to produce more crops for people, and we could reduce land degradation because we would not need so much land dedicated to intensive crop production (Hamblin, 2017). In addition, many scientific studies show that excess meat consumption increases people's risk of heart disease, stroke, type 2 diabetes, obesity, and some types of cancers (Harvard School of Public Health, 2012; Johns Hopkins School of Public Health, 2017). This means we could have healthier people and landscapes if we shifted our diets (Carrington, 2016; Marlow et al., 2009; McMichael et al., 2007; Springmann et al., 2017).

These are ethical issues. Do citizens of wealthy countries have the right to use their agricultural system to produce crops that will be used indirectly and inefficiently for animal production when 815 million people in the world do not have enough food to eat (THP 2018)? There are also issues of large agricultural corporations that earn huge profits from the current animal production system. Agricultural subsidies also heavily favor meat and dairy production at the expense of vegetable and grain crops for direct human consumption.

Animal advocates claim that the meat and dairy industries also hide the truth of the animal cruelty that is part of this system—from extremely overcrowded conditions (eight chickens held in cages the size of a shoebox) to unnecessary cruelty in slaughterhouses (cattle cut and hided before death). These corporations have even made it illegal to report from within a slaughterhouse because they claim they need to protect company secrets (Humane Society, 2018; Solotaroff, 2013).

In addition to land-use choices concerning animal agriculture, we see that meat and dairy production negatively affect our use of resources in terms of deforestation, grassland destruction, freshwater depletion, air pollution, waste disposal, energy consumption, and especially global warming (FAO, 2006b; Smith and Bustamante, 2014; WWI, 2017).

DIET AND CLIMATE CHANGE

As noted earlier, human actions causing deforestation have had a huge impact on the Earth, but as we look to the future, our impacts are shifting. In the last two decades, deforestation has slowed as an impact on greenhouse gas (GHG) emissions, but "agriculture emissions have continued to grow, at roughly 1% annually, and remained larger than the land-use sector" due to increased production of animal products (Tubiello et al., 2015). This means that our choice of crops and our use of lands for agriculture have a growing influence on climate change. Animal agriculture produces several types of emissions, which are potent greenhouse gases. Of course, everybody hears about carbon dioxide (CO_2), but methane (CH_4) is 25 times more potent than CO_2 and is produced at huge levels by livestock, particularly cattle (Smith and Bustamante, 2014; WWI, 2017).

Indeed, animal agriculture is responsible for over 20% of GHG emissions, more than the combined exhaust from all transportation (FAO, 2006b; University of Minnesota, 2017; Vermeulen et al., 2012). So diet is an important factor driving GHG emissions in the food supply chain (Bellarby et al., 2012; Garnett, 2011). Research shows that plant-based diets produce significantly lower GHG emissions than animal-based diets (Bellarby et al., 2012; Berners-Lee et al., 2012; Carlsson-Kanyama and González, 2009; Hamblin, 2017; Pathak et al., 2010).

Stehfest et al. (2009) evaluated effects of dietary changes on CO_2, methane (CH_4), and nitrous oxide (N_2O) emissions and projected it forward for five years. In contrast to the current typical animal-based diets, they evaluated no ruminant meat, no meat, and a diet without any animal products. Dietary changes resulted in GHG emission savings of 34 to 64% compared to the "business-as-usual" meat-based scenario (Stehfest et al., 2009).

POLLUTION

Over time, humans have caused vast land-use change. More recently, people and their consumer lifestyles are causing increasing amounts of pollution, which is defined as "environmental contamination with man-made waste" (Merriam-Webster, 2018). Specific pollutants are unwanted byproducts of our activities or things people have made, used, or thrown away. There are many categories of pollution, including air, water, land, noise, and even light pollution caused by human activities. Pollution can contribute to human health problems; for example, air pollution can lead to asthma or even lung cancer. Some pollutants in water can lead to reproductive problems like birth defects, and other pollution can even cause early death. Epidemiologists study the patterns, causes, and effects of health and disease on populations in defined areas. This is a geographic area of study, because location is very crucial to understanding public health, pollution, and the environment.

Several complex factors make it difficult to solve pollution problems. First, it is often nonlinear, which means that the relationship between the input (cause) and the output (effect) is not clear. We might not be sure that a certain chemical is causing an illness because it can occur in one place but the effects are felt far away, so it is hard to prove it. If a smokestack emits a toxic chemical high above the factory, the source of this pollution might not be obvious if somebody miles away downwind gets sick.

Second, pollution is often caused at one time, but the effects are not felt until many years later. This is called lag time, and it makes it hard to prove what pollutants are most dangerous. For example, if a chemical is ejected into the ground, and then 20 years later the people who live above the area get sick, it is difficult to prove the cause.

Third, there are feedbacks to any environmental system, which can multiply the negative effects and make it difficult to stop pollution. For example, A causes B, and B causes C, but then C increases the situation and makes A worse—and the process is intensified. An example of this would be warming climates that force people to turn on their air conditioner more often, which demands more electricity, which causes more GHG emission, which further increases temperatures, thus causing greater use of air conditioning.

Finally, when we consider pollution, we need to be aware of "tipping points," or irreversible consequences. Scientists are keenly aware of the need to identify a situation when it reaches a level of extreme concern so that we can implement change before it is too late. The problem is that people do not really know when a tipping point is reached. For example, what is the maximum human population that Earth can support? Or what is the maximum level of CO_2 the Earth's atmosphere can withstand without being forever altered and made inhospitable to human life? Nobody, not even the smartest scientist, knows the answers to these questions.

An important concept related to these issues is the "precautionary principle," which is a policy established and followed by the European Union (EU) countries since 2000. The precautionary principle is a way of managing risk: if a given action might cause harm to people or the environment, this action should not be taken, even if there is no scientific consensus on the issue. Once the scientific information becomes available, then the action will be reviewed. The precautionary principle can only be implemented when an event of real potential risk is present. This is related to the EU's environmental policy, which aims to preserve and protect environmental quality and human health, which is based on precautionary and preventative actions (EU, 2018).

The United States does not follow a precautionary principle, but rather takes a reactive approach, whereby action is taken after a problem is proven. Considering these complex factors of nonlinearity, lag time, feedback, and tipping points, it is obvious that addressing pollution risks is highly complex. Society makes pollution due to people's growing demand for consumer goods, transportation, food, and housing. How pollution is managed has varied over time, with an increasing level of environmental protection occurring when environmental laws began to be implemented in the 1970s in North America and much of Europe.

ENVIRONMENTAL POLICY

It is hard to imagine life without basic environmental policies, but the fact is that these only began in the 1970s. Prior to these policies, a factory could dump all chemicals into their local river and spew any chemical out their smokestack with no oversight and no penalty. And nobody knew what was being dumped, even if it affected their neighborhood.

Famous examples exist, where in 1969 the Cuyahoga River in Cleveland, Ohio, caught on fire. Imagine—a river (in other words, flowing water) was so full of chemicals that it caught fire! And in the 1960s the air pollution in Los Angeles was so severe that they had warnings where they told children to stay indoors. Unregulated pollution from factories and automobile exhaust directly affected people's lives and they began to force policy action.

Environmental policies are important statements about people's relationship with the environment. The fact that people have enacted policies to clean up their environment and regulate some types of industry says that they do care about the Earth and realize their future is closely linked to a healthy environment.

Each nation enacts its own environmental policies, and there is a great deal of variation among countries. What is allowed in one country may not be allowed in other nations. For example, the herbicide atrazine is a commonly used weed killer in the United States, but has been banned in Europe since 2003 because it is a dangerous water pollutant (Sass and Colangelo, 2006). The United States has had strict rules about air pollution since the Clean Air Act of 1970, but China just implemented a similar law in 2014, and there are still questions about how it will be enforced (Pettit, 2015; Zhang and Cao, 2015). Bangladesh has environmental regulations on the books, but tragically, many of their industries dump toxic chemicals, and the government ignores severe water pollution due to the need for economic growth (Yardley, 2013). Costa Rica collects taxes on fuel, cars, and energy to pay for environmental protection and managing nature preserves (Watts, 2010). Every country has its own environmental laws and decides how it will regulate its natural resource use.

Globally, the United Nations Environmental Programme (UNEP) was founded in 1972 as an agency within the United Nations with the mission to "provide leadership and encourage partnership in caring for the environment by inspiring, informing, and enabling nations and peoples to improve their quality of life without compromising that of future generations" (UNEP, 2018). This is accomplished through assessing national and global environmental conditions and assisting institutions in their environmental management. So the UNEP coordinates national efforts and keeps everyone informed of global trends.

Clearly, humans have a significant impact on the global environment. Environmental policies are important statements that indicate international cooperation and national goals for reducing the negative effects of human activities on the environment. Going from global to national is not really far enough. Think local! What environmental actions, policies, or rules are important to a local community? This can motivate individuals to get involved in groups that work to promote a better environment in their hometown.

CASE STUDIES 1-5

Within this context of local to global concern, the Case Studies and Key Concepts for this section focus on how humans affect the environment at each geographic scale. Environmental geographers are concerned about location and place, so these case studies provide thought-provoking examples of causes and effects that are often geographically separate, which makes the problems so very difficult to tackle. Individual decisions, national policies, and global environmental systems must all be addressed to find solutions. This section presents five Case Studies that detail humans' negative impact on the environment.

Case Study 1 focuses on what is common in daily life for many people: reliance on a car. This specifically describes how and why the United States is currently so dependent on the automobile and if there are any ways to change this reliance on fossil fuels and individual transportation. Cars exemplify how society can set up what becomes "normal" often without people fully realizing that their daily lives have such a negative environmental impact.

Case Study 2 is about the Great Barrier Reef located off the northeastern coast of Australia. It is one of the great wonders of the world, but it is currently in dramatic decline, with vast bleaching events that indicate that the reef is dying. It is linked to global issues, particularly the warming oceans, and so local efforts to improve the health of the reef are somewhat hopeless without global action to

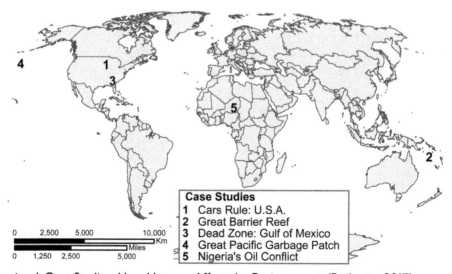

Section 1 Case Studies: How Humans Affect the Environment. (Bathgate, 2017)

combat climate change. This case describes a fascinating, wondrous habitat and the complexity of local to global environmental problems.

Case Study 3 provides an overview of marine hypoxia, with specific reference to the gigantic dead zone in the Gulf of Mexico, at the mouth of the Mississippi River. This case is an example of how the cause of a problem (agricultural production in the U.S. Midwest) may be far away upstream and thus extremely difficult to tackle.

Case Study 4 is also a gigantic problem: the Great Pacific Garbage Patch is estimated to be twice the size of Texas. It is an area of the north Pacific that has billions of tiny pieces of plastic floating like a vast soup of garbage. This exemplifies the idea of "out of sight, out of mind," meaning that anybody who throws away plastic is somewhat part of the problem because this plastic breaks down into tiny bits, but never goes away. Much of this waste ends up in the oceans and harms marine wildlife.

Case Study 5 describes the devastating environmental and social impacts of the oil industry in Nigeria. This is both an example of extreme ecological degradation and an abuse of basic human rights of the local people. In this case, outside petroleum corporations have been active in Nigeria for over 50 years, causing destruction through pipeline leaks, water pollution, and air pollution in the manufacturing process. Anytime local people tried to protest, they were brutally suppressed by their national government, which was only concerned about oil profits. Global dependence on fossil fuels causes this situation, which is so starkly felt at the local level in Nigeria.

Overall these five Case Studies paint a picture of how people continue to affect the environment, both through local choices and global systems. The Key Concepts continue along this theme, providing details about pollution, hazardous waste, fossil fuels, mining, agriculture, and land-use change. In addition, several alternatives are described: renewable energy, conservation, and alternative agriculture.

FURTHER READING

Bellarby, J., R. Tirado, A. Leip, F. Weiss, J. P. Lesschen, and P. Smith. 2012. "Livestock Greenhouse Gas Emissions and Mitigation Potential in Europe." *Global Change Biology* 19(1): 3–18.

Berners-Lee, M., C. Hoolohan, H. Cammack, and C. N. Hewitt. 2012. "The Relative Greenhouse Gas Impacts of Realistic Dietary Choices." *Energy Policy* 43: 184–190.

Billington, C., V. Kapos, M. Edwards, S. Blyth, and S. Iremonger. 1996. *Estimated Original Forest Cover Map: A First Attempt.* Cambridge, UK: World Conservation Monitoring Centre.

Carlsson-Kanyama A., and A. D. González. 2009. "Potential Contributions of Food Consumption Patterns to Climate Change." *The American Journal of Clinical Nutrition* 89: 1704S–1709S.

Carrington, Damian. 2016. "Tax Meat and Dairy to Cut Emissions and Save Lives, Study Urges." *The Guardian.* November 7. www.theguardian.com/environment/2016/nov/07/tax-meat-and-dairy-to-cut-emissions-and-save-lives-study-urges.

Cassidy, E. S., P. C. West, J. S. Gerber, and J. A. Foley. 2013. "Redefining Agricultural Yields: From Tonnes to People Nourished per Hectare." *Environmental Research Letters* 8(3):8 pp.

Center for Biological Diversity. 2018. "The Extinction Crisis." www.biologicaldiversity.org/programs/biodiversity/elements_of_biodiversity/extinction_crisis/.

Chivian, E. and A. Bernstein, eds., 2008. *Sustaining Life: How Human Health Depends on Biodiversity.* London: Oxford University Press.

EU. 2018. "Precautionary Principle." European Union. Access to European Union Law. Treaty on the Functioning of the European Union. Article 191. http://eur-lex.europa.eu/summary/glossary/precautionary_principle.html.

FAO. 2006a. "Livestock's Long Shadow." Food and Agriculture Organization. http://www.fao.org/docrep/010/a0701e/a0701e00.HTM.

FAO. 2006b. "Livestock a Major Threat to Environment." Food and Agriculture Organization. http://www.fao.org/Newsroom/en/news/2006/1000448/index.html.

FAO. 2010. "World Deforestation Decreases, But Remains Alarming in Many Countries." Food and Agriculture Organization. www.fao.org/news/story/en/item/40893/.

FAO. 2012. "Planted Forests." Food and Agriculture Organization. http://www.fao.org/forestry/plantedforests/en/.

FAO. 2015. "World Deforestation Slows Down as More Forests Are Better Managed." Food and Agriculture Organization. http://www.fao.org/news/story/en/item/326911/icode/.

FAOSTAT. 2017. "Data." Food and Agriculture Organization. http://faostat.fao.org/site/339/default.aspx.

Foley, Jonathan. 2014. "Five Step Plan to Feed the World." *National Geographic.* May. http://www.nationalgeographic.com/foodfeatures/feeding-9-billion/.

French, Howard. 2011. "E. O. Wilson's Theory of Everything." *The Atlantic.* www.theatlantic.com/magazine/archive/2011/11/e-o-wilsons-theory-of-everything/308686/.

Garnett T. 2011. "Where Are the Best Opportunities for Reducing Greenhouse Gas Emissions in the Food System (Including the Food Chain)?" *Food Policy* 36(Supp 1): 23–32.

Giller, Geoffrey. 2014. How Many Species Are There on Earth?" *Scientific American.* April 8. https://www.scientificamerican.com/article/are-we-any-closer-to-knowing-how-many-species-there-are-on-earth/.

Hamblin, James. 2017. "If Everyone Ate Beans Instead of Beef." *The Atlantic.* August 2. https://www.theatlantic.com/health/archive/2017/08/if-everyone-ate-beans-instead-of-beef/535536/.

Harvard School of Public Health. 2012. "Red Meat Consumption Linked to Increased Risk of Total, Cardiovascular, and Cancer Mortality." https://www.hsph.harvard.edu/news/press-releases/red-meat-consumption-linked-to-increased-risk-of-total-cardiovascular-and-cancer-mortality/.

Herrero, Mario and Philip K. Thornton. 2013. "Livestock and Global Change: Emerging Issues for Sustainable Food Systems." *Proceedings of the National Academy of Science* (PNAS). 110(52): 20878–20881.

Humane Society. 2018. "Farm Animal Welfare: Science and Research." The Humane Society of the United States. http://www.humanesociety.org/news/publications/whitepapers/farm_animal_welfare.html.

Johns Hopkins School of Public Health. 2017. "Health and Environmental Implications of U.S. Meat Consumption and Production." Johns Hopkins Public Health Magazine. http://www.jhsph.edu/research/centers-and-institutes/johns-hopkins-center-for-a-livable-future/projects/meatless_monday/resources/meat_consumption.html.

Kline, Benjamin. 2011. *First Along the River: A Brief History of the U.S. Environmental Movement.* Lanham, MD: Rowman and Littlefield.

Marlow H. J., W. K. Hayes, S. Soret, R. L. Carter, E. R. Schwab, and J. Sabaté. 2009. "Diet and the Environment: Does What You Eat Matter?" *The American Journal of Clinical Nutrition* 89(5): 1699S–1703S.

Marsh, George Perkins. 1864. *Man and Nature: Physical Geography as Modified by Human Action.* New York: Charles Scribner.

McMichael, Anthony J., J. W. Powles, C. D. Butler, and R. Uauy. 2007. "Food, Livestock Production, Energy, Climate Change, and Health." *The Lancet* 370: 1253–1263.

Merriam-Webster. 2018. "Definition of Pollution." *Merriam-Webster Dictionary.* https://www.merriam-webster.com/dictionary/pollution.

Muir, John. 1901. *Our National Parks.* Boston: Houghton, Mifflin and Company. http://vault.sierraclub.org/john_muir_exhibit/writings/our_national_parks/.

Muir, John. 1912. *The Yosemite.* New York: The Century Company. http://vault.sierraclub.org/john_muir_exhibit/writings/the_yosemite/.

NPS. 2018. "Theodore Roosevelt Quotes." Theodore Roosevelt National Park: North Dakota. United States Department of the Interior. National Parks Service. https://www.nps.gov/thro/learn/historyculture/theodore-roosevelt-quotes.htm.

Pappas, Stephanie. 2016. "There Might Be 1 Trillion Species on Earth." *Live Science.* May 5. https://www.livescience.com/54660-1-trillion-species-on-earth.html.

Pathak H., N. Jain, A. Bhatia, J. Patel, and P. K. Aggarwal. 2010. "Carbon Footprints of Indian Food Items." *Agriculture, Ecosystems & Environment* 139, 66–73.

Pettit, David. 2015. "Environmental Law Progress in China." Natural Resources Defense Council. April 16. https://www.nrdc.org/experts/david-pettit/environmental-law-progress-china.

Sass, J. B. and A. Colangelo. 2006. "European Union Bans Atrazine, While the United States Negotiates Continued Use." *International Journal of Occupational and Environmental Health* 12(3):260–267.

Smith, Pete and Mercedes Bustamante. 2014. "Agriculture, Forestry and Other Land Use (AFOLU)." In: *Climate Change 2014: Mitigation of Climate Change. Contribution of Working Group III to the Fifth Assessment Report of the Intergovernmental Panel on Climate Change.* Intergovernmental Panel on Climate Change. New York: Cambridge University Press.

Solotaroff, Paul. 2013. "In the Belly of the Beast." *Rolling Stone.* http://www.rollingstone.com/feature/belly-beast-meat-factory-farms-animal-activists.

Springmann, Marco, D. Mason-D'Croz, S. Robinson, K. Wiebe, H.C.J. Godfray, M. Rayner, and P. Scarborough. 2017. "Mitigation Potential and Global Health Impacts from Emissions Pricing of Food Commodities." *Nature Climate Change* 7: 69–74.

Stehfest Elke, L. Bouwman, D. P. Vuuren, M.G.J. Elzen, B. Eickhout, and P. Kabat. 2009. "Climate Benefits of Changing Diet." *Climatic Change* 95: 83–102.

THP. 2018. "World Hunger." The Hunger Project. http://www.thp.org/knowledge-center/know-your-world-facts-about-hunger-poverty/.

Thoreau, Henry David. 1854. *Walden; or, Life in the Woods.* Boston: Ticknor and Fields.

Tubiello, F. N., M. Salvatore, A. F. Ferrara, J. House, S., Federici, S., Rossi, . . . P. Smith. 2015. "The Contribution of Agriculture, Forestry and Other Land Use Activities to Global Warming, 1990–2012." *Global Change Biology* 21: 2655–2660.

UNEP. 2001. "Original Forest Cover." Forest Programme: United Nations Environmental Programme Monitoring Centre. http://www1.biologie.uni-hamburg.de/b-online/afrika/africa_forest/www.unep_wcmc.org/forest/original.htm.

UNEP. 2018. "About UN Environment." United Nations Environmental Programme. https://www.unenvironment.org/about-un-environment.

University of Minnesota. 2017. "How Does Agriculture Change Our Environment?" Institute on the Environment, University of Minnesota. http://www.environmentreports.com/how-does-agriculture-change/.

USFS. 2007. "The Greatest Good: A Forest Service Centennial Film." U.S. Forest Service. https://www.fs.fed.us/greatestgood/.

Vermeulen, Sonja J., Bruce M. Campbell, and John S. I. Ingram. 2012. "Climate Change and Food Systems." *Annual Review of Environment and Resources* 37:195–222.

Watts, Jonathan. 2010. "Costa Rica Recognised for Biodiversity Protection." *The Guardian.* October 25. https://www.theguardian.com/environment/2010/oct/25/costa-rica-biodiversity.
World Bank. 2017. "Terrestrial Protected Areas (% of total land area)." http://data.worldbank.org/indicator/ER.LND.PTLD.ZS.
WWF. 2018. "Deforestation." World Wildlife Fund. www.worldwildlife.org/threats/deforestation.
WWI. 2017. "Is Meat Sustainable?" World Watch Institute. http://www.worldwatch.org/node/549.
Yardley, Jim. 2013. "Bangladesh Pollution, Told in Colors and Smells." *The New York Times.* July 14. http://www.nytimes.com/2013/07/15/world/asia/bangladesh-pollution-told-in-colors-and-smells.html.
Zhang, Bo and Cong Cao. 2015. "Policy: Four Gaps in China's New Environmental Law." *Nature.* January 21. http://www.nature.com/news/policy-four-gaps-in-china-s-new-environmental-law-1.16736.

Case Studies

Case Study 1: Cars Rule—American Dependence on the Automobile

American society is built around the automobile, which claims to offer incredible mobility and convenience. Indeed, it is commonly assumed that Americans love cars. But why? Is there a choice to live without a car in this country?

In the United States, the impacts of automobile dependence include suburban sprawl, environmental degradation, and social disparities. Overall, this is an unsustainable transportation system in terms of ecological, social, and economic factors.

This case study explains how the car came to determine U.S. land-use patterns and the ecological and social ramifications of this domination. Environmental geographers can provide suggestions for future actions that would allow people to take back their communities from the automobile.

AMERICAN DRIVING HABITS

In the United States, there are 253 million cars and trucks on the roads, with ownership rates at 800 per 1,000 people. This contrasts with most European countries, which have 580 motor vehicles per 1,000 people. As context, in China there are about 128 cars per 1,000 people, and many countries in Africa and Asia match Bangladesh, with less than 3 cars per 1,000 people (Energy.gov, 2014).

There are also significant variations in how much Americans drive compared to people in other countries. Often, this is explained as an issue of distance: greater distances in the United States mean that Americans must drive more. But this does not ring true.

Americans drive 85% of their daily trips by car, and about 30% of these trips are shorter than 1 mile in distance. Amazingly, even for a trip that is under 1 mile, Americans will drive it by car 70% of the time! This contrasts with Europeans who make 70% of their 1-mile trips by bike, on foot, or with public transit like bus or streetcar (Buehler, 2014). This is a significant variation that has both social and environmental impacts.

ENVIRONMENTAL IMPACTS OF CARS

Many environmental impacts are associated with cars. Life Cycle Assessment is an approach to analyze the environmental impacts of a product throughout its "life," beginning with extraction of raw materials and ending with the disposal of waste (ISO, 2006). For an automobile, this involves the following stages: vehicle development, sourcing raw materials, vehicle manufacture, vehicle use, and vehicle end of life (Maclean and Lave, 2003).

Manufacturing a car demands steel, rubber, glass, plastics, electronics, paints, and more. Each of these things requires raw materials, the sourcing of which may cause environmental degradation. Fuels are used to transport these natural resources to a central location—a factory—where vehicles are assembled. Then completed cars demand fuel in order to distribute them to car dealers across the country and the world.

Not only is car production resource intensive up-front, but there are significant environmental costs at the end of a car's life, too. Many of the components remain in the environment for a long time; toxic batteries, paints, and plastics do not decompose. Luckily many parts can be recycled. It's estimated that about 75% of a car can be recycled (National Geographic, 2018), but the recycling rate varies greatly. Thus, because of Americans' supposed love affair with automobiles, junkyards must still take a huge amount of waste from worn-out old cars that are no longer on the road.

Although the entire cycle of a car's life has consequences, overall the bulk of its environmental impacts come from its use. Cars use chemicals that can be harmful to both ecosystems and humans: windshield wiper fluid, antifreeze, and motor oil can be poisonous when they leak and run off into water supplies. Fuel consumption and the emission of air pollution and greenhouse gases (GHG) cause about 90% of a car's negative environmental effects (National Geographic, 2018). But the good news is that each driver then has a great deal of control over these impacts.

Cars use fuel. Petroleum products must be extracted from the Earth, and this process is very energy intensive. Local environmental degradation can occur because of broken pipelines.

Likewise fuel must be moved from one location (pumping well) to another (oil refinery) (EIA, 2017). As society demands increasing amounts of oil, the ecological impacts of petroleum withdrawal will continue to intensify, as more risky methods of extraction will be necessary (EESI, 2018). This is why fuel efficiency is so important. Each individual driver has a significant impact if they drive a more fuel-efficient car that gets higher miles per gallon.

Cars and trucks have a huge negative impact on air quality. This is particularly of concern because cars pollute right where humans breathe: a car's tailpipe is at ground level. The emissions from cars and trucks include nitrogen oxides (NOx), hydrocarbons (HC), sulfur dioxide (SO_2), particulate matter, carbon monoxide

(CO), and several GHGs, including carbon dioxide (CO_2). Each of these pollutants causes distinct problems, ranging from respiratory diseases and cancer to changing the chemical composition of the Earth's atmosphere and accelerating climate change. In fact, about one-third of all greenhouse gas emissions that cause climate change are produced by the transportation sector. The word "smog" is a combination of the words smoke and fog. It is caused when particulate matter, ozone, and carbon monoxide, all found in large amounts in car exhaust, interact with sunlight. Smog causes numerous health problems, including lung and heart diseases; it can aggravate asthma, bronchitis, and other illnesses (UCS, 2018).

Luckily, in 1970 the Clean Air Act gave the newly created Environmental Protection Agency the authority to regulate pollution from cars and other modes of transportation. This historic legislation has profoundly reduced vehicle pollution in the United States. Thanks to these continued regulations, cars today run cleaner than cars did in the 1960s (EPA Air, 2018).

But the number of vehicles on the road continues to increase, thus making the need for air pollution regulations—indeed, all environmental regulations related to the life cycle of an automobile—ever more important.

WHY IS THE UNITED STATES SO DEPENDENT ON THE AUTOMOBILE?

It is not pure chance or the result of innovation that the United States is so dependent on automobiles; rather there are at least 10 reasons:

1. Historically, the U.S. government has greatly subsidized the car. Only about 60 % of necessary money for roadways actually comes from gas taxes, registration, and tolls paid. The rest of our car-based infrastructure comes from property tax, income tax, and other taxes not related to transportation. So our government made a decision to promote the automobile and its use. These many subsidies act to reduce the cost and increase people's demand for driving their cars (Cortright, 2015.)

2. In the United States, government support for public transit decreased significantly from the 1940s on, so in most cities the transit systems were privately owned and were forced to cut services and increase fares. Most of these went out of business, and for the few that were bought up by city governments, it was often too late (Buehler, 2014). For example, street cars used to run in every city, but these systems disappeared almost entirely by the late 1960s (UM-Dearborn, 2010).

3. The National Interstate Highway System started in the 1950s. Eventually 75,400 km (46,800 miles) of highways were built across America. Land away from city centers was cheaper, so developers bought it up and put in huge tracts of suburban houses. So Americans moved to the suburban housing and used their cars to get to work in distant cities (Chase, 2010).

4. The U.S. federal government paid 90% of the construction of this vast network of highways that stretched across the country and crisscrossed most cities (Buehler, 2014). Many urban neighborhoods were cut off by large exit

ramps and busy highways. Urban planning at this time emphasized the development of suburbs, which were subsidized by U.S. mortgage policies. It is worth noting that the interstate highways could have been planned to link cities rather than cut through them.

5. American city planners and engineers quickly adapted city infrastructures to the car: traffic lights, bridges, parking lots and garages, tunnels, and more. These concrete surfaces cause cities to be "heat islands" with higher temperatures (Giuliano and Hanson, 2017). In addition, U.S. city planners have always required a high number of parking places around businesses, which indeed makes it easier to drive, but also creates many unpleasant concrete parking lots that makes the city even more spread out (UM-Dearborn, 2010).

6. The high level of car ownership in America has triggered even more car ownership, because roads and towns are set up for cars and not other forms of transit. Now over 95% of Americans own at least one car (Chase, 2010).

7. Taxes on cars are very low in the United States, both in terms of ownership and use (gas tax). From the beginning the few taxes that were levied on car owners were used specifically for roadways, thus again building the car infrastructure early on (Giuliano and Hanson, 2017; Rodrigue, 2017).

8. In the United States, problems with cars focus on technological fixes rather than changing behaviors. For example, the problem of air pollution was addressed by adding catalytic converters to reduce pollution from leaded fuel, rather than a campaign encouraging people to drive less. Beyond that, the United States does not acknowledge the link automobile dependence, land-use, and poor health. In fact, there are clear correlations, as neighborhoods built for cars lead to adverse health outcomes (Dannenberg et al., 2003).

9. A few American cities began to implement bike- and pedestrian-friendly policies in the 1970s, but these are mostly underdeveloped with few bike lanes and signals. In addition, U.S. traffic laws tend to protect car drivers over pedestrians and cyclists. So it can be scary to ride a bike or walk to the store, because you are often forced to cross four-lane roads with cars zooming by at high speeds (VTPI, 2017).

10. Zoning in U.S. cities keeps land-uses very separated. So people live in residential areas—large neighborhoods or subdivisions—that are zoned only for single-family homes. By forbidding mixed uses of land, Americans cannot ride their bike from home to go to the doctor, grocery store, or school. These land-uses are all kept very separate, so the distance between home and the grocery store are often too far to walk or bike (Buehler, 2014). And crossing a four-lane road, with traffic zooming by, is not pleasant or safe. Indeed, 35,000 people are killed in U.S. car accidents each year (Gonzales, 2016).

Tragically, this is the ultimate example of cars dominating people in American society. Overall, there are clear examples of social problems and ecological damage that are directly attributable to automobiles. The issue is whether Americans have the will to change their addiction to cars and the land use that is devoted to them.

Case Study: Cars Rule

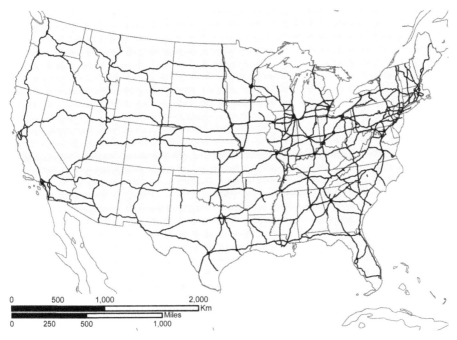

The U.S. Interstate System. (Bathgate, 2017)

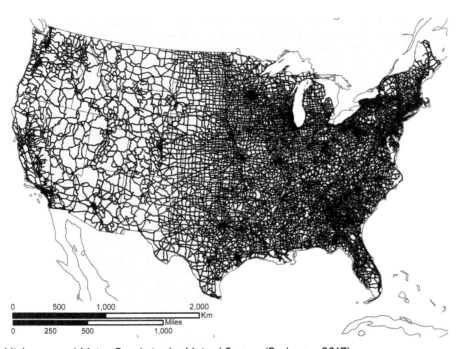

Highways and Major Roads in the United States. (Bathgate, 2017)

FINDING SOLUTIONS

Individuals have the power to change their personal actions, to vote for new policies, and to push for a new approach to transportation. Each person has a choice in their mode of transportation, so bike to work or take the bus. In order to make these options more feasible, people can vote for city leaders who make public transit, good sidewalks, and safe bike networks a priority.

Land-use and transportation policies can encourage healthy communities (Dannenberg et al., 2003). Smart growth is an urban planning approach that focuses on affordable, healthy neighborhoods (EPA, 2018). At a time when society must address and mitigate climate change, new approaches that focus on transportation systems that favor people over cars provide an opportunity for innovation (UN, 2018). People can demand these new transit amenities, borrowing from successful examples and tailoring them to their own region. It is a choice: Will cars rule America or will people?

LOOKING AHEAD

How can people reduce their dependence on cars? Several options exist, such as creating car-free zones, reducing the amount of car parking available, reducing speed limits, and paying for public transportation. Overall, this requires government action and a full-out behavior shift (Newman and Kenworthy, 2015).

New York City provides an interesting example: car ownership is low; just 50% of households own a car. The city has begun to build bike lanes that are separated from car traffic in the city; the city also closed some roads to make them pedestrian malls. Public transit is convenient and frequent (Chase, 2010).

In New York City and in other dense urban areas, car sharing programs make a lot of sense. Car sharing is a program where companies own cars, vans, and pickups parked throughout the city. Members rent the vehicles by the hour or day. This greatly reduces the number of automobiles within the city; sometimes 1 shared car takes the place of 20 individually owned cars (Chase, 2010).

Bike sharing programs are similar. One good example is found in Portland, Oregon: BIKETOWN's motto is join, unlock, ride, lock; and that summarizes the program well. There are 1,000 bikes at 100 stations throughout downtown and in several neighborhoods. BIKETOWN is intended to be convenient, fun, and affordable (Biketown, 2017).

Even cell phones can help people shift away from car dependence. Smart phone users can find public transit schedules in real time, compare cost and carbon emissions, and decide the best mode of transportation. If society demands it, the future may allow people, not cars, to rule our cities.

See also: Section 1: Air Pollution, Urbanization; *Section 3:* Assessments and "Footprints," Electric Cars, Sustainable Cities

FURTHER READING

Biketown. 2017. "A Whole New Way to Get Around." Portland, Oregon Bike Share Program. https://www.biketownpdx.com/how-it-works.

Buehler, Ralph. 2014. "9 Reasons the U.S. Ended Up So Much More Car-Dependent Than Europe." *The Atlantic*. February 4. http://www.citylab.com/commute/2014/02/9-reasons-us-ended-so-much-more-car-dependent-europe/8226/.

Chase, Robin. 2010. "You Asked: Does Everyone in America Own a Car?" U.S. Embassy. http://iipdigital.usembassy.gov/st/english/pamphlet/2012/05/201205165791.html#ixzz4XxazNW16.

Cortright, Joe. 2015. "The True Costs of Driving." *The Atlantic*. October 25. https://www.theatlantic.com/business/archive/2015/10/driving-true-costs/412237/.

Dannenberg, Andrew L., Richard J. Jackson, Howard Frumkin, Richard A. Schieber, Michael Pratt, Chris Kochtitzky, and Hugh H. Tilson. 2003. "The Impact of Community Design and Land-Use Choices on Public Health: A Scientific Research Agenda." *American Journal of Public Health* 93(9): 1500–1508. https://www.ncbi.nlm.nih.gov/pmc/articles/PMC1448000/.

EESI. 2018. "Fossil Fuels." Environmental and Energy Study Institute. http://www.eesi.org/topics/fossil-fuels.

EIA. 2017. "Crude Oil and Petroleum Products Explained." U.S. Energy Information Administration. https://www.eia.gov/energyexplained.

Energy.gov. 2014. "Fact #841: October 6, 2014 Vehicles per Thousand People: U.S. vs. Other World Regions." Office of Energy Efficiency & Renewable Energy. https://energy.gov/eere/vehicles/fact-841-october-6-2014-vehicles-thousand-people-us-vs-other-world-regions.

EPA. 2018. "Smart Growth and Affordable Housing." U.S. Environmental Protection Agency. https://www.epa.gov/smartgrowth/smart-growth-and-affordable-housing.

EPA Air. 2018 "History of Reducing Air Pollution from Transportation in the United States (U.S.)." U.S. Environmental Protection Agency. https://www.epa.gov/air-pollution-transportation/accomplishments-and-success-air-pollution-transportation.

Giuliano, Genevieve and Susan Hanson. 2017. *The Geography of Urban Transportation*, 4th ed. New York: Guilford.

Gonzales, Richard. 2016. "Traffic Deaths in 2015 Climb by Largest Increase in Decades." NPR. August 29. http://www.npr.org/sections/thetwo-way/2016/08/29/491854557/traffic-deaths-climb-by-largest-increase-in-decades.

ISO. 2006. "Life Cycle Assessment." International Organization for Standardization. https://www.iso.org/standard/37456.html.

Maclean, Heather and Lester B. Lave. 2003. "Life Cycle Assessment of Automobile/Fuel Options." *Environmental Science & Technology* 37: 5445–5452.

National Geographic. 2018. "Car Buying Guide." http://environment.nationalgeographic.com/environment/green-guide/buying-guides/car/environmental-impact/.

Newman, Peter and Jeffrey Kenworthy. 2015. *The End of Automobile Dependence: How Cities Are Moving Beyond Car-Based Planning*. Washington, D.C.: Island Press.

Rodrigue, Jean-Paul. 2017. *The Geography of Transport Systems*. New York: Routledge.

UCS. 2018. "Cars, Trucks, and Air Pollution." Union of Concerned Scientists. https://www.ucsusa.org/clean-vehicles/vehicles-air-pollution-and-human-health/cars-trucks-air-pollution#.

UM-Dearborn. 2010. "Automobile in American Life and Society." University of Michigan-Dearborn and Benson Ford Research Center. http://www.autolife.umd.umich.edu/.

UN. 2018. "7.d. Paris Agreement. Chapter XXVII: ENVIRONMENT." United National Treaty Collection. https://treaties.un.org/pages/ViewDetails.aspx?src=TREATY&mtdsg_no=XXVII-7-d&chapter=27&clang=_en.

VTPI. 2017. "Land Use Impacts on Transport: How Land Use Patterns Affect Travel Behavior." Victoria Transport Policy Institute. http://www.vtpi.org/tdm/tdm20.htm.

Case Study 2: Great Barrier Reef—Human Activities Endanger Coral Reefs

People are a real threat to coral reefs. They have overfished, polluted, changed ocean chemistry, introduced invasive species, and warmed the oceans through climate change. All of these things have a negative impact on coral reefs, which have been completely destroyed in some places. Overall on Earth, coral reefs are gravely endangered today. There is increasing urgency because each year more and more coral is dying (Amos, 2018; Daley, 2017; Griffith, 2016).

Coral reefs provide significant biodiversity to our oceans and help humans in numerous ways, but are endangered due to human activities (Worland, 2018). Environmental geographers are concerned about reefs, particularly because they are a common resource that we all affect with our activities. This means that environmental management and policy also require broader geographic collaboration. This case study of the Great Barrier Reef describes what coral reefs are, why they are important, the problems humans are causing, and the controversy of trying to solve these problems.

CORALS

Corals are fascinating animals. Found in fossils dating back 400 million years, these ancient creatures evolved about 25 million years ago into the reef-building species we know today. Coral reefs are the largest biologically built structures on Earth. There are over 6,000 known species of corals, but the stony corals (scleractinians) are the ones that build the reef structures (NOAA, 2014).

These corals live in colonies composed of up to hundreds of thousands of individuals, called polyps, which are very small, usually with a diameter of less than half an inch (1 cm). Each tiny polyp builds a protective crust of calcium carbonate around its body, and these crusts accumulate to create a reef. Inside each polyp are tiny single-celled algae that give food and oxygen and create reefs' beautiful colors (Ocean Portal Team, 2017).

There are corals in many of the Earth's oceans, but reef-building corals, like the Great Barrier Reef, only live in shallow tropical and subtropical waters. They need sunlight for photosynthesis and they thrive in water temperatures between 70 and 85°F (22 and 29°C) (Ocean Portal Team, 2017). In your mind, follow the equator around the Earth and think about all the island and ocean coasts in this area. This is likely where coral reefs are native: the clear shallow waters around tropic islands. Because they live close to shore, coral is especially vulnerable to human impacts.

Coral reefs only take up 2% of the ocean bottom, but they are very important to ocean ecosystems, with approximately 25% of all ocean life depending on reefs for food and shelter. Coral reefs are also extremely important to people for protection of shorelines, providing food, tourism money, and even medicines (Ocean Portal Team, 2017).

But these colorful treasures may not survive in the future. There are both local and global threats.

THREATS TO CORALS

Coral reefs grow slowly and are sensitive to disruptions in their local area. Because coral reefs are near the shore, pollution from human activities is often nearby. Oil and chemical spills and simply runoff from urban and industrial areas are hazardous. Water pollution can block sunlight or even settle on them, kill off their polyps, and then the coral reefs die. In addition, overfishing removes too many herbivore fish species, allowing excess seaweed to grow, which smothers the coral and kills it (National Ocean Service, 2017; Py-Lieberman, 2008).

Globally, the biggest threats to coral reefs are climate change and acidification of the oceans. We all know that humans have created massive amounts of CO_2 that has heated our atmosphere and oceans. High water temperatures lead to coral bleaching. This is caused by warmer water killing the coral polyps' algae that produce the corals' food. Coral bleaching can kill vast areas of reefs or leave them more vulnerable to other threats (UNEP, 2016).

Ocean acidification is related to ocean temperature. As humans continue to burn more fossil fuels and our emissions have increased, oceans have helped slow the Earth's atmospheric warming by removing about 50% of excess CO_2 since the year 1800. So now about 2.5 billion metric tons of additional carbon from human activities enter the ocean in the form of carbon dioxide (NOAA, 2018). But the oceans covert this CO_2 into carbonic acid (H_2CO_3) in the seawater, which has increased the acidity of ocean surface water by about 30% since 1850 (Royal Society, 2005). This makes the building and maintaining of calcium carbonate structures like reefs and shells increasingly difficult for coral and shellfish.

THE GREAT BARRIER REEF

The Great Barrier Reef is located along the northeastern coast of Australia. It is 2,300 km (1,430 miles) long, so huge that it is visible from outer space. It takes up an area of 344,400 km² (132,974 mi²) or about the size of Germany or half the size of Texas (GBRMPA, 2018). It is the largest living thing on Earth (Lewis, 2016).

The Great Barrier Reef ecosystem comprises thousands of reefs and hundreds of islands. It is also home to diverse sea life, including 600 types of soft and hard corals, 100 species of jellyfish, 3,000 varieties of mollusks, 1,625 types of fish, 133 varieties of sharks and rays, and more than 30 species of whales and dolphins (GBRMPA, 2018).

Most of the reef is part of the Great Barrier Reef Marine Park, which helps to protect the area and manage fishing and tourism. The reef was declared a World Heritage Site by the United Nations Educational, Scientific and Cultural Organization (UNESCO) in 1981. This means that the Great Barrier Reef is part of the global UNESCO program that encourages the identification, protection, and preservation of natural areas and cultural sites around the world "considered to be of outstanding value to humanity" (UNESCO, 2018).

Despite its ecological importance and recognized value to both people and the planet, the Great Barrier Reef is dying (Ritter, 2016).

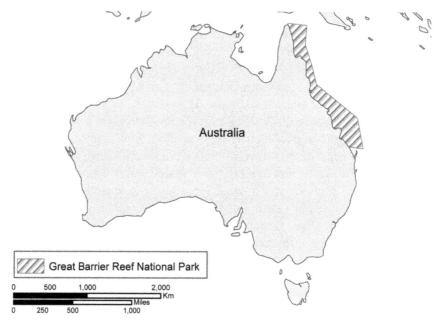

Great Barrier Reef National Park, Australia. (Bathgate, 2017)

The Great Barrier Reef Marine Park Authority conducts surveys to assess the extent of the reef's degradation, and the news is not good. Approximately 93% of the reef is affected by bleaching and 22% of coral on the reef is dead from the worst mass coral bleaching ever experienced (Slezak, 2016a, 2016b). The northern portion of the reef is particularly hard hit (GBRMPA, 2017; Griffith, 2016).

CONTROVERSY OVER PROTECTION OF THE REEF

There is a huge controversy around the protection of and possible methods to save the reef. In 2015, the Australian government released the Reef 2050 sustainability plan for the region, but many environmental scientists say that it does not address the biggest problem: climate change (Norman et al., 2015). The Reef 2050 plan suggests cutting back on dredging and waste near the reef, but does not act to reduce the mining and use of fossil fuels, particularly coal, which is so destructive and a significant cause of global warming (Australian Government, 2015).

In addition, there is concern that the Australian government has been covering up the true level of reef destruction in order to maintain the tourism money that the reef ensures (Griffith, 2016; Selzak, 2016b). Critics point to the fact that the Australian government requested that a United Nations (UN) report on climate change remove a chapter about the destruction of the Great Barrier Reef (Ritter, 2016).

Speaking of the global crisis, well-known marine biologist Nancy Knowlton said: "If people don't change the way they're doing things, reefs as we know them will be gone by the year 2050. It's actually really depressingly unbelievable" (Py-Lieberman, 2008).

Specifically in regard to the Great Barrier Reef, Professor Terry Hughes, a well-respected researcher at the Centre of Excellence for Coral Reef Studies at James Cook University in Queensland, Australia, is a critic of the Australian Reef 2050 Plan. He stated that the "biggest omission in the plan is that it virtually ignores climate change, which is clearly the major ongoing threat to the reef" (Ritter, 2016). He and other experts says that Australia, and really the world, has a choice: "Either we take action to adequately protect reefs and transition away from fossil fuels, or we abandon the reef and develop the world's largest thermal coal mines. We can't possibly do both" (Norman et al., 2015). This exemplifies the conflict between economic and ecological sustainability, and now society must step in and decide which aspect they value more. Will it be a healthy and vibrant reef or continued use of coal and other fossil fuels (Norman et al., 2015)? This is also an example of short-term gain versus long-term sustainability. We have an energy system in place that is built on fossil fuel extraction and use; we need to change this to be sustainable into the future. Such a shift is monumental and requires significant economic and political change, and thus most people are still scared to make the change to clean energy.

LOOKING AHEAD

Recently, scientists have undertaken significant efforts to better understand coral reef ecosystems. The U.S. government's National Oceanic and Atmospheric Administration (NOAA) has an important Ocean Acidification Program (OAP) with the mission "to better prepare society to respond to changing ocean conditions and resources by expanding understanding of ocean acidification, through interdisciplinary partnerships, nationally and internationally." One part of this is the Coral Reef Ocean Acidification Monitoring Portfolio (CROAMP). This is a collaborative effort to monitor chemical and ecological changes in Atlantic and Pacific Ocean coral reefs (NOAA, 2018).

Beginning in 2012, with significant funding from an insurance company and passion for photography, the XL Catlin Seaview Survey began documenting the Great Barrier Reef (Walsh, 2013). Now the Catlin Survey is creating a "baseline record of the world's coral reefs, in high-resolution 360-degree panoramic vision." (University of Queensland, 2018). This breathtaking website takes viewers to the depths of the ocean, as if we are each explorers on a reef survey mission (XL Catlin Seaview Survey, 2018). Monitoring the health of reefs and documenting their decline are important, but ultimately action must be taken.

The controversy around the health of the Great Barrier Reef exemplifies the complexity of trying to communicate science while confronting strong varying viewpoints. An article in *Outdoor Magazine* proclaimed: "The Great Barrier Reef of Australia passed away in 2016 after a long illness. It was 25 million years old" (Jacobson, 2016). The author describes the plight of the reef, explains that it is endangered, places some blame on the Australian government, and notes that we are not doing enough to save this site of global importance. A creative approach was used by the author to present the article as an obituary of the reef. But at the same time, this article sent many people, as displayed on social media, reeling in hopelessness (Lewis, 2016).

Is it really too late to do anything? Will we be able to save the Great Barrier Reef? Scientists warn that conditions are declining, as major bleaching events are occurring more often (Amos, 2018). But it is not too late. Still, we must take action immediately. We must cut emissions to slow global warming and decrease ocean acidification. This will help us save our coral reefs. Shouting out sensationalized headlines may lead to misunderstanding, but relying on a calm rational approach may be ignored. This is often the dilemma for environmental geographers; it is difficult to educate people about environmental problems and solutions while emphasizing the importance of immediate action that does not coincide with existing economic and political goals.

See also: Section 1: Climate Change; *Section 2:* Anthropocene, Ecotourism, Human Modification of Ecosystems, Protected Areas and National Parks

FURTHER READING

Amos, Jonathan. 2018. "Coral Reefs Head for Knock-Out Punch." BBC. January 4. http://www.bbc.com/news/science-environment-42571484.
Australian Government. 2015. "The Reef 2050 Plan." Department of Environment and Energy. http://www.environment.gov.au/marine/gbr/long-term-sustainability-plan.
Daley, Jason. 2017. "Great Barrier Reef Braces for Another Massive Bleaching Event." *Smithsonian Magazine.* February 27. http://www.smithsonianmag.com/smartnews/great-barrier-reef-braces-another-massive-bleaching-event-180962305/.
GBRMPA. 2017. "Reef Health." Great Barrier Reef Marine Park Authority. http://www.gbrmpa.gov.au/about-the-reef/reef-health.
GBRMPA. 2018. "Facts about the Great Barrier Reef." Great Barrier Reef Marine Park Authority. http://www.gbrmpa.gov.au/about-the-reef/facts-about-the-great-barrier-reef.
Griffith, Hywel. 2016. "Great Barrier Reef Suffered Worst Bleaching on Record in 2016, Report Finds." BBC. November 28. http://www.bbc.com/news/world-australia-38127320.
Jacobsen, Rowan. 2016. "Obituary: Great Barrier Reef (25 Million BC-2016)." *Outside Magazine.* October 11. https://www.outsideonline.com/2112086/obituary-great-barrier-reef-25-million-bc-2016.
Lewis, Sophie. 2016. "The Great Barrier Reef Is Not Actually Dead." CNN. October 14. http://www.cnn.com/2016/10/14/us/barrier-reef-obit-trnd/.
National Ocean Service. 2017. "How Does Land-Based Pollution Threaten Coral Reefs?" National Oceanic and Atmospheric Administration. October 10. http://oceanservice.noaa.gov/facts/coral-pollution.html.
NOAA. 2014. "What Are Coral Reefs?" National Oceanic and Atmospheric Administration. December 16. http://www.coris.noaa.gov/about/what_are/.
NOAA. 2018. "Ocean Acidification Program." National Oceanic and Atmospheric Administration. http://oceanacidification.noaa.gov.
Norman, Barbara, Iain McCalman, and Terry Hughes. 2015. "Government Unveils 2050 Great Barrier Reef Plan: Experts React." The Conversation: Academic Rigor,

Journalistic Flair. March 23. https://theconversation.com/government-unveils-2050-great-barrier-reef-plan-experts-react-39172.
Ocean Portal Team. 2017. "Corals and Coral Reefs." Smithsonian National Museum of Natural History. http://ocean.si.edu/corals-and-coral-reefs.
Py-Lieberman, Beth. 2008. "Nancy Knowlton: Reef Biologist." *Smithsonian Magazine.* http://www.smithsonianmag.com/arts-culture/nancy-knowlton-7330452/?no-ist.
Ritter, David. 2016. "Great Barrier Reef: Why Are Government and Business Perpetuating the Big Lie?" *The Guardian.* October 31. https://www.theguardian.com/sustainable-business/2016/nov/01/great-barrier-reef-why-are-government-and-business-perpetuating-the-big-lie.
Royal Society. 2005. "Ocean Acidification Due to Increasing Atmospheric Carbon Dioxide." https://royalsociety.org/~/media/Royal_Society_Content/policy/publications/2005/9634.pdf.
Slezak, Michael. 2016a. "Great Barrier Reef: 93% of Reefs Hit by Coral Bleaching." *The Guardian.* April 19. https://www.theguardian.com/environment/2016/apr/19/great-barrier-reef-93-of-reefs-hit-by-coral-bleaching.
Slezak, Michael. 2016b. "Agencies Say 22% of Barrier Reef Coral Is Dead, Correcting 'Misinterpretation.'" *The Guardian.* June 3. https://www.theguardian.com/environment/2016/jun/03/agencies-say-22-of-barrier-reef-coral-is-dead-correcting-misinterpretation.
UNEP. 2016. "Urgent Need for Sustainable Management of Coral Reefs." United Nations Environmental Programme. May 16. http://web.unep.org/stories/story/urgent-need-sustainable-management-coral-reefs.
UNESCO. 2018. "Great Barrier Reef." United Nations Educational, Scientific, and Cultural Organization. http://whc.unesco.org/en/list/154.
University of Queensland. 2018. "XL Catlin Seaview Survey." Global Change Institute. http://www.gci.uq.edu.au/xl-catlin-seaview-survey.
Walsh, Bryan. 2013. "Breaking the Waves: Catlin Seaview Survey Digitizes the Endangered Oceans." *Time.* July 31. http://science.time.com/2013/07/31/breaking-the-waves-catlin-seaview-survey-digitizes-the-endangered-oceans/.
Worland, Justin. 2018. "A Most Beautiful Death." *Time.* http://time.com/coral/.
XL Catlin Seaview Survey. 2018. "Our Oceans Are Changing." http://catlinseaviewsurvey.com/.

Case Study 3: Dead Zones—The Gulf of Mexico

A dead zone is a more popular term for hypoxia, which means lacking oxygen. This occurs when the level of oxygen in the water is greatly reduced to the point where aquatic life cannot survive (National Ocean Service, 2018).

In the United States and Europe, hypoxia is mainly caused by agricultural techniques that lead to runoff. Farmers add lots of nitrogen and phosphorus to fertilize their crops, but rain washes away the excess. These nutrients pollute nearby streams and then run into rivers that flow into the ocean. At the mouth of these rivers, such as the Mississippi River, the surplus nitrogen and phosphorous leads to excessive algae blooms (MRC, 2018). When this algae decomposes, it depletes underwater oxygen. This lack of oxygen in the deep water means that fish cannot survive. The area then becomes a dead zone that is void of aquatic life (National Ocean Service, 2018). As with many environmental geography issues, the overuse of agricultural chemicals is a complicated problem to solve because the sources of pollution are far away and it is difficult to prove responsibility.

Dead zones have doubled in frequency every decade since the 1960s, due to the increased use of synthetic fertilizers and the associated runoff (Minogue, 2014). Over 90% of current dead zones are in regions that are expected to warm at least 2°C (3.6°F) (Zielinski, 2014) and these warmer waters will intensify algae growth, thus creating more dead zones and making existing ones expand more quickly.

WORLDWIDE DEAD ZONES

On Earth, there are more than 400 dead zones, ranging in size from just 1 square kilometer (0.4 mi²) to tens of thousands of square kilometers (or square miles) (Scientific American, 2018; NASA Earth Observatory, 2010). The size and location of dead zones vary by season as more fertilizer is applied in the spring and then washes downstream. This, combined with warmer summer temperatures, causes the algae to grow more rapidly in summer months, thus leading to hypoxia during this season.

Dead zones are found worldwide. Of particular note are zones in the Baltic Sea in northern Europe; the Black Sea in central Europe; the Sea of Japan and the Yellow Sea off the coasts of China, South Korea, and Japan; the Atlantic Ocean along the eastern coast of the United States; and most notably in the Gulf of Mexico near the coast of Louisiana, Mississippi, and parts of Texas (NASA Earth Observatory, 2010).

It is estimated that there are approximately 200 dead zones in U.S. waterways alone, mostly within coastal waters (On Earth, 2014). These sites have increased

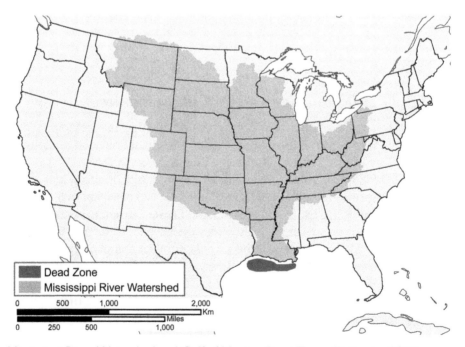

Mississippi River Watershed and Gulf of Mexico Dead Zone. (Bathgate, 2017)

and intensified as fertilizer use on U.S. farms creates the nutrient-rich setting for hypoxia.

The Gulf of Mexico dead zone varies depending on the season and the year, but is typically about 15,500 square kilometers (6,000 square miles), or the size of the state of Connecticut (Gulf Restoration Network, 2018; NOAA, 2016).

The U.S. Gulf Coast is known as the northern portion of the Gulf of Mexico ecosystem by scientists and ecologists. This region holds nearly half of all U.S. coastal wetlands and supports a $3 billion a year commercial and recreational fishing industry (NCCOS, 2017). For communities along the Gulf Coast, fishing is the economic basis of their livelihoods. Shrimp, for example, had been a $500 million per year industry (UPI, 2009). But this area changed a great deal due to pollution from the massive Mississippi River flowing into the gulf. The pollution is from nutrients like nitrogen and phosphorus that are washed downstream from intensively farmed fields throughout the U.S. Midwest agricultural belt.

THE MIDWEST'S CONTRIBUTION TO HYPOXIA IN THE GULF

The U.S. Midwest produces huge harvests of corn and soybean crops, most of which supports the production of meat and dairy at relatively low prices to consumers. Approximately $49 billion dollars' worth of corn is produced on this rich farmland every year. The United States produces 40% of the world's corn in the top corn-producing states of Iowa, Illinois, Nebraska, and Minnesota (USDA-ERS, 2017). Corn requires high nitrogen to grow, so farmers often add chemical fertilizers to their field because the crop removes the nutrients each year. Corn is a popular crop for feeding cattle and other animals for slaughter; consequently, farmers are trying to grow more and more corn, so they often add more fertilizer. There are only voluntary programs aimed at encouraging farmers to reduce their chemical usage (Marder, 2011). If the soil is not healthy with good natural microbial activity and drainage, the fertilizer will run off the crop fields. A watershed is an intricate system of small creeks, medium streams, and big rivers that all join together and flow into one outlet. In the case of the U.S. Midwestern agricultural region, this is the Mississippi River, which flows into the Gulf of Mexico.

The Mississippi River is approximately 2,350 miles long, starting as a trickle in northern Minnesota and gathering tributaries until it flows into the Gulf of Mexico in New Orleans at a rate of 4,493,400 gallons per second. The Mississippi River watershed is one of the largest in the world, extending from the Rocky Mountains in the west to the Allegheny Mountains in the east. It covers about 40% of the contiguous 45 states, including all or part of 31 states and two Canadian provinces (NPS, 2017).

Given the size and location of the Mississippi River watershed, millions of people and many diverse activities affect the health of the water. In the case of the dead zone, nutrient loads, greatly increased in the last 50 years of industrial farming that has pushed up fertilizer usage, have had catastrophic effects on the river and the ocean it flows into (Gulf Restoration Network, 2018).

A DIFFICULT PROBLEM TO SOLVE

First, think about how complex it is to manage a resource that is as vast as the Mississippi River watershed. Each individual who lives in any of the 31 states could play a role in water quality, and in this case, agriculture has a special role to play. Second, think about how environmental rules vary from state to state. And the national rules, although important, cannot always overcome state laws. Finally, the way in which any sort of cleanup must occur is actually far upstream from the problem. This geographic separation between where a problem is seen and where it is caused often makes a solution more challenging.

Mississippi River pollution is mostly nonpoint source, meaning that it flows into the river from overland sources. This is different from point-source pollution, which is from a single obvious location, like a wastewater pipe from a factory. The U.S. Clean Water Act of 1972 (and amendments in later years) primarily addresses our waterways by regulating point-source pollution. Nonpoint-source pollution is very difficult to regulate, and this has mostly been attempted through incentives like grants for programs that set up water and land management plans. State rules for nonpoint-source water pollution vary, and this does not help the huge and diverse Mississippi watershed. Because of privacy laws, it is difficult to monitor what flows from a farmer's fields, so one farmer can blame another, without anybody ever taking responsibility for the pollution.

This is a common problem in environmental regulation: negative externalities. When a person or company causes pollution but does not have to pay the cost of this pollution, it is a negative externality. It is external to or outside the structures of society, so the thing being produced creates a cost to society that is bigger than the cost the consumer is paying for it (IMF, 2017). In this case, the corn produced in Iowa causes pollution that flows down the Mississippi and causes the dead zone, but the price we pay for meat (as a corn product) does not include the cost of cleaning up the hypoxic zone. Indeed, the externalities affect economic, social, and environmental segments of society. Economic impacts include a decline in gulf fishing and tourism dollars lost from lack of recreation in the region. Social impacts are human health problems from the pollution and cultural decline in the fishing communities. Environmental impacts are the losses to biological diversity experienced in the ecosystems that are decimated by the dead zones.

Overcoming the complex problems that have caused the Gulf of Mexico dead zone will require new strategies and collaborations. Certainly the U.S. national government can take action by restructuring agricultural subsidies that encourage higher chemical use (Gulf Restoration Network, 2018.) Researchers and organizations can help by providing information and assistance through scientific advising and technology. The biggest hurdle is to overcome individuals' unwillingness to change their current behaviors, but this is a necessity.

LOOKING AHEAD

Dead zones can be brought back to life. A good example is the Black Sea, which had one of the largest hypoxic areas in the world, estimated at 39,000 square kilometers (15,000 square miles). But when the USSR broke apart in the 1990s, the subsidies for cheap fertilizer ended, and the fertilizer runoff decreased by 50%. After about three years, the rivers began to recover and through careful management, the Black Sea has come back to life (On Earth, 2014).

The United States and other countries could adopt new agricultural policies that would help reduce the amount of fertilizer that flows off farm fields. A large amount of agricultural research shows how specific techniques can reduce runoff, but much of this is ignored because fertilizer is relatively cheap and farmers constantly need to increase their crop yields to survive in the super-competitive realm of industrialized agriculture.

The interagency Mississippi River/Gulf of Mexico Hypoxia Task Force was established in 1997 to help assess and develop options for addressing the dead zone in the Gulf of Mexico. Their 2001 Action Plan included 11 specific actions, and an updated 2008 Action Plan stated that a 45% reduction in nitrogen and phosphorus load is needed to reach the goal of reducing the hypoxic zone to 5,000 square kilometers (1,931 square miles) (NCCOS, 2017). The task force website provides a great deal of useful information for people wishing to educate themselves about the problem of dead zones, and the Gulf of Mexico dead zone, in particular. They also note examples of successful projects and programs that help reduce nutrients in the waterways and improve water quality in the Mississippi River watershed (EPA, 2017).

Although these collaborative groups gather information and provide opportunities for education, there is no unified effort to change the land-uses and agricultural practices that cause the dead zones. The United States (in the case of the Gulf dead zone) and the world (in the case of the hundreds of other dead zones) must take action to change the dominant agricultural system to promote sustainable farming methods that do not externalize the costs of fertilizer on the environment. Until this occurs, these ecological problems will continue. There is no simple fix, as society must be willing to rethink their demand for cheap crops that feed our cattle and dairy cows. Agricultural policies must prioritize sustainable farms, with healthy soils that hold nutrients, rather than allowing them to wash away downstream.

See also: Section 1: Agriculture, Animal Agriculture, Water Pollution; *Section 2:* Ecosystem Services; *Section 3:* Alternative Agriculture, Environmental Policy, Sustainable Diet

FURTHER READING

EPA. 2017. "Mississippi River/Gulf of Mexico Hypoxia Task Force." U.S. Environmental Protection Agency. https://www.epa.gov/ms-htf.
Gulf Restoration Network. 2018. "The Dead Zone and Mississippi River." Our Work. http://healthygulf.org/our-work/water-wetlands/dead-zone-and-mississippi-river.
IMF. 2017. "Externalities: Prices Do Not Capture All Costs." International Monetary Fund. July 29. http://www.imf.org/external/pubs/ft/fandd/basics/external.htm.
Marder, Jenny. 2011. "Farm Runoff in Mississippi River Floodwater Fuels Dead Zone in Gulf." *PBS News Hour.* May 18. http://www.pbs.org/newshour/rundown/the-gulf-of-mexico-has/.

Minogue, Kristen. 2014. "Climate Change Expected to Expand Majority of Ocean Dead Zones." Smithsonian Insider. November 10. http://insider.si.edu/2014/11/climate-change-expected-expand-majority-ocean-dead-zones/.
MRC. 2018. "What Is the Gulf of Mexico 'Dead Zone'"? Mississippi River Collaborative. http://www.msrivercollab.org/focus-areas/dead-zone/.
NASA Earth Observatory. 2010. "Aquatic Dead Zones." National Aeronautics and Space Administration. http://earthobservatory.nasa.gov/IOTD/view.php?id=44677.
National Ocean Service. 2018. "What Is a Dead Zone?" National Oceanic and Atmospheric Administration. http://oceanservice.noaa.gov/facts/deadzone.html.
NCCOS. 2017. "NCCOS-Supported Research Provides Foundation for Management of the 'Dead Zone' in the Northern Gulf of Mexico." National Centers for Coastal Ocean Science. https://coastalscience.noaa.gov/research/stressor-impacts-mitigation/habhrca/dead-zone/.
NCCOS. 2018. "Gulf of Mexico Ecosystems & Hypoxia Assessment (NGOMEX) Program Highlights." National Centers for Coastal Ocean Science. https://coastalscience.noaa.gov/research/stressor-impacts-mitigation/habhrca/ngomex/.
NOAA. 2016. "Average 'Dead Zone' for Gulf of Mexico Predicted." National Oceanic and Atmospheric Administration. June 9 http://www.noaa.gov/media-release/average-dead-zone-for-gulf-of-mexico-predicted.
NPS. 2017. "Mississippi River Facts." National Park Service. https://www.nps.gov/miss/riverfacts.htm.
On Earth. 2014. "Devil in the Deep Blue Sea." Natural Resources Defense Council. November 11. https://www.nrdc.org/onearth/devil-deep-blue-sea.
Scientific American. 2018. "What Causes Ocean "Dead Zones?" Earth Talk. www.scientificamerican.com/article/ocean-dead-zones/.
UPI. 2009. "Shrimp industry at Risk from Dead Zone." United Press International. October 30. http://www.upi.com/Science_News/2009/10/30/Shrimp-industry-at-risk-from-dead-zone/UPI-98381256912603.
USDA-ERS. 2017. "Background." United States Department of Agriculture. Economic Research Service. September 14. https://www.ers.usda.gov/topics/crops/corn/background/.
Zielinski, Sarah. 2014. "Ocean Dead Zones Are Getting Worse Globally Due to Climate Change." Smithsonian Magazine. November 10. http://www. smithsonianmag.com/science-nature/ocean-dead-zones-are-getting-worse-globally-due-climate-change-180953282/.

Case Study 4: Great Pacific Garbage Patch

The expression "out of sight, out of mind" is a dangerous concept in environmental geography. Although people may not see pollution or degradation that is far away, its negative consequences are still very real. Humans have dumped waste into the oceans for thousands of years, but since plastics were developed in the 1900s, the problem is exacerbated due to the fact that these materials never break down. They never decompose, so they never go away. We might think that our trash is gone once it goes into our trashcan and is hauled away by a garbage truck each week. But that is far from the truth. Some of this waste ends up in the ocean and is called marine debris. Because of ocean currents, garbage follows patterns in the ocean and it flows and accumulates in certain areas. This case study describes one area of the ocean where our plastic trash accumulates and harms sea life: the Great Pacific Garbage Patch.

MARINE POLLUTION

According to the United Nations Environment Programme: "Marine litter poses a vast and growing threat to the marine and coastal environment. It is found in all sea and ocean areas of the world" (UNEP, 2005). It is mind boggling to think about how all this garbage got in our oceans. About 80% of marine debris comes from land-based sources and 20% is from ocean-based activities (CCC, 2018). That means most of it is from coastal activities: litter left by beach visitors or garbage overflows that carry trash from storm drains and sewers, including garbage and street litter. To a lesser extent the trash is from fishing debris or shipping waste that boats accidentally or deliberately dump overboard.

In the last 40 years, this trash is increasingly composed of plastics, which are found in nearly all consumer goods because they are cheap and durable. And plastic, in particular, causes problems because the wave action and sun breaks it down into tiny pieces (microplastics), which never go away (Kostigen, 2008; *Plastic Paradise*, 2013).

There are known garbage patches made up of these plastics in many oceans on Earth. "While the total amount in the ocean is unknown, plastic is found all over the world including in the polar regions, far from its source" (UNEP, 2014: 49).

The garbage patches, composed of these tiny microplastics, are not always visible. Instead the water looks murky or cloudy and even satellite imagery does not always show the garbage patches. This is why people only discovered the huge floating patches of plastic waste within the past few decades (NOAA GPGP 2018).

THE GREAT PACIFIC GARBAGE PATCH

The Great Pacific Garbage Patch, discovered by Captain Charles Moore, is in the northern Pacific Ocean, north of the Hawaiian Islands. It covers hundreds of square miles of ocean, and some say it is twice the size of Texas. But it is not a solid mass; instead, the tiny pieces of plastic are like "pepper in soup" both at the surface and down into the deeper water column. Thus its true size is hard to quantify.

Moore is an expert sailor, who often sailed his large catamaran between Hawaii and California. In 1997, he veered off his typical route and went through a remote region known as the North Pacific Gyre. "I was confronted, as far as the eye could see, with the sight of plastic," he stated in a *Natural History* article (Moore, 2003). He was shocked that he never found a clear spot, day or night, during the week-long crossing: "plastic debris was floating everywhere: bottles, bottle caps, wrappers, fragments." There were larger items, as well, including a full drum of hazardous chemicals, an inflated volleyball, an inflated truck tire mounted on a steel rim, a tangled mess of polypropylene net lines, and more. Moore and his research team published some preliminary findings indicating that for every pound of naturally floating zooplankton in the Great Pacific Garbage Patch, there are six pounds of plastic waste (Moore, 2003).

The Great Pacific Garbage Patch is also known as the North Pacific Trash Vortex, and this provides a good image of the circular motion at the core of this

Marine researcher Charles Moore holds a tray of debris collected on a beach in Hawaii, washed ashore from the Great Pacific Garbage Patch. There are at least five massive floating garbage dumps across the planet as gyres are created by ocean currents. (Jonathan Alcorn/Bloomberg via Getty Images)

phenomenon. Ocean water from Japan to the west coast of North America is part of this system, influenced by the North Pacific Subtropical Gyre. A gyre is a large system of rotating ocean currents created by the rotation of the Earth, which go in a clockwise direction in the Northern Hemisphere and counterclockwise in the Southern Hemisphere.

Scientists are making new, tragic discoveries. These gyres truly act as funnels and garbage is gathering in all of them. Islands that are located near the gyres are particularly hard-hit. For example, it was recently discovered that Marshall Island, an uninhabited island located in a remote region of the South Pacific on the edge of the South Pacific Gyre, now has vast amounts of plastic waste on its shores. This occurred recently, because only in 1988, Marshall Island was named a UNESCO World Heritage Site for its unique, rare ecology that was "practically untouched by a human presence" (UNESCO, 2018). Now it has some of the highest amounts of plastic garbage seen anywhere in the world (Lavers and Bond, 2017; Ramzymay, 2017.

This case study focuses on the North Pacific garbage patch because it has been studied for a few years. The North Pacific Subtropical Gyre is created by the interaction of currents that move in a clockwise direction around an area of about 20 million square kilometers (7.7 million square miles). This is about the size of three times the area of the continental United States. The circular motion of the gyre pulls debris into its center, which has calm, stable waters. So a water bottle tossed in a California beach may be pulled south, then west, and then north in a huge 8,000-mile loop before, after a few years, getting drawn into the Great Pacific Garbage Patch (NGS, 2018).

Scientists have found that most debris comes from plastic bags, bottle caps, plastic water bottles, and Styrofoam cups. They have collected up to 750,000 bits of microplastic in a single square kilometer (1.9 million bits per square mile) of the

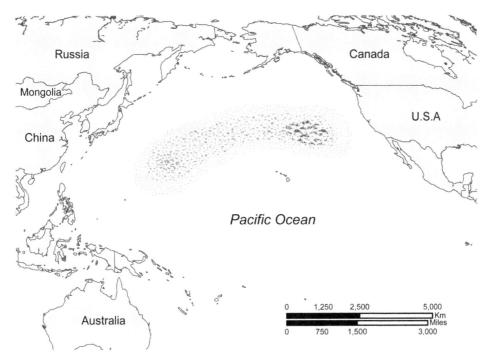

Great Pacific Garbage Patch. (Bathgate, 2017)

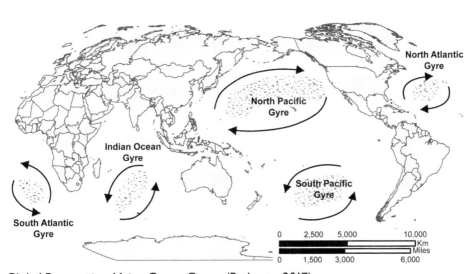

Global Perspective: Major Ocean Gyres. (Bathgate, 2017)

Great Pacific Garbage Patch. Oceanographers and marine ecologists believe that about 70% of marine debris actually sinks to the bottom of the ocean (NGS, 2018). So it is possible that the seafloor beneath the Great Pacific Garbage Patch may also be covered with piles of our trash.

IMPACT OF THE GREAT PACIFIC GARBAGE PATCH

This garbage has greatly affected hundreds of different marine species that have been found to have ingested or been entangled in marine debris. This includes seabirds, turtles, seals, sea lions, whales, and fish. In the garbage patch, for example, sea turtles die because they mistake plastic bags for their major food source, jellyfish. Several bird species confuse plastic pellets for fish eggs and feed them to their chicks, which then starve to death or die from internal bleeding. Seals and sea mammals are curious, so are especially affected, as they get tangled in plastic fishing nets and drown. Fish species are affected, as thousands die in remote discarded nets, a tragic phenomenon known as "ghost fishing" (NOAA, 2018).

In addition to these direct impacts on sea life, the plastic garbage patch creates ecological disruptions that have long-term consequences on the health of the ocean. Plastics leach out harmful chemicals, such as bisphenol A (BPA), which can get into the food chain and harm fish and mammals. Further, when the microplastics and trash gather on the surface of the water, they block sunlight, which kills off algae and plankton. This changes the entire ecosystem, as species like turtles and fish that depend on algae and plankton will have less food and thus decline in numbers, which in turn means less food for sharks, tuna, and whales. In the long term, this will lead to less seafood available to humans.

WORKING TOWARD A SOLUTION

Many people contribute to the Great Pacific Garbage Patch due to consumption and American's "throw-away" culture. Any nonbiodegradable items that get tossed in the trash or left at the beach add to the garbage that enters the ocean. Several organizations, government agencies, and researchers are working to document the garbage patches and educate the public.

Captain Charles Moore, who discovered the patch, continues to educate others through the Algalita Marine Research Foundation. He has built a research team and has gone on more than 11 sea expeditions in five oceans, often using drones to document and inform people around the world (Algalita, 2018).

The United States National Oceanic and Atmospheric Association (NOAA) Marine Debris Program was authorized by the U.S. Congress in the Marine Debris Act of 2006. Their mission is to investigate and prevent the adverse impacts of marine debris. Through their strategic plan, they conduct research and fund others to do research that should help educate people on this important issue.

The United Nations Environment Programme (UNEP) oversees the Regional Seas Programme, which seeks to develop collaborative management activities related to marine litter (UNEP, 2018). This U.S. organization notes that plastic garbage is a growing threat to coastal and marine environments because it degrades slowly and there is so very much waste. There are negative trends with marine debris getting worse, so action is urgently needed.

One very interesting recent development is seen in the work of Boyan Slat, a young man who gave a famous YouTube TED talk presenting his solution for ocean cleanup (YouTube, 2012). Now his website shows sophisticated designs for a floating platform that gathers, filters, and sorts the plastic debris (Ocean Cleanup, 2018). Slat has done an amazing job obtaining financial support through a crowdfunding campaign and is moving forward with prototypes. In fact, his Ocean Cleanup company now has over 30 employees (McCoy, 2016). This is clearly something to watch for in the future as we seek solutions to this plastic pollution.

ANALYZING THE PROBLEM

Developed, industrialized countries create a lot of waste, including plastics, which get into the Earth's oceans. In the future, it is anticipated that as more countries industrialize, they will likely also produce more plastics and waste, which will increase the problem of marine debris and garbage patches. This is "a problem that pervades the entire globe" (Allsopp et al., 2006); however, no nation will take responsibility or pledge money for a cleanup because now the garbage patch is far from every country's coast. Like so many complex problems studied by environmental geographers, ocean garbage is a global problem that requires local and regional action.

There is a long history of humans dumping their waste in the Earth's waterways and oceans. In fact, it was not until 1972 that any international action was taken on this topic. That year, the Convention on the Prevention of Marine Pollution by Dumping of Wastes and Other Matter, known as the London Convention, was the first international agreement to protect oceans from garbage dumping. Even now just 64 countries have signed on to this agreement, which controls or bans the deliberate disposal of wastes at sea (EPA, 2017). Only since 1988 has the International Convention for the Prevention of Pollution (MARPOL) sought to ban ships from dumping garbage and plastics at sea. Approximately 120 countries have ratified this treaty, but research does not show that it has had much impact. This is due to lack of enforcement and the fact that most marine debris (80%) is from land sources.

Cleaning up the Great Pacific Garbage Patch is indeed a momentous task. The microplastics are often the same size as small sea animals, so any attempts to scoop up the debris would also catch and kill millions of tiny ocean species. If this hurdle could be overcome, the sheer size of the patch means that it would take a very long time to completely eliminate it.

One thing that might halt the growth of the Great Pacific Garbage Patch is for humans to shift and use only biodegradable polymers rather than the synthetic plastics used so ubiquitously in daily life (Kubota et al., 2005). But our current consumer society depends on petroleum-based plastics, and retooling to use compostable plastics (made from corn, for example) is a huge barrier. In addition, there is confusion about what biodegradable plastic truly means. Such items cannot break down in a landfill because the trash there is in deep piles without oxygen. Instead, these bioplastics need to be recycled and taken to a composting facility. There is no consensus about how long it takes for biodegradable plastic to break down; thus, it may still end up in our oceans.

LOOKING AHEAD

So while we try to figure out a better system relying on biodegradable plastic, the only real solution to stop the patch from increasing is to stop our use of plastics completely. This brings us to the important term: zero-waste. This concept means waste reduction, reuse of items, recycling of goods, ecodesign, and producer responsibility—but these actions require massive education and infrastructure along with significant government action to manage and tax waste.

In a practical sense, zero-waste means that all industrial processes are retooled so that every material is reused or recycled. All consumer goods would be created, from the beginning, with consideration for their end use, so that "trash" becomes the new raw materials for manufacturing all consumer items (Kaufman, 2009; ZWIA, 2015;). If globally our manufacturing would adopt a zero-waste approach, over time we could reduce the amount of garbage that becomes marine debris.

One scientist with expertise in marine ecology warns: "[T]hese items that we call 'disposable' or 'single-use' are neither of those things . . . items that were constructed decades ago are still floating around there in the ocean today, and for decades to come" (Ramzymay, 2017).

Prevention of waste is the key to reducing the size of the garbage patches. Specifically, every purchase, use, and disposal of plastic items must be reduced, and people must implement a comprehensive recycling system that strives for zero-waste. If the microplastics entering the oceans could be reduced, this would give society more time to develop the technological solutions needed to actually clean up the existing garbage patches.

See also: Section 1: Solid Waste, Water Pollution; *Section 3:* Recycling

FURTHER READING

Algalita, 2018. "We Envision a World Free of Plastic Pollution." Algalita: Marine Research and Education. http://www.algalita.org/.

Allsopp, Michelle, Adam Walters, David Santillo, and Paul Johnston. 2006. "Plastic Debris in the World's Oceans." Greenpeace International. http://www.greenpeace.org/international/Global/international/planet-2/report/2007/8/plastic_ocean_report.pdf.

CCC. 2018. "The Problem with Marine Debris." California Coastal Commission. https://www.coastal.ca.gov/publiced/marinedebris.html.

EPA. 2017. "Ocean Dumping: International Treaties." United States Environmental Protection Agency. November 6. https://www.epa.gov/ocean-dumping/ocean-dumping-international-treaties.

Kaufman, Leslie. 2009. "Nudging Recycling from Less Waste to None." *New York Times.* October 19. http://www.nytimes.com/2009/10/20/science/earth/20trash.html.

Kostigen, Thomas M. 2008. "The World's Largest Dump: The Great Pacific Garbage Patch: In the Central North Pacific, Plastic Outweighs Surface Zooplankton 6 to 1." *Discover Magazine.* July 10. http://discovermagazine.com/2008/jul/10-the-worlds-largest-dump.

Kubota, M., K. Takayama, and D. Namimoto. 2005. "Pleading for the Use of Biodegradable Polymers in Favor of Marine Environments and to Avoid an Asbestos-Like Problem in the Future." *Applied Microbiology and Biotechnology*, 67(4): 469–476.

Lavers, Jennifer L. and Alexander L. Bond. 2017. "Exceptional and Rapid Accumulation of Anthropogenic Debris on One of the World's Most Remote and Pristine Islands." *Proceedings of the National Academy of Sciences of the U.S.* 224(23): 6052–6055.

McCoy, Terrence. 2016. "Can the 'Largest Cleanup in History' Save the Ocean?" *The Washington Post.* February 1. www.washingtonpost.com/news/inspired-life/wp/2016/02/01/can-the-largest-cleanup-in-history-save-the-ocean/.

Moore, Charles. 2003. "Trashed: Across the Pacific Ocean, Plastics, Plastics, Everywhere." Natural History. http://www.naturalhistorymag.com/htmlsite/master.html?http://www.naturalhistorymag.com/htmlsite/1103/1103_feature.html.

NGS. 2018."Great Pacific Garbage Patch: Pacific Trash Vortex." National Geographic Society. http://nationalgeographic.org/encyclopedia/great-pacific-garbage-patch/.

NOAA. 2018. "Discover the Issue." National Oceanic and Atmospheric Administration. https://marinedebris.noaa.gov/discover-issue.

NOAA GPGP. 2018. "Great Pacific Garbage Patch." National Oceanic and Atmospheric Administration. February 6. https://marinedebris.noaa.gov/info/patch.html.

Ocean Cleanup. 2018. "Who We Are." Ocean Cleanup Foundation. https://www.theoceancleanup.com/.

Plastic Paradise. 2013. Film. Angela Sun, Director.

Ramzymay, Austin. 2017. "A Remote Pacific Island Awash in Tons of Trash." *New York Times.* May 16. https://www.nytimes.com/2017/05/16/world/australia/henderson-island-plastic-debris-south-pacific.html.

Ryan, Peter G., Charles J. Moore, Jan A. van Franeker, and Coleen L. Moloney. 2009. "Monitoring the Abundance of Plastic Debris in the Marine Environment." *Philosophical Transactions B: Biological Science.* 364(1526): 1999–2012.

UNEP. 2005. "Marine Litter, an Analytical Overview." United Nations Environmental Programme. http://www.cep.unep.org/content/about-cep/amep/marine-litter-an-analytical-overview/view.

UNEP. 2014. "Emerging Issues Update: Plastic Debris in the Ocean." United Nations Environmental Programme. http://staging.unep.org/yearbook/2014/PDF/chapt8.pdf.

UNEP. 2018. "Marine Litter." United Nations Environmental Programme. http://web.unep.org/regionalseas/what-we-do/marine-litter.

UNESCO. 2018. "Marshall Island." United Nations Educational, Scientific and Cultural Organization. http://whc.unesco.org/en/list/487.

YouTube. 2012. "How the Oceans Can Clean Themselves: Boyan Slat at TEDxDelft." TEDx Talks. October 24. www.youtube.com/watch?v=ROW9F-c0kIQ.

ZWIA. 2015. "Standards and Policies." Zero Waste International Alliance. www.zwia.org.

Case Study 5: Nigeria's Oil Causes Human Rights Abuses and Environmental Degradation

The world is dependent on oil. It fuels cars, heats buildings, and creates plastic products. Obtaining this oil has economic, social, and environmental consequences. Since the 1960s, international oil companies have been active in the Niger Delta and caused significant environmental degradation and human rights violations. All local protests have been met with severe violence (HRW, 2018). Oil corporations, such as Shell, Mobil, and Chevron, have teamed up with crooked government officials who sought to make themselves wealthy (Shah, 2010). After decades of environmental degradation, political corruption, and ever-worsening conditions for local people, militant groups emerged in Nigeria. Violence between government forces and militants escalates to severe conflict in some places (Iaccino, 2015). This case study shows how humans' dependence on fossil fuels causes the destruction of both the environment and human rights.

POVERTY DESPITE OIL PROFITS

Nigeria is located at the inner corner of the Gulf of Guinea on the West Coast of Africa, and it has an area of 923,768 square kilometers (356,669 square miles), or more than twice the size of California. Throughout history the Niger Delta had excellent, fertile agricultural land so the area was densely populated by several different tribes, each of which had smaller groups. It is estimated that over 30 million people representing more than 40 ethnic groups live in the area (CIA, 2018).

Given that oil corporations have been drilling here for over 50 years, one would expect that some profits would have trickled in to benefit the Nigerian people, but this is not the case (*The Guardian*, 2015). Indeed, life expectancy is just 53 years, literacy rates are only 60%, and only 70% of Nigerians have access to safe drinking water (CIA, 2018). Many people in the Niger Delta live in abject poverty, despite the wealth of oil that lies beneath them. Clearly, the oil profits have not been used to improve the lives of local people.

Additional disparity is seen in the fact that the lifestyle supported by petroleum products is mostly not accessible to the majority of Nigerians. The oil pumped from their lands and seas by multinational corporations is refined in other countries. Oil is very expensive back in Nigeria, so most people cannot afford it or the lifestyle of cars and material items that accompanies it.

Why have oil corporations been active in Nigeria for decades, yet so little economic development has occurred here? A combination of corrupt government officials, placeless multinational corporations, lax environmental regulations, ethnic discrimination, and greed have led to relentless violence. Sadly, the continued human rights abuses have fueled terrorist extremists, thus increasing violence and making life even more difficult for Nigerian people (HRW, 2018). This is a case where sustainable development is desperately needed, but is mostly unattained.

EFFECTS OF OIL PRODUCTION IN THE NIGER DELTA

In the Niger Delta, several ethnic groups, including the Ijaw and the Ogoni, live in villages with no schools, medical clinics, or clean drinking water (Junger, 2007; Shah, 2010). There are very few paying jobs, so people scratch out a living by fishing in oil-polluted waters. At the same time, they are surrounded by oil wells that pump billions of dollars of oil each year. This setting provides a glimpse into the complicated forces that create environmental injustice.

The effect of oil production in this region is extreme. The Nigerian government allowed companies to pump oil-drilling waste directly into the river. This waste, dumped into the Niger River Delta, has killed most of the fish, which had been people's major source of food (Eboh and Onuah, 2011). The groundwater is

Case Study: Nigeria's Oil

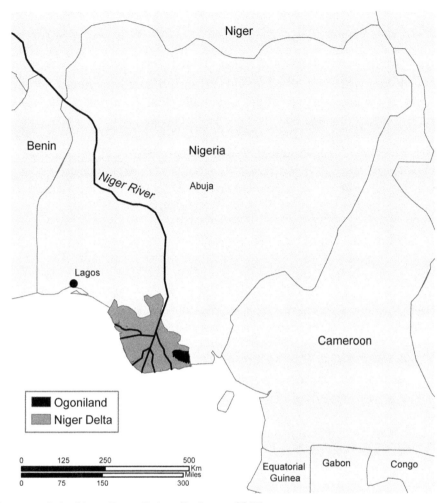

Nigeria and the Niger River Delta. (Bathgate, 2017)

contaminated, and even rainwater falls as acid rain, so drinking water sources are scarce. The air pollution is severe due to the oil drilling, refineries, and petrochemical complex in the area (UNEP, 2018). Drilling for oil creates natural gas as a by-product, and this is burned off in several flaring stations near villages and homes. This process involves collecting batches of gas and creating flares (sudden flames) that cause loud explosions and add more pollution to the air. This poor air quality is linked to cancer, asthma, and other diseases. In terms of air quality, three of the ten most polluted cities in the world in 2016 were in Nigeria (World Economic Forum, 2016).

FIGHTING INJUSTICE

The Ogoniland region has been particularly hard hit by the oil industry. In 1990, poet Ken Saro-Wiwa formed the Movement for the Survival of Ogoni People

(MOSOP). MOSOP demanded environmental justice and fair allocation of oil profits to local people (Junger, 2007). Because Saro-Wiwa was known internationally, media attention came to the issue. Shell funded the national Nigerian government, which helped control protesters in order to keep the oil profits flowing. Even with repression, the MOSOP organized numerous rallies, some with hundreds of thousands of protesters. One night, four government officials were killed. Although there was no evidence that he had anything to do with these killings, Saro-Wiwa and eight other MOSOP leaders were arrested and charged with murder. The Nigerian government executed the nine men in 1995, despite an international outcry demanding justice for Saro-Wiwa (Amnesty International, 2015). None of the goals of the MOSOP were met. Shell decided that the political unrest and negative press were not worth it, so they pulled out of Ogoniland, but remain in Nigeria.

In addition to the Ogoni, other local people have attempted to draw attention to the environmental degradation and the plight of ethnic groups. The Ijaw, for example, initiated the Ijaw Youth Movement (IYM) and began a direct action campaign they called Operation Climate Change in 1998, through which they pledged to struggle for freedom and ecological justice. This was met with two warships and over 10,000 Nigerian troops who occupied the Delta states, opened fire, and killed the rebel leader. The Nigerian military has often declared a state of emergency, imposed curfews, and banned meetings. Soldiers from a Chevron oil facility attacked and set fire to several villages, and when one leader attempted to negotiate, he was shot dead. Still the IYM actions continued and did disrupt Nigerian oil supplies in the early 2000s. More recently, Ijaws continue to resist with both nonviolent and violent actions against the oil companies, including taking foreign oil workers hostage (Junger, 2007).

Throughout the last 50 years, all protests have been met with military repression, often with the assistance of the oil companies (Shah, 2010). Even in the present day, Shell and the other oil companies point to increased violence in Nigeria as a reason they don't come and fix broken pipelines, which makes the pollution even worse. It is true that "widespread violence, kidnappings for ransom and clashes between militant groups and the army since the early 2000s" continues (Iaccino, 2015). Several armed militant groups are causing this violence at the same time the ecological degradation continues to ruin the daily lives of local people. Nigeria is currently so dangerous that both the UK and United States warn against travel to most regions of the country (Gov.UK, 2018; U.S. State Department, 2018).

ENVIRONMENTAL ASSESSMENT

In 2011, the United Nations Environment Programme (UNEP) completed a comprehensive scientific assessment of the effects of oil operations on Ogoniland. Data were gathered from medical records and public meetings. Soil and water contamination was investigated with over 4,000 samples analyzed across 69 sites. Key findings included the overall assessment that cleanup of the Ogoniland area will take 25 to 30 years (UNEP, 2011).

But there are immediate actions that must be taken in specific areas—for example, the community of Nisisioken Ogale has drinking water wells with over 900 times the acceptable level of the known carcinogen benzene. This well is near a Nigerian National Petroleum Company pipeline, which had reported an oil spill six years earlier. Yet a thick layer of refined oil still covers the groundwater that supplies the community's wells (Eboh and Onuah, 2011).

Other findings include oil up to 1 centimeter thick on mangrove roots, which had once provided fishing and natural pollution filtration. Agriculture is affected—fires often occur after an oil spill and this kills all vegetation, leaving a crust of ash and tar for decades. UNEP data also showed oil contamination of soil down to 5 meters (16 feet) deep. Surface water throughout Ogoniland is contaminated with oil—much of it visible as a thick black layer. Due to habitat destruction, fish populations have decreased significantly. Ogoni people are exposed to oil and its by-products daily through many routes, including food and water (Eboh and Onuah, 2011; UNEP, 2011).

Authors of the UNEP report stated that it would offer a scientific basis to guide the cleanup and help other countries avoid such catastrophes in the future. UNEP noted that a fund of at least $1 billion be established for the necessary environmental restoration projects, and this should be paid by both the oil industry and the Nigerian government. However, neither the oil companies nor the Nigerian government were willing to take responsibility for the cleanup.

Indeed, several years after the UNEP report was published, a group of organizations, including Friends of the Earth and Amnesty International, investigated the follow-up to the situation in Ogoniland. Sadly, few of the UNEP's recommendations had been implemented (Amnesty International, 2015).

A court case in 2015 culminated with Shell agreeing to pay $84 million (£55 million) in compensation. Still, critics wonder why this amount will be given by Shell, when the total cost of cleanup was estimated at over $1 billion (PRI, 2016). Given the historical ties between the oil industry and the Nigerian government, many are skeptical that the cleanup will occur. But in mid-2016, Nigerian President Muhammadu Buhari stated that the $1 billion cleanup would begin, with full remediation work starting about 18 months later (Adenle, 2016). All of this probably sounds like hollow promises for the local people in the Delta region, as they continue to suffer the consequences of environmental disaster and violence in one of the world's most oil-polluted regions.

WHO IS RESPONSIBLE?

Corrupt governments, with the support of oil corporations, have riled up ethnic divisions and allowed unregulated environmental destruction all because they want to profit from fossil fuel exploitation.

This cycle of violence is complex. Certainly, terrorism from within the local region is horribly violent. At the same time, according to Human Rights Watch, multinational oil companies allow or even encourage the violent acts committed by the Nigerian military and police (HRW, 2018).

In addition to the corrupt politicians and corporations, there are many other players in this conflict. Nigeria exports oil mainly to the United States, India, Brazil, Spain, France, and the Netherlands. Ultimately, these nations create the market for profit that is obtained from oil. But if these countries didn't buy their oil from Nigeria, other nations would. Overall, the global demand for petroleum products has led to cases of corruption that harm both people and the environment.

Many brave protesters have stood up to the abuse they see in the Niger Delta. There are others whose frustration has turned to inexcusable violence; numerous militant groups undertake piracy, kidnappings, and violence. Still others sit by idly (HRW, 2018).

LOOKING AHEAD

There are other options—sustainable development—as noted by the United Nations (UN). The world could build on the three dimensions of sustainability—social, economic, and environmental—while also helping alleviate poverty. One UN goal is to provide access to affordable, reliable, sustainable, and modern energy for all (UN, 2018). This would make the example of the Niger Delta oil conflict obsolete, as the value of oil would no longer trump the value of human life and ecological health.

In 1948, the United Nations passed the Universal Declaration of Human Rights as a united statement about the common fundamental human rights for all peoples that should be protected. This includes Article 3, which states: "Everyone has the right to life, liberty and security of person" (UN, 1948). When environmental degradation harms one's health and violence shakes one's security, their human rights have been lost. Sustainable development must be implemented to secure communities and protect individuals' rights in Nigeria.

In the meantime, as humans attempt to deal with the shattered lives and divisions within the region, the environment continues to be destroyed. The Niger Delta must be cleaned up and the livelihoods of millions of local Nigerians must be restored.

The lesson of the Niger Delta oil conflict is both tragic and ominous. It could happen anywhere that corruption and greed win over sustainable development. If the oil industry had worked with honest government officials to safeguard the environment and equitably distribute oil profits among local people, the region could have flourished. Instead, the people and the environment continue to suffer and this desperate situation fuels continued conflict.

See also: Section 1: Air Pollution, Energy, Water Pollution; *Section 2:* Global North and Global South; *Section 3:* Environmental Justice, Renewable Energy, Sustainable Development

FURTHER READING

Adenle, Ademola. 2016. "How Nigeria's Buhari Can Clean Up Ogoniland's Oil Spills." *Newsweek.* July 1. http://www.newsweek.com/how-nigerias-buhari-can-clean-ogonilands-oil-spills-476654.

Amnesty International. 2015. "Niger Delta: Shell's Manifestly False Claims about Oil Pollution Exposed, Again." Amnesty International. Business and Human Rights News. November 3. www.amnesty.org/en/latest/news/2015/11/shell-false-claims-about-oil-pollution-exposed/.

CIA. 2018. "Nigeria." U.S. Central Intelligence Agency. January 23. https://www.cia.gov/library/publications/resources/the-world-factbook/geos/ni.html.

Eboh, Camillus and Felix Onuah. 2011. "U.N. Slams Shell as Nigeria Needs Biggest Ever Oil Clean-Up." August 4. http://www.reuters.com/article/us-nigeria-ogoniland-idUSTRE7734MQ20110804.

Gov.UK. 2018. "Foreign Travel Advice: Nigeria." https://www.gov.uk/foreign-travel-advice/nigeria.

The Guardian. 2015. "The UNEP REPORT On Ogoniland." August 23. https://guardian.ng/opinion/the-unep-report-on-ogoniland/.

HRW. 2018. "Nigeria." Human Rights Watch. https://www.hrw.org/africa/nigeria.

Iaccino, Ludovica. 2015. "Nigeria's Oil War: Who Are the Niger Delta Militants?" *International Business Times*. http://www.ibtimes.co.uk/nigerias-oil-war-who-are-niger-delta-militants-1520580.

Junger, Sebastian. 2007. "Blood Oil." *Vanity Fair*. http://www.vanityfair.com/news/2007/02/junger200702.

PRI. 2016. "Shell Oil Faces a Lawsuit in the UK over Oil Spills in Nigeria." Public Radio International. March 27. https://www.pri.org/stories/2016-03-27/shell-oil-faces-lawsuit-uk-over-oil-spills-nigeria.

Shah, Anup. 2010. "Nigeria and Oil." Global Issues. June 10. http://www.globalissues.org/article/86/nigeria-and-oil.

UN. 1948. "The Universal Declaration of Human Rights (UDHR)." United Nations. http://www.un.org/en/universal-declaration-human-rights/.

UN. 2018. "Goal 7: Ensure Access to Affordable, Reliable, Sustainable and Modern Energy for All." United Nations Sustainable Development Goals. www.un.org/sustainabledevelopment/energy/.

UNEP. 2011. "Ogoniland Oil Assessment Reveals Extent of Environmental Contamination and Threats to Human Health." United Nations Environmental Programme. August 4. http://web.unep.org/disastersandconflicts/news/unep-ogoniland-oil-assessment-reveals-extent-environmental-contamination-and-threats-human.

UNEP. 2018. "Disasters & Conflicts: Nigeria." United Nations Environmental Programme. http://web.unep.org/disastersandconflicts/where-we-work/nigeria.

U.S. State Department. 2018. "Nigeria Travel Warning." https://travel.state.gov/content/travel/en/international-travel/International-Travel-Country-Information-Pages/Nigeria.html.

World Economic Forum. 2016. "Which Are the World's Most Polluted Cities?" https://www.weforum.org/agenda/2016/05/which-are-the-world-s-most-polluted-cities/.

Key Concepts

Agriculture

Agriculture is the intentional cultivation and management of plants and animals for human consumption. Initially people were nomadic and relied on a hunter-gatherer form of society that was dependent on seasonal movement to obtain enough food. Then about 11,000 years ago, people realized they could save seeds, plant them, and remain in that location to tend the plants and obtain the harvested crops. This allowed settlement in one geographic region, and some folks could focus on agriculture, whereas others could build things or develop other tools. It is thought that agriculture had a profound impact on human society because it allowed people to focus on specialized tasks, which developed into more complex economic relationships (Balter, 2005). Agriculture has had a profound impact on the Earth's environment, as the clearing of land for crops has decreased forests, changed watersheds, and influenced the animals and plants across the landscape.

In these ancient times, agricultural production continuously improved because sensible farmers saved and reused the best seeds each season. Farmers knew nature and the local geography very well because they had to depend on it for survival: if crops failed, they would starve. So farmers worked within the nature system and eventually began to cross-pollinate the best varieties of plants. For example, if they saved the seeds of two really big wheat shocks and plant them the next year, they could cross-pollinate with the help of the bees and the wheat crop might be even bigger and more hearty the next year. With better crops, human populations grew. Still, most people remained on the land, and the vast majority of people on Earth were rural and involved in food production.

This changed in the 1800s as industry pulled many people from the rural farmlands into the cities to work in factories. With fewer people on the farm and more food needed in the cities, the industrialization of agriculture began. After World War II, the chemical companies that made nerve gas from a family of chemicals known as organophosphates discovered that these chemicals could also kill bugs that destroy food crops (Biddle, 1984).

Food Miles

How far does food travel from farm to fork? Food miles is the distance between where a crop is grown and where it is consumed. In the United States, average food miles are the vast distance of 1,500 miles (Pirog et al., 2001).

Food miles have increased significantly as national and global distribution is possible due to relatively cheap transportation costs. The true cost of fossil fuels (e.g., air pollution, climate change) is not built into the price of food that people buy at the grocery store, so the negative environmental impacts are external and will have to be paid by society later.

Several important themes are related to food miles. First, advocates of local food say that buying directly from a local farmer is beneficial to the local economy and the environment. A dollar spent on local food is thought to multiply eight times within the community (from the consumer, to the farmer, on to other businesses that the farmer interacts with).

Second, food miles represent the concept of how "big" agriculture has become. But distance alone is not the most important factor, as only 11% of the greenhouse gases (GHG) associated with agriculture are from transport. Actually, if the main concern is lowering GHG emissions, the best approach is to shift away from meat and dairy (Webber and Matthews, 2008).

Third, life cycle assessment (LCA) provides a more complete view of the impact of food because it considers both the direct emissions (from crop production and transport) and indirect emissions (from manufacturing all agricultural inputs, such as fertilizer, pesticides, and fuel used in production) (Edwards-Jones et al., 2008).

Fourth, transportation method is important. For example, fuel use and CO_2 emissions are much lower for produce transported by sea than by air, even if the distances are double or triple.

Fifth, larger, more industrial methods of food production can cause more concentrated environmental impacts. For example large dairy or livestock operations produce a great deal of nitrogen and phosphorus waste from manure in one location.

Overall, "food miles" is an important concept because it encourages the geographic question: Where does your food come from? If it is from far away, a consumer does not know much about it. If food is local, the consumer may know the farmer and learn about how their food is grown.

Further Reading

Edwards-Jones, Gareth, Llorenç Milà i Canals, Natalia Hounsome, Monica Truninger, Georgia Koerber, Barry Hounsome,... D. Jones. 2008. "Testing the Assertion that 'Local Food Is Best': The Challenges of an Evidence-Based Approach." *Trends in Food Science and Technology* 19(5): 265–274.

Martinez, Steve, M. Hand, M. DaPra, S. Pollack, K. Ralston, T. Smith,... C. Newman. 2010. "Local Food Systems: Concepts, Impacts, and Issues." Economic Research Service. https://www.ers.usda.gov/webdocs/publications/46393/7054_err97_1_.pdf?v=42265.

Pirog, Rich S., T. Van Pelt, K. Enshayan, and E. Cook. 2001. "Food, Fuel, and Freeways: An Iowa Perspective on How Far Food Travels, Fuel Usage, and Greenhouse Gas Emissions." Leopold Center Pubs and Papers, Iowa State University. http://lib.dr.iastate.edu/leopold_pubspapers/3.

Weber, Christopher L. and Scott Matthews. 2008. "Food-Miles and the Relative Climate Impacts of Food Choices in the United States." *Environmental Science & Technology* 42(10): 3508–3513.

Agricultural chemicals can be toxic for humans, just as they are toxic to bugs and weeds. For example, Monsanto was a wartime government contractor that produced Agent Orange, a chemical used during the Vietnam War to defoliate the jungle so that U.S. soldiers could see the enemy (Monsanto, 2017). Later many veterans had health problems they blamed on Agent Orange. Monsanto is also one of the largest agricultural pesticide producers in the world, and this industry has grown significantly in the last 50 years.

Using chemicals, both pesticides and fossil fuel–based fertilizers, meant that more crops could be grown to feed the world's expanding populations. Fewer farmers were needed because big farm machinery and these chemical "inputs" could substitute for farm labor. Today most U.S. farmland is used for industrial agriculture, with intensive production dependent on chemicals and fossil fuel usage. These enormous farming operations and animal production facilities have left many farmers in debt because they always need to buy more off-farm supplies, equipment, and inputs in order to compete with larger and larger industrial farms.

According to the U.S. Department of Agriculture: "American agriculture and rural life underwent a tremendous transformation in the 20th century. Early 20th century agriculture was labor intensive, and it took place on many small, diversified farms in rural areas where more than half the U.S. population lived. Agricultural production in the 21st century, on the other hand, is concentrated on a small number of large, specialized farms in rural areas where less than a fourth of the U.S. population lives" (USDA, 2017).

The benefits of industrial agriculture are the high production of food per acre. But the costs are soil degradation, water pollution and overuse, public health concerns, and decline of rural communities. Some people question whether this type of agriculture will be sustainable into the future, as it is so dependent on fossil fuels, intensive use of soils, and chemical inputs that cause pollution downstream and downwind (UCS, 2017). The history and geography of industrial agriculture depict the changing relationship of people to their local environment and the pros and cons that go along with these developments.

See also: Section 1: Case Study 2; *Section 3:* Alternative Agriculture, Case Study 14

FURTHER READING

Balter, Michael. 2005. "The Seeds of Civilization." *Smithsonian Magazine.* May. www.smithsonianmag.com/history/the-seeds-of-civilization-78015429/#KA5j6aBiuoXzyFi6.99.

Biddle, Wayne. 1984. "Nerve Gases and Pesticides: Links Are Close." *New York Times.* March 30. http://www.nytimes.com/1984/03/30/world/nerve-gases-and-pesticides-links-are-close.html.

Monsanto. 2017. "Agent Orange: Background on Monsanto's Involvement." April 7. https://monsanto.com/company/media/statements/agent-orange-background/.

UCS. 2017. "Industrial Agriculture." Union of Concerned Scientists. http://www.ucsusa.org/our-work/food-agriculture/our-failing-food-system/industrial-agriculture#.WUr-4WjyuUk.

USDA. 2017. "Farming and Farm Income." U.S. Department of Agriculture. February 7. https://www.ers.usda.gov/data-products/ag-and-food-statistics-charting-the-essentials/farming-and-farm-income/.

Air Pollution

Clean air is composed of nitrogen (78.04%), oxygen (21%), and other trace gases (0.96%) (Smithsonian, 2018). Air pollution occurs when a substance is introduced into the air and has damaging effects, including harm to human health. Pollutants are found throughout the United States and the world. Globally approximately 3.3 million people die prematurely each year due to air pollution (Zielinski, 2015).

Air pollution can be either from the release of solid particles or gases into the air. Examples of particulate pollution include car emissions, chemicals from factories, dust, pollen, and mold spores. On the other hand, smog—composed of ground-level ozone—is a gas that is a major air pollutant in cities. Pollution is mostly related to burning fossil fuels, so industry, motor vehicles, and electricity generation are considered the primary sources of air pollution.

Air pollution regulations vary by region. For example, the European Union has health-based standards for 12 distinct air pollutants (EU, 2017). The Clean Air Act of 1970 authorized the U.S. Environmental Protection Agency (EPA) to set limits called national ambient air quality standards (NAAQS) for six of the most common air pollutants known as "criteria air pollutants": particulates, ozone (O_3), carbon

Geography and the high number of vehicles on the road in Los Angeles contribute to a form of air pollution known as smog. (PhotoDisc, Inc.)

monoxide (CO), lead (Pb), sulfur dioxide (SO_2), and nitrogen dioxide (NO_2) (EPA, 2017; NRDC, 2018). These criteria air pollutants can harm human health, create environmental degradation, and cause property damage. In turn, each EPA criteria air pollutant has its own sources, problems, and cumulative effects.

Particulates (or PM for particulate matter) are solid particles such as dust, dirt, soot, or smoke. Some are so tiny they can only be seen through an electron microscope, but others can be readily seen as smog or haze in our surroundings. Most PM is formed by complex reactions of chemicals such as sulfur dioxide and nitrogen oxides, which are pollutants emitted from power plants, industries, and automobiles. PM contains microscopic solids, so when inhaled they cause serious health problems "because they can get deep into your lungs, and some may even get into your bloodstream" (EPA, 2017). Known health problems from PM include eye irritation, lung and throat irritation, trouble breathing, lung cancer, and other health problems.

Ozone (O_3) is "good" in the upper atmosphere, as it contributes to the protective layer of gases that shield the Earth from most of the sun's strong ultraviolet rays. But ground-level ozone is a dangerous pollutant that is created by chemical reactions between oxides of nitrogen (NOx) and volatile organic compounds (VOC) in the presence of sunlight. The major sources of NOx and VOC are emissions from industries, electric power plants, vehicle exhaust, and use of various chemicals. Ground-level ozone is harmful to most plants, animals, and ecosystems. Breathing ozone can trigger human health problems. Overall, ground-level ozone and PM in the air can cause heart disease, strokes, respiratory illnesses, and lung cancer. Global data on this key type of pollution have been lacking because air quality is not monitored in many parts of the world.

Carbon monoxide (CO) is an odorless and colorless gas that is released when something is burned. Outdoors, the major sources of CO are vehicles that burn fossil fuels, whereas indoor sources are unvented gas stoves and leaky furnaces. Breathing air with high CO concentrations is harmful because this reduces the amount of oxygen in a person's bloodstream, causing confusion, dizziness, or even death (Medline Plus, 2017).

Lead (Pb) can enter the air through metal processing, waste incinerators, electricity power plants, and lead-acid battery manufacturing. Fortunately, the U.S. EPA banned leaded gasoline, so the levels of lead in the air decreased 98% between 1980 and 2014. Still, lead air pollution from factories is hazardous, as it leads to significant health problems in humans because it is taken into the bloodstream and deposited in the bones. Large exposures to lead can damage people's brains and kidneys, and even in small amounts it can affect children's IQ and ability to learn (NRDC, 2018). Lead enters ecosystems as it settles out of air pollution into the soil and waterways. In the environment, lead can cause neurological problems in animals and decreased growth and reproductive rates in plants and animals.

Sulfur dioxide (SO_2) emissions usually occur with other sulfur oxides (SOx), so by regulating SO_2, many related pollutants are also addressed. SOx can react with other compounds in the atmosphere to form even smaller particles. This PM pollution can penetrate deeply into the lungs and cause many health problems. Likewise SO_2 can affect the environment by damaging trees and plants and slowing their growth. The largest source of SO_2 air pollution is the burning of fossil fuels by power plants and other industrial facilities.

Nitrogen dioxide (NO_2) is used as an indicator for the larger group of nitrogen oxides (NOx) that are highly reactive gases. The main sources of NO_2 air pollution are from the burning of fossil fuels, meaning the emissions from electric power plants and motor vehicles. Breathing NO_2-polluted air can irritate the respiratory system and aggravate or cause asthma and other breathing problems. NO_2 and NOx react with other chemicals in the air to form both PM and ground-level ozone, which are harmful to humans and ecosystems. NOx and SO_2 interact with water vapor and other chemicals in the atmosphere to form acid rain, which harms lakes, forests, and even buildings (EPA, 2017).

Under the Clean Air Act, the EPA sets primary and secondary limits called national ambient air quality standards (NAAQS) for these six criteria air pollutants. Primary NAAQS are set to protect public health, and secondary NAAQS are to protect public welfare (for example, reduced visibility or damage to animals, crops, and buildings). The Clean Air Act states that the EPA must intermittently conduct a comprehensive review of scientific research on the health and welfare effects of exposure to the criteria air pollutants. These assessments are the basis for EPA rules for air pollution standards that note the allowable concentration of each pollutant in the ambient air. NAAQS are not at a zero-risk level, but instead at a level that sufficiently reduces risk to protect public health, with attention given to at-risk populations, such as the elderly or children (EPA, 2017).

See also: Section 1: Case Study 1, Case Study 5; *Section 3:* Renewable Energy, Sustainable Cities

FURTHER READING

EPA. 2017. "Criteria Air Pollutants." U.S. Environmental Protection Agency. https://www.epa.gov/criteria-air-pollutants.
EU. 2017. "Air Quality Standards." European Union. http://ec.europa.eu/environment/air/quality/standards.htm.
Medline Plus. 2017. "Air Pollution." U.S. National Library of Medicine. https://medlineplus.gov/airpollution.html.
NRDC. 2018. "Air Pollution: Everything You Need to Know." Natural Resources Defense Council. https://www.nrdc.org/stories/air-pollution-everything-you-need-know.
Smithsonian. 2018. "The Air We Breathe." Smithsonian Environmental Research Center. http://forces.si.edu/Atmosphere/02_01_02.html.
Zielinski, Sarah. 2015. "Air Pollution Kills More Than 3 Million People Every Year." *Smithsonian Magazine*. September 16. http://www.smithsonianmag.com/science-nature/air-pollution-kills-more-3-million-people-every-year-180956638/#gRYR7Lj7BjeXWUVV.99.

Animal Agriculture

Land-use change due to animal agriculture has had a significant impact on the Earth. Over time, with an increasing emphasis on meat consumption, livestock are consuming more and more of the Earth's land. Livestock animals far outnumber humans on Earth. In fact, there are 19 billion chickens, 1.4 billion cattle, and 1 billion pigs according to the United Nation's Food and Agriculture Organization (*The Economist*, 2011). Twenty-six percent of the Earth's ice-free land is used for livestock grazing, but that's not all—33% of all croplands are used for livestock feed production and 40% of all crops grown on Earth are fed to livestock. These livestock animals cause over 7% of greenhouse gas emissions due to their enteric fermentation (burping) and manure (FAO, 2012).

Keeping livestock can cause environmental problems. Those issues, however, vary depending on the location of the livestock—in the Global North or in the Global South. In the Global North, industrial production is striving for efficiency; thus, huge animal production facilities exist. In the Global South, a large percentage of the population gains their livelihoods from pastoralism and livestock cover large areas of land.

In the United States, which is in the Global North, there are approximately 450,000 confined feeding operations (CAFOs). The Environmental Protection Agency (EPA) defines CAFOs as agricultural enterprises where animals are kept and raised in confined situations. Specifically, a CAFO has more than 1,000 head of beef cattle, 700 dairy cows, 2,500 swine, 125,000 broiler chickens, or 82,000 laying hens confined on a site for over 45 days during one year (USDA-NRCS, 2018). At these facilities there is a concentration of live animals, animal feed, manure, urine, dead animals, and water demands on a small area of land. In this setting, feed is shipped in and animals are confined to a very small space, rather than grazing in pastures or fields. As an example, chickens are held in cages that are 645 square centimeters (99 square inches) with six birds (Pohle and Cheng, 2009). Like an industry, CAFOs are regulated by the Clean Water Act through amendments added in 2003 and 2008, as these production facilities became more prominent. In contrast to the EPA, the "USDA's goal is for AFO/CAFO owners and operators to take voluntary actions to minimize potential air and water pollutants from storage facilities, confinement areas, and land application areas." Thus, this agency seeks to provide technical and financial assistance to facility operators to help them adopt "practices that will protect our natural resources" (USDA-NRCS, 2018).

The situation in the Global South shows a sharp contrast to this industrial system of livestock production, yet it is economically extremely important. In the Global South, shepherds (also called pastoralists) are typically poor and depend on livestock for their food and economic well-being. Livestock is the basis of 40% of the economy (gross domestic product) in many regions of sub-Saharan Africa and South Asia (FAO, 2012). Demand for animal products continues to surge as incomes rise and human populations grow in the developing world. Livestock are a key driver of land-use change on the landscape. Problems of desertification and degradation of soils are intensified by livestock production. Over 13 billion hectares (32 billion acres) of forest are destroyed to develop pastures or cropland for livestock feed each year, which results in environmental degradation. Water quality and quantity, soil fertility, biodiversity, and climate change are all linked to these shifts in land-use and will continue to intensify as livestock density increases on available land (FAO, 2012).

See also: Section 1: Agriculture, Case Study 3, Genetically Modified Crops; *Section 3:* Assessments and Footprints, Sustainable Diet

FURTHER READING

The Economist. 2011. "Global Livestock Counts." July 27. http://www.economist.com/blogs/dailychart/2011/07/global-livestock-counts.
FAO. 2012. "Livestock and Landscapes." Food and Agriculture Organization of the United Nations. http://www.fao.org/docrep/018/ar591e/ar591e.pdf.
Pohle, K. and H. Cheng. 2009. "Furnished Cage System and Hen Well-Being: Comparative Effects of Furnished Cages and Battery Cages on Behavioral Exhibitions in White Leghorn Chickens." *Poultry Science* 88(8): 1559–1564.
USDA-NRCS. 2018. "Animal Feeding Operations." U.S. Department of Agriculture- Natural Resources Conservation Service. https://www.nrcs.usda.gov/wps/portal/nrcs/main/national/plantsanimals/livestock/afo/.

Biodiversity Loss

Biodiversity is the variety of life on Earth. It is plants and animals, from the very tiny to the complexity of entire ecosystems. In fact, there are three integrated levels: genetic diversity, species diversity, and ecosystem diversity that interact to create the complexity of life on Earth.

Seen here from an aerial view is the Great Barrier Reef, located off the northeastern coast of Australia. (Dreamstime.com)

Agrobiodiversity Loss

Grandma ate different apples! In the last 100 years, over 85% of the apple varieties in the United States have gone commercially extinct. That means hundreds of varieties of apples that were hearty and locally adapted were abandoned. The apple industry decided on a few varieties to promote worldwide, and the rest were forgotten. Now, over 40% of apples sold are Red Delicious, but the Harrison and Wolf River, to name a few older varieties, cannot be found in a grocery store (Nuwer, 2014).

Apples are not alone. Globally the Food and Agriculture Organization of the United Nations estimates that "[m]ore than 90% of crop varieties have disappeared from farmers' fields; half of the breeds of many domestic animals have been lost" (FAO, 2004).

Agrobiodiversity is the variety of species and the variability within species of plants and animals used for food and agriculture, including crops, livestock, fisheries, and forests.

As crop varieties vanish, genetic information is lost that could help in developing food crops that are more resistant to pests, diseases, drought, and salinity. Such knowledge will be necessary to address food shortages related to climate change. This is particularly important for food crops, as they are already at a disadvantage—people have selected and tailored them for profitable traits such as high yield and durability—but not adaptability. Those old varieties of apples that Grandma ate were hardier and could more easily adapt to changes in temperature and precipitation (Nuwer, 2014).

This narrowing of the food supply occurred because farmers started using just a few varieties in order to get fast-growing, same-sized fruits, vegetables, or animals, so they would fit on an assembly line for picking, transporting, and processing. But people may need to rethink this approach in the future.

The Svalbard Global Seed Vault in a remote area in the far north of Norway holds the seeds of over 825,000 crop plants. This facility is also called the Doomsday Vault, as it is supposed to provide safety against extinction of food crops. The far northern location was originally thought to protect the seeds in case of electricity failure. Very recently, however, temperatures are rising faster than expected due to climate change (Carrington, 2017). The Vault may be in danger of heating up and will need to rely on electricity in the future. Still it provides some defense against the loss of food diversity.

According to experts, agrodiversity "is the most effective, easiest, cheapest and most sustainable way to help agriculture adapt to change" (Nuwer, 2014).

Further Reading

Carrington, Damian. 2017. "Arctic Stronghold of World's Seeds Flooded after Permafrost Melts." *The Guardian*. May 19. https://www.theguardian.com/environment/2017/may/19/arctic-stronghold-of-worlds-seeds-flooded-after-permafrost-melts?CMP=fb_gu.

FAO. 2004. "Agrobiodiversity Fact Sheet." Food and Agriculture Organization of the United Nations. http://www.fao.org/docrep/007/y5609e/y5609e02.htm.

Nuwer, Rachel. 2014. "The World's Most Endangered Food." BBC. April 1. http://www.bbc.com/future/story/20140401-the-worlds-most-endangered-food.

Genetic diversity refers to variations within a species. Genes are the basic building blocks on Earth. Because individuals each have their own genetic makeup, and a species is composed of many individuals, a species may be quite diverse with distinct populations. For example, dogs are all one species, but due to genetic diversity, they may be a Saint Bernard or a poodle. To conserve genetic diversity, different populations of a species must be conserved. This helps preserve variations in size, color, and characteristics of plants and animals on Earth.

Species diversity refers to variations of species within a region. Some habitats, such as a healthy forest, have many. Others, such as a polluted river, have few. Another consideration is whether a species is endemic, or only existing naturally in one area. For example, in Australia, more than 80% of plant and animal species are endemic, meaning they only occur naturally there. This concept raises the notion of ecosystems and larger interrelationships among species. Indeed, species are grouped together into families according to shared characteristics. In Australia, whole families of animals and plants are endemic because species evolved into specific niches due to geographic isolation (Australian Museum, 2018).

Ecosystem diversity refers to variations in ecosystems in a geographic region. Ecosystems are communities of interacting organisms and their physical environment. Ecosystems range from very large—like the Great Barrier Reef—to quite small—like a pond in a local park. Because ecosystems are complex, self-regulating communities of animals and plants interacting with each other and with their environment, they are vulnerable to outside modification. One change, like the introduction of a pollutant, can upset the whole balance. Many ecosystems are at risk, and some have already been lost.

The best way to conserve biodiversity is to save habitats and ecosystems rather than trying to save a single species. Although it is easier to increase awareness about individual species—"Save the Whale!"—scientists emphasize that no single species can live in isolation. If one species is endangered, then most likely its entire habitat is probably under threat too.

"Researchers have estimated that there are between 3–30 million species on Earth, with a few studies predicting that there may be over 100 million species on Earth" (NWF, 2018). Since only 1.7 million species have been identified by humans, there is a lot of information about biodiversity that remains to be discovered on Earth! It is thought that tropical rainforests have the most diversity of any other place on the planet. Protecting these areas could have significant impact on future generations seeking the benefits of biodiversity. Biodiversity hotspots are known areas of high biodiversity.

There are numerous benefits from biodiversity, including those related to health, economics, and sustainability. Many medical discoveries that have helped cure diseases were based on genetic research of plants and animals. Diversity of pollinators, plants, and soils is the basis of the human food supply. Biodiversity makes Earth livable for people: ecosystems work to clean our water, absorb chemicals, and decrease the effects of flooding. Plants make the oxygen people breathe!

Biodiversity is now threatened like never before in human history. The main threats to biodiversity are human population growth, urban sprawl, pollution, climate change, and overexploitation of species. Human population growth decreases biodiversity as people destroy natural habitats for more cropland and grazing lands. Urban sprawl is the process whereby cities extend ever farther into the countryside, replacing natural habitats with roads and concrete. Human activities have drastically changed the landscape; for example, an estimated 80% of the original forest that covered the Earth 8,000 years ago has been cleared. Pollution affects animal and plant health, as well as human health, so environmental disasters can have negative effects on biodiversity. Climate change is changing native habitats and could affect up to half the world's species by 2050 (EU, 2004). Globally, overexploitation is decreasing biodiversity. In Europe, 80% of fish species are facing extinction or collapse due to overfishing (EU, 2004). And in North America, birds have declined 61% between 1966 and 2014 due to land-use changes and destruction of habitat (NWF, 2018).

See also: Section 1: Biodiversity Loss, Deforestation, Overfishing; *Section 3:* Case Study 13

FURTHER READING

Australian Museum. 2018. "What Is Biodiversity?" https://australianmuseum.net.au/what-is-biodiversity.
EU. 2004. "Biodiversity Loss: Facts and Figures." European Union. February 9. http://europa.eu/rapid/press-release_MEMO-04-27_en.htm.
IUCN. 2018. International Union for Conservation of Nature. www.iucn.org.
NWF. 2018. "What Is Biodiversity?" National Wildlife Federation. https://www.nwf.org/Wildlife/Wildlife-Conservation/Biodiversity.aspx.

Climate Change

Human activities, mainly burning fossil fuels and destroying forests, have caused higher temperatures on Earth—this is global warming. To be clear, the greenhouse effect itself is a good thing, as it is a "blanket" of gases around the Earth that help keep temperatures moderate and livable (NASA, 2018). The problem is that since the Industrial Revolution of the 1800s, humans have quickly added vast amounts of greenhouse gases, particularly carbon dioxide, methane, and nitrous oxide, to the Earth's atmosphere in a short amount of time (BBC, 2017). The Earth's ability to regulate the atmosphere cannot handle this rapid onslaught of "extra" gases. For example, vast tons of CO_2 that was held in the ground for millions of years has been emitted in a very short time through the burning of fossil fuels. The Earth now has more CO_2 in the atmosphere than at any time in the last 20,000 years (Cook, 2018). This is causing the Earth to warm, which in turn disrupts atmospheric processes and weather patterns—this is climate change.

Over the billions of years of Earth's history, the climate has changed many times. But today, these natural fluctuations in atmospheric processes are being superseded by the rapid changes caused by human-induced warming. Climate change has the capacity to upset the stability of the Earth's climate. "Scientific evidence for warming of the climate system is unequivocal," according to the international experts on the Intergovernmental Panel on Climate Change (NASA, 2018). Indeed, climate science research shows consensus that human-induced climate change is occurring and has dire consequences for the Earth's people and environment (NASA Consensus, 2018).

The National Climate Assessment (NCA) is a summary of the impacts of climate change on the United States. It is written by a team of more than 300 experts guided by a 60-member Federal Advisory Committee. It undergoes extensive review by scientists and the public, including members from the U.S. National Academy of Sciences. The NCA indicates that in addition to direct increases in temperature, the changing patterns may greatly affect people's lives. For example, long periods of high temperatures and especially high nighttime temperatures will affect people, livestock, and wildlife because there will be no break from the heat. Longer periods of high temperatures can also lead to bigger wildfires and longer fire seasons. Extreme heat is becoming more common, whereas extreme cold is less common. Human influence on the climate has already doubled the likelihood of extreme heat events, and these events are likely to increase in the future (NCA, 2014; USGCRP, 2018).

The World Health Organization outlines many impacts of climate change on human health. Extreme weather may cause increasing harm. Already over 1 million deaths are attributed to weather-related natural disasters in the last 20 years, mostly in developing countries. Higher temperatures cause an increase in pollen and other allergens in the air, which triggers respiratory problems and asthma. Rising sea levels will cause more flooding, and with half the world's population now living within 60 kilometers (37 miles) of coastline, more people will be displaced. Floods also increase the rise of infections from water-borne diseases. Climate change affects disease vectors; for example, with changing precipitation there will be an increase in mosquitoes that spread malaria. This is already a public health problem, as over 1 million people die of malaria each year, mostly in Africa. Finally, climate change will affect water and food supplies, which may intensify current shortages. Water scarcity is already a problem for nearly 40% of the world's population. With variable rainfall, water supplies may be unpredictable, and lack of clean water directly affects human health. In the lesser developed countries, an estimated 2.2 million die each year from diarrhea caused by dirty drinking water. Malnutrition may increase across the world, because climate change will bring higher temperatures and variable rainfall, which will reduce crop yields, particularly in tropical developing countries where food insecurity is already a problem (WHO, 2018).

National Geographic clearly outlines key facts about climate change that people should consider. Record temperatures keep getting broken: in 2016, the average global surface temperature set a record high, which had been set in 2015, which had just broken the record of 2014. This is comprehensive global data from thousands of weather stations, ocean buoys, ships, and even atmospheric satellites. CO_2 has increased rapidly and significantly since the 1870s, but especially since 1960, as the Earth's atmosphere can no longer keep up with the level of greenhouse gas emissions. Rising temperatures are causing sea ice to melt. In addition, melting of land ice (glaciers) has caused the seas to rise eight to nine inches since 1900. Ocean acidification due to climate change is killing coral reefs that help protect shorelines. Weather-related disasters have tripled since 1980; this includes intensified heat waves, hurricanes, heavy precipitation, and forest fires (caused by lightning strikes from increased severe storms). The Earth's plants and animals cannot adapt to such a rapidly changing environment, so extinctions will increase (National Geographic, 2017).

There are thousands, if not millions, of scientific websites, articles, and research findings that describe the causes and effects of climate change. To make informed policy and personal choices, every citizen of Earth has the opportunity to become familiar with the facts of climate change. With 16 of the hottest years on record occurring since 2000, the Earth is undergoing changes that may have permanent impacts on humans and the environment (NASA, 2018).

See also: Section 1: Case Study 3; *Section 2:* Case Study 6, Case Study 7

FURTHER READING

BBC. 2017. "What Is Climate Change?" BBC. November 14. http://www.bbc.com/news/science-environment-24021772.
Cook, John. 2018. "Explaining Climate Change Science & Rebutting Global Warming Misinformation." Skeptical Science. https://skepticalscience.com/.
NASA. 2018. "Climate Change: How Do We Know?" National Aeronautics and Space Administration. https://climate.nasa.gov/evidence/.
NASA Consensus. 2018. "Scientific Consensus: Earth's Climate Is Warming." https://climate.nasa.gov/scientific-consensus/.
NCA. 2014. "National Climate Assessment." U.S. Global Change Research Program. http://nca2014.globalchange.gov/.
National Geographic. 2017. "Seven Things to Know about Climate Change." National Geographic. http://www.nationalgeographic.com/environment/climate-change/.
USGCRP. 2018. "Fourth National Climate Assessment." U.S. Global Change Research Program. www.globalchange.gov.
WHO. 2018. "10 Facts on Climate Change and Health." World Health Organization. http://www.who.int/features/factfiles/climate_change/facts/en/.

Climate Change Policies

The Intergovernmental Panel on Climate Change (IPCC) was established by the United Nations in 1988 to allow scientists from across the globe to share data and collaborate in wide-ranging research on climate change. The mission of the IPCC is to assess "the scientific, technical and socioeconomic information relevant for the understanding of the risk of human-induced climate change" (IPCC-organization, 2017). Thousands of scientists have contributed to, reviewed, and edited the IPCC reports that were published in 1990, 1995, 2001, 2007, and 2014 (IPCC, 2018). Each of these huge reports documents climate changes that have already occurred and impacts that are currently underway and provides a basis for policy actions. The IPCC is so important and well respected internationally that they were awarded the Nobel Peace Prize in 2007.

The 1994 United Nations Framework Convention on Climate Change (UNFCCC) established the framework for global cooperation on climate change, and now nearly every country on Earth (195 at a recent count) has ratified this document. Every country is thus a "party" to the convention, so every international climate meeting is called a Convention of the Parties (COP), and they bring together many politicians and experts who try to plan cooperative actions to combat climate change. In fact, there have been over 60 COP meetings from the late 1990s to the present, but most people just hear about the major ones, such as Kyoto (1997), Montreal (2005), Copenhagen (2009), and Paris (2015), because these were meetings that attempted to negotiate significant agreements to reduce greenhouse gas emissions.

IPCC reports provide up-to-date scientific and technical data for the COP meetings. The interrelationship between these two groups is important and underscores the critical interaction between science and decisions negotiated by policy makers. "Preventing 'dangerous' human interference with the climate system is the ultimate aim of the UNFCCC," and this requires scientific understanding and informed international cooperation and agreement (UNFCCC, 2018).

The 2014 IPCC Assessment provided the following key points of synthesis: "1) Human influence on the climate system is clear; 2) The more we disrupt our climate, the more we risk severe, pervasive, and irreversible impacts; and 3) We have the means to limit climate change and build a more prosperous, sustainable future" (United Nations, 2014).

Taking this IPCC report to heart, the representatives from nearly 200 countries worked for months to negotiate the Paris Agreement at the 2015 COP meeting. This document set up limits to greenhouse gas emissions, mitigation approaches including reduction of fossil fuel use, and global agreements for funding these activities. The overall stated goal of the Paris Agreement is to limit global warming to 1.5° to 2°C (2.7° to 3.6°F). In November 2016, the Paris Agreement went into effect, and all nations are bound to these goals (IPCC, 2018).

However, in June 2017, U.S. President Trump formally announced that he had decided to pull the United States out of the Paris Agreement. There are no legal ramifications to his decision, as the agreement was not legally binding. The main fallout will be angering countries like Germany, Japan, and Canada that the United States traditionally counted as allies and the loss of jobs in renewable energy as China is taking over this employment sector. Even more notable, the United States is the world's second largest GHG emitter, so if the United States abandons the Paris Agreement, then meeting global climate mitigation goals and target reductions in GHG emissions will not be possible (UN Dispatch, 2017).

See also: Section 1: Climate Change; *Section 2:* Case Study 6, Case Study 7

FURTHER READING

IPCC-organization. 2017. "Organization: Intergovernmental Panel on Climate Change." United Nations Environment Programme (UNEP) and World Meteorological Organization (WMO). https://www.ipcc.ch/organization/organization.shtml.
IPCC. 2018. Intergovernmental Panel on Climate Change. http://www.ipcc.ch/.
NASA Consensus. 2018. "Scientific Consensus: Earth's Climate Is Warming." https://climate.nasa.gov/scientific-consensus/.
UN Dispatch. 2017. "Trump Is Pulling Out of the Paris Agreement? Here's What Will Happen." United Nations. May 31. https://www.undispatch.com/trump-pulling-paris-agreement-heres-will-happen/.
UNFCCC. 2018. "Meetings." United Nations Framework Convention on Climate Change http://unfccc.int/meetings/items/6240.php.
United Nations. 2014. "IPCC Launches Complete Synthesis Report." http://www.un.org/climatechange/blog/2015/03/ipcc-launches-complete-synthesis-report/.

Deforestation

"Forests play key roles in the water cycle, soil conservation, carbon sequestration, and habitat protection" (FAO, 2016). Forest ecosystems are important to the Earth's natural systems, and forest products are important economic drivers in some communities. Forest management is complex and involves timber, wildlife habitat, water quality, recreation, and employment. Forestry is the science of conserving and managing forests, tree plantations, and associated resources. The field of forestry has a wide range of specializations, including forest ecology, genetics, wildfire science, climate change, watershed management, tree pathogens, entomology, and geographic information systems (GIS) to model landscape changes.

Trends in forest land-use vary by region. In the temperate forests of North America and Europe the net forested area has increased in recent years, but there was a net forest loss of 7 million hectares (27 million square miles) per year in tropical countries during the decade from 2000 to 2010. During this time there was a net gain of 6 million hectares (23 million square miles) per year in agricultural land. This is caused by both international corporate interests and local population pressures. Large-scale commercial agriculture caused about 40% of tropical deforestation, whereas local subsistence agriculture was responsible for 33%, infrastructure for 10%, urban expansion for 10%, and mining for 7% (FAO, 2016).

Although most countries have forest regulations, the global forest products industry earns an estimated $150 billion per year and is geographically integrated. "A tree may be cut in Indonesia, manufactured into a table in China, sold to a retailer in New York, and bought by a business in Florida" (TNC, 2018). So individual countries' forest regulations have a limited impact globally if other countries have lax rules.

Illegal logging is cutting, transporting, processing, buying, or selling that is in violation of national or regional laws. This happens because of the high demand for specialty timber, paper, and related forest products. Clearing forests is such a lucrative business that nearly 1,000 people have been murdered for trying to protect their forests in the last decade—these forest defenders are local activists, government rangers, and indigenous people (WWF, 2017). Illegal clear-cutting of forest lands also occurs because high demand for palm oil (for use in personal products such as shampoos and food products such as processed foods) pushes people to destroy natural forests to plant palm oil plantations.

The Forest Stewardship Council (FSC) is an organization with the mission of promoting environmentally sound, socially beneficial, and economically prosperous management of the world's forests. They provide one of the most highly regarded certification systems to carry out this mission. FSC standards are set and then independent (third-party) certifiers do the auditing and verification. FSC has developed a set of 10 principles and 57 criteria that apply to FSC-certified forests around the world (FSC, 2018).

The key principles include the following: 1) benefits from the forest include products, environmental improvements, and social gains; 2) long-term land tenure and use rights must be documented; 3) indigenous peoples' rights must be upheld; 4) workers' rights and community relations must be maintained; 5) compliance with all laws and FSC principles must be met; 6) environmental impacts must be minimized so that biodiversity and ecological integrity are maintained; 7) management plans must be appropriate in scale and intensity to the specific forest; 8) forest activities must be monitored and assessed for environmental and social impacts; 9) precautionary approaches must be taken, especially in high-conservation-value forests; and 10) tree plantations must be planned and managed to reduce the pressures on conservation of natural forests.

For people who want to get involved with local forestry organizations, the Arbor Day Foundation and American Forests both provide opportunities. The Arbor Day Foundation is a nonprofit conservation and education organization with the mission: "We inspire people to plant, nurture, and celebrate trees." It was founded in 1972, 100 years after the observance of the first Arbor Day to celebrate trees. Now the group has over a million members with a focus on three areas: replanting forests to help damaged areas, urban forest management programs to help communities manage trees, and programs to help kids learn about nature and trees (Arbor Day Foundation, 2018).

American Forests is a conservation organization that is over 140 years old. Since 1990 alone, this group has planted more than 50 million trees to help restore forests in all 50 states and nearly 50 countries. Their vision is "a world in which forests are thriving and valued for their significant environmental, societal and economic benefits" (American Forests, 2018). Thus they are active in protecting and restoring threatened forest ecosystems, promoting and expanding urban forests, and increasing understanding of the importance of forests.

The Nature Conservancy (TNC), founded in 1951, also protects forests, especially through scientific staff and partners from government and local communities. TNC's mission is "to conserve the lands and waters on which all life depends." They employ more than 600 scientists who study and have a positive impact on conservation in 69 countries (TNC, 2018). TNC uses a nonconfrontational, collaborative approach that seeks mutual benefit to nature and the local communities.

At the global level, the World Wildlife Fund for Nature (called the World Wide Fund in the United States and Canada) has an extensive program to promote biodiversity and reduce deforestation. WWF's mission is "to stop the degradation of the planet's natural environment and to build a future in which humans live in harmony with nature, by: conserving the world's biological diversity; ensuring that the use of renewable natural resources is sustainable; and promoting the reduction of pollution and wasteful consumption." They have invested $10 billion in more than 13,000 conservation projects in more than 150 countries since they were established in 1961 (WWF, 2017).

See also: Section 1: Biodiversity Loss; *Section 2:* Protected Areas and National Parks; *Section 3: Case Study 15*

FURTHER READING

American Forests. 2018. "Our Story." http://www.americanforests.org/discover-american-forests/our-story/.

Arbor Day Foundation. 2018. "About the Arbor Day Foundation." www.arborday.org.
AUFSC. 2018. "What Is Forestry?" Association of University Forestry Schools of Canada. http://www.aefuc-aufsc.ca/what-is-forestry/.
FAO. 2016. "State of the World's Forests 2016." Food and Agriculture Organization. http://www.fao.org/publications/sofo/2016/en/.
FSC. 2018. "News." Forest Stewardship Council. www.us.fsc.org.
TNC. 2018. "Forests: for People, Water and Wildlife." The Nature Conservancy. https://www.nature.org/ourinitiatives/urgentissues/land-conservation/forests/.
WWF. 2017. "Deforestation." World Wildlife Fund. http://wwf.panda.org/about_our_earth/deforestation.

Endangered Species

Extinction is a natural evolutionary process through which some species may gradually die out due to natural environmental changes over thousands of years. But today, species extinction is occurring at a rapid rate due to non-natural environmental changes caused by humans: spread of non-native species and diseases, habitat loss due to land-use change, overharvesting of some species, climate change, and pollution. Experts believe species are going extinct at a rate that is 1,000 to 10,000 times higher than the natural rate would be (EU, 2004). Some experts say the Earth is now experiencing a mass extinction so severe they are calling it a "biological annihilation" (Gunaratna, 2017).

The International Union for Conservation of Nature (IUCN) was founded in 1948 with the mission: "Influence, encourage and assist societies throughout the world to conserve the integrity and diversity of nature and to ensure that any use of natural resources is equitable and ecologically sustainable" (IUCN, 2018). With over 1,600 organizational members (governments and environmental groups) and 16,000 experts, IUCN is well known for their Red List of Endangered Species, which is a comprehensive global evaluation of the conservation status of animal and plant species (IUCN Red List, 2018). The Red List is based on quantitative data of detailed measurable criteria, including population size, number of mature individuals, generation length, fragmentation, extent, and location. The IUCN also has classification schemes to describe the habitat within which species occur, which helps plan environmental conservation strategies. These topics include threats, stresses, actions needed, research needed, trade, ecosystem services, and livelihoods.

Globally, about one-third of all known species are threatened with extinction (IUCN, 2018). Specifically, this works out as 29% of all amphibians, 21% of all mammals, and 12% of all birds. This rate continues to accelerate, with vertebrate species (mammals, birds, reptiles, amphibians, and fish) declining by 52% over just the last 40 years (WWF, 2018). Activities that humans take for granted, such as farming, can have a huge impact. For example, as large-scale agriculture has replaced smaller diversified farms, wildlife species have been negatively affected. A study of common farmland birds monitored in 18 European countries show a decline in numbers by 71% between 1980 and 2002 (EU, 2004). Other impacts such as tropical deforestation push species to extinction more rapidly, as these ecosystems are the site of high biodiversity.

If these extinctions continue, humans could be facing widespread environmental problems with extreme consequences to human health and life on Earth. Many environmental groups are concerned about species extinction. For example, the World Wildlife Fund (WWF) is active in over 100 countries. Their biggest goal is to save wildlife. They seek to achieve this by focusing on populations of the most ecologically, economically, and culturally important species in the wild. They call these the WWF flagship species, and they represent "the iconic animals that serve as ambassadors for conservation" (WWF, 2018). They also work to help wildlife threatened by human actions such as overfishing.

Globally, the IUCN notes that over 19,817 species are threatened with extinction. Among those species, 3,947 are classified as "critically endangered," 5,766 as "endangered," and more than 10,000 species are listed as "vulnerable" (IUCN, 2018). The 1973 Convention on International Trade in Endangered Species of Wild Fauna and Flora (CITES) is an international treaty that seeks to ensure that international trade in wild animal and plant species does not threaten any species' survival. Other than this international agreement, laws and regulations regarding endangered species are implemented at the national level, and these vary in funding and implementation.

In the United States, the Endangered Species Act (ESA) of 1973 is the national law that seeks to protect endangered and threatened species through conservation efforts that seek to prevent the extinction of native plants and animals. The act defines an endangered species as "an animal or plant in danger of extinction within the foreseeable future throughout all or a significant portion of its range." According to the act, threatened species are those that are "likely to become an endangered species within the foreseeable future throughout all or a significant portion of its range." Rather than establishing a law that protects ecosystems themselves, the ESA addresses individual species, which is not necessarily the best ecological approach. Still, once a species is listed as endangered or threatened, it receives specific protection from the U.S. government. For example, the animal cannot be traded or sold—this is known as the "Take" clause from Section 3(18) of the Endangered Species Act: "The term 'take' means to harass, harm, pursue, hunt, shoot, wound, kill, trap, capture, or collect, or to attempt to engage in any such conduct" (USFWS, 2015). There are over 1,300 endangered or threatened species in the United States today.

See also: Section 1: Biodiversity Loss; *Section 2:* Case Study 8

FURTHER READING

EU, 2004. "Biodiversity Loss: Facts and Figures." European Union. February 9. http://europa.eu/rapid/press-release_MEMO-04-27_en.htm.

Gunaratna, Shanika. 2017. "Earth Faces 'Biological Annihilation' in Sixth Mass Extinction, Scientists Warn." CBS News. July 10. http://www.cbsnews.com/news/sixth-mass-extinction-biological-annihilation/.

IUCN. 2018. International Union for Conservation of Nature (IUCN). www.iucn.org.

IUCN Red List. 2018. "Overview of the IUCN Red List." International Union for Conservation of Nature. http://www.iucnredlist.org/about/introduction.

USFWS. 2015. "Glossary." U.S. Fish and Wildlife Service. April 14. https://www.fws.gov/midwest/endangered/glossary/index.html.
WWF. 2018. "Wildlife Conservation." World Wildlife Fund. https://www.worldwildlife.org/initiatives/wildlife-conservation.

Energy

As defined by the U.S. Energy Information Administration, there are several main types of energy sources used by people: petroleum, natural gas, coal, nuclear, and renewable energy. These are called primary energy sources. Electricity is a secondary energy source that is generated from one of these primary sources (EIA, 2017).

Fossil fuels account for 81% of world energy production. Renewable sources are expanding at rapid rates each year (wind +11%, solar thermal +7%, solar photovoltaic [PV] +35%, and geothermal +8%) but still only comprise 1% of total world energy production. In terms of economic sectors, industry has the largest consumption (37%), followed the transportation sector (28%), and residential (23%). Energy consumption is somewhat concentrated, with nearly half of world consumption by just the top four countries: China, United States India, and Russia (IEA, 2018). Americans are only 4.3% of the world's population, but consume nearly 20% of its oil (Scientific American, 2017).

Patterns of energy consumption have changed over time, as new sources of energy were discovered and prices shifted. In the United States, wood was the dominant form of energy through the mid-to-late 1800s, with some use of water mills or early industrial growth. Then coal became the primary source of energy in the late 1800s. Petroleum and then natural gas came to dominate in the mid-1900s. Since then, coal again increased, as it is used in electric power generation, and nuclear electric power also makes a contribution (EIA, 2011). Patterns of U.S. energy sources are fairly stable, with the nation's primary energy production composed of natural gas 33%, petroleum (crude oil and natural gas plant liquids) 28%, coal 17%, renewable energy 12%, and nuclear electric power 10% (EIA, 2017).

People don't realize that right under our homes and streets is a huge network of pipelines. The United States has more fuel pipelines than any country in the world: 2.6 million miles. These pipelines deliver huge volumes of natural gas and liquid oil products to American industrial and residential consumers. The pipelines truly provide a transportation system that could not be accomplished through just trucks and trains. Pipeline systems have three main parts: gathering pipelines collect products from the source (well or offshore rig) and move fuel to storage and processing. Transmission pipelines move huge volumes of fuel great distances to power plants or industries. Finally, distribution lines are the mains and smaller service lines that go through cities and into homes. In addition to the pipelines, there are pump stations and distribution facilities to manage the product's movement. The Pipeline and Hazardous Materials Safety Administration is part of the U.S. Department of Transportation, and it oversees pipeline safety. Layers of data, including pipelines, power plants, and refineries, can be seen on the U.S. Department of Energy interactive maps: https://www.eia.gov/state/maps.php.

Corn Ethanol from Mining the Soil

When fueling up in the United States, drivers commonly see "E10" or "E15" on the pump, which indicates the percentage of ethanol that is blended into the gasoline. Ethanol is considered a renewable fuel source because it is made from plants—in the United States it is primarily from corn. The process of creating ethanol actually requires a lot of resources: inputs for growing the corn crops (water, fuels, fertilizers, etc.), transportation of this biomass to a facility that uses energy to convert the plant biomass into ethanol, and transportation to a fuel terminal where ethanol is blended with gasoline. Ethanol contains about 30% less energy per gallon than gasoline (U.S. Department of Energy, 2017).

The use of ethanol has expanded greatly over the last 15 years, so that now approximately 143,000 square kilometers (55,000 square miles) of farmland in the U.S. Midwest are dedicated to producing corn for this fuel. The U.S. government and many states have laws that require gas manufacturers to use huge amounts of ethanol and blend it into the nation's fuel supply (Charles, 2016).

Production of crops for fuel takes away farmland from food production. This is both an ethical concern, with fuel crops taking farmland away from food crops that people need, and an economic concern because fuel production drives up the price of farmland and agricultural production. There is even an international dimension: ethanol production causes higher demand and increased prices for corn. So farmers grow more corn and less of other crops, such as soybeans, which drives up prices of these other crops. This prompts farmers and corporations in other countries to tear down rainforests in order to plant corn and soybeans—destruction of vast swaths of Brazil's Amazon rainforest are a recent example of this (Grunwald, 2008).

Corn-based biofuels raise several environmental concerns, including the overuse of farmland and degradation of soils. Fossil fuel–based synthetic pesticides and fertilizers are needed to produce the biofuel crops, and these are finite resources. Water use is a concern, as corn demands high levels of irrigation, often from unsustainable sources. Finally, the reason that corn is inexpensive and available to the oil companies for ethanol is partially due to the U.S. government subsidizing corn production, which is not sustainable. There is disagreement as to whether ethanol actually reduces greenhouse gas emissions—it burns cleaner, but a lot of fossil fuel is burned to create it.

People say that corn ethanol is a renewable fuel, but is it really? If soils and farmland are abused, they cannot rejuvenate in the short term. Essentially biofuels are like drilling for oil, except in this case soils are mined, water resources depleted, and land removed from food production.

Further Reading

Charles, Dan. 2016. "The Shocking Truth about America's Ethanol Law: It Doesn't Matter (For Now)." *The Salt*, National Public Radio. February 10. http://www.npr.org/sections/thesalt/2016/02/10/466010209/the-shocking-truth-about-americas-ethanol-law-it-doesnt-matter-for-now.

Grunwald, Michael. 2008. "The Clean Energy Scam." *Time*. March 27. http://content.time.com/time/subscriber/article/0,33009,1725975,00.html.

U.S. Department of Energy. 2017. "Ethanol Fuel Basics." Alternative Fuels Data Center. https://www.afdc.energy.gov/fuels/ethanol_fuel_basics.html.

Imports and exports of fossil fuels also show interesting patterns. The United States imported about 10.1 million barrels per day (MMb/d) of petroleum in 2016. The vast majority of this (78%) was crude oil, but other products included natural gas plant liquids, liquefied refinery gases, and refined petroleum products such as gasoline. These imports came from 70 countries, but the top 5 (comprising 69% of the total) were Canada, Saudi Arabia, Venezuela, Mexico, and Colombia. On the other hand, the United States exported about 5.2 MMb/d of petroleum to 101 countries, with the top 5 being Mexico, Canada, the Netherlands, Brazil, and Japan. Most of the exports were petroleum products. The United States was thus a net importer in 2016 (meaning the monetary value of imports was greater than the value of exports) of petroleum, at 4.9 MMb/d (EIA, 2018). Looking forward, the new president seeks to greatly increase the level of U.S. energy exports. Given the importance of fossil fuel energy on the economic well-being of most nations, diplomatic relationships are often developed to support energy imports and exports.

There is geographic variation in energy sources and usage between the Global North and Global South. In developing countries, approximately 1.3 billion people do not have access to electricity, and over 3 billion people cook and heat their homes using open fires and simple stoves burning coal or biomass (wood, animal manure, and crop waste). This causes poor air quality within the home and in the community, resulting in illness and over 4 million premature deaths each year. There are new initiatives to increase renewable energy technologies in the Global South. For example, low-cost solar cookers have been developed that are based on a solar concentrator that could be developed for other applications (Oxford, 2018). But many developing nations still rely on fossil fuels, and demand for gasoline cars and electric appliances is increasing as nations become wealthier.

The world is highly dependent on fossil fuels, and each year there is an increase in fossil fuel consumption. This means that greenhouse gas emissions are also increasing each year. Despite projections that fossil fuel dependence may decrease from its current 82% to 75% by 2050, significant amounts of greenhouse gases will continue to be emitted, pushing up global temperatures well into the future (IEA, 2018).

See also: Section 1: Case Study 5; *Section 3:* Green Buildings, Renewable Energy

FURTHER READING

EIA. 2011. "History of Energy Consumption in the United States, 1775–2009." Energy Information Administration. February 9. https://www.eia.gov/todayinenergy/detail.php?id=10.
EIA. 2018. "Energy Explained." Energy Information Administration. May 19. https://www.eia.gov/energyexplained/?page=us_energy_home.
IEA FAQs. 2018. "Energy and Climate Change." International Energy Agency. https://www.iea.org/publications/freepublications/publication/WEO2015SpecialReportonEnergyandClimateChange.pdf
Oxford. 2018. "Energy in Developing Countries." ONE: Oxford Networks for the Environment. http://www.energy.ox.ac.uk/wordpress/energy-in-developing-countries/.
PHMSA 2013. "General Pipeline FAQs." Pipeline and Hazardous Materials Safety Administration. https://www.phmsa.dot.gov/faqs/general-faqs.
Scientific American. 2018. "Use It and Lose It: The Outsize Effect of U.S. Consumption on the Environment." https://www.scientificamerican.com/article/american-consumption-habits/.

E-waste

E-waste is the popular term for electronic items that are near the end of their use by a consumer. This includes things like phones, computers, tablets, televisions, light-emitting diodes (LEDs), music players, and copiers.

The average American family owns 28 electronic products, such as personal computers, mobile phones, televisions, and tablets. Americans generated 3.05 million kilograms (3.36 million tons) of obsolete electronic products in 2014 (EPA, 2017).

Why do people buy new electronics so often? There are several reasons: convenience, trendiness, and planned obsolescence. Older items run slower, and technology changes quickly, so new gadgets are perceived as better. Companies themselves promote this constant purchasing by upgrading operating systems so that older devices don't work. Planned obsolescence is when companies build their products to break within a specific number of years.

The huge and ever-growing demand for electronics has several negative effects on the environment. First, it means that more mining of metals and minerals must occur. Second, ever-growing piles of electronics that are discarded create a huge amount of e-waste, some of which is toxic in landfills.

The United Nations set up a partnership of organizations, governments, and science organizations called the Solving the E-Waste Problem (STEP) Initiative, which predicts e-waste generation will continue to increase by 33% in the next five years. The group says that EEE, or electrical and electronic equipment—anything with a battery or a cord—will continue to grow rapidly, thus emphasizing the need to address the issue of e-waste immediately. Their most recent data indicate that 56.5 million metric tons (mmt) (125 million pounds) of EEE comes on the market each year, which equals 8.2 kilograms (18 pounds) for each of the 7+ billion people

An E-waste drop-off location collects electronic waste, but there are few programs that actually recycle the components for reuse. Instead, e-waste often ends up in a landfill or is exported to other countries. (imging/Shutterstock)

on Earth. And each year 41.8 mmt (92 million pounds) of e-waste is generated, which equals 5.9 kg (13 pounds) per person (STEP, 2018). STEP has created an online map of the world's e-waste, which shows global and country-specific data. For example, each person in the United States produces 22 kg (49 pounds) of e-waste each year (STEP, 2018). Indeed, U.S. consumers spend about $200 billion on electronics each year and less than 30% of these are recycled (Ahmed, 2016). But for every million cell phones we recycle, 35,000 pounds of copper, 772 pounds of silver, 75 pounds of gold, and 33 pounds of palladium can be recovered (EPA Recycling, 2017).

Even before the recycling stage, consumers can think about reusing and refurbishing items. First, consumers can reduce their generation of e-waste by making smart purchases and keeping good maintenance of their electronics. Second, consumers can reuse electronics that are still functioning by donating or selling it to someone who would still use it.

Finally, electronics can be recycled by specific e-waste recycling programs. In the European Union, e-waste recycling has been available for over a decade, but in other countries, it has been slower to develop. In the United States, for example, 25 states have electronics recycling laws, which began in 2003 with California, and now also includes Maine, Washington, Connecticut, Minnesota, Oregon, Texas, North Carolina, New Jersey, Oklahoma, Virginia, West Virginia, Missouri, Hawaii, Rhode Island, Illinois, Michigan, Indiana, Wisconsin, Vermont, South Carolina, New York, Pennsylvania, and Utah (ETC, 2018).

National e-waste policies are found in the U.S. National Strategy for Electronics Stewardship, which laid out four major goals under President Barack Obama: 1) the federal government will lead by example, by reusing electronics and tracking items through their life cycle; 2) federal incentives should focus on greener electronics to enhance science, research, and technology; 3) to increase the safe handling of used electronics, the government will establish voluntary partnerships with the electronics industry; and 4) the United States must reduce the harm of its e-waste exports to developing countries by helping them better manage used electronics.

This final point is extremely important, as e-waste is a significant health hazard and environmental problem in many developing nations. With few economic options, people work at the toxic e-waste dump sites to retrieve small bits of metal and minerals that earn them a small wage.

An example of one site is the Agbogbloshie e-waste processing area in Ghana, West Africa, which imports 215,000 tons (474 million pounds) of used consumer electronics from overseas each year. This facility is planned to double in size by 2020. The environmental degradation is already severe, with high levels of lead and other toxins in the soil and air. For example, lead levels in the soil were found to be over 18,000 parts per million (ppm), which is significantly higher than the 400 ppm allowable soil standard noted by the U.S. EPA. Workers have high levels of aluminum, copper, iron, and lead in their bloodstream, and all nearby neighborhoods are at risk (Green Cross, 2016).

Consumers in the wealthier nations are part of the problem because they buy electronics and throw them away without realizing the environmental and human health impacts. This is a geographic example of variation between the Global North creating the e-waste and the Global South dealing with its consequences.

See also: Section 1: Hazardous Waste, Solid Waste; *Section 3:* Sustainable Cities, Sustainable Development

FURTHER READING

Ahmed, Syed Faraz. 2016. "The Global Cost of Electronic Waste." *The Atlantic.* September 29. https://www.theatlantic.com/technology/archive/2016/09/the-global-cost-of-electronic-waste/502019/.
EPA. 2017. "Basic Information about Electronics Stewardship." U.S. Environmental Protection Agency. https://www.epa.gov/smm-electronics/basic-information-about-electronics-stewardship.
EPA Recycling. 2017. "Electronics Donation and Recycling." U.S. Environmental Protection Agency. https://www.epa.gov/recycle/electronics-donation-and-recycling.
ETC. 2018. "State Legislation." Electronics Takeback Coalition. http://www.electronicstakeback.com/promote-good-laws/state-legislation/.
Green Cross. 2016. "The World's Worst Pollution Problems 2016: The Toxics beneath Our Feet." http://www.worstpolluted.org/.
ITFES. 2011. "National Strategy for Electronics Stewardship." Interagency Task Force on Electronics Stewardship. July 20. https://www.epa.gov/sites/production/files/2015-09/documents/national_strategy_for_electronic_stewardship_0.pdf.
Parameswaran, Siva. 2013. "Toxic Waste 'Major Global Threat.'" BBC. November 20. http://www.bbc.com/news/science-environment-24994209.
STEP. 2018. "World Overview of E-Waste Related Information." United Nations University. http://www.step-initiative.org/overview-world.html.

Fracking

Hydraulic fracturing, or fracking, is a method used to extract gas and oil from shale rock deep underground. Companies buy the mineral rights within a certain area and then set up rigs that enable drilling an average of 2.4 kilometers (1.5 miles) down into the Earth. Some drilling is horizontal to the rock layer. Then a high-pressure mixture of chemicals, sand, and water is injected into the rock, causing it to crack (fracture) and release the flow of gas into the well (EPA, 2017).

Fracking has "revolutionized" the energy industry in the United States in the last decade, but it has prompted environmental concerns. These concerns revolve around water, chemical contamination, and earthquakes (BBC, 2015).

Fracking uses a lot of water—up to 7 million gallons for each well drilled. Depending on the location, the water must be trucked to the site, which has negative environmental costs of fossil fuel use and emissions. In water-deficient areas, this intensifies existing water scarcity.

Fracking involves injecting chemicals into the ground that act to fracture the rock. Community members and environmentalist are concerned that these chemicals could be carcinogenic and could contaminate the groundwater. Geologists state that contamination is unlikely due to the separation of deep shale deposits from freshwater aquifers; however, spills and blowouts at fracking sites have led to pollution. But the EPA is investigating shallow fracking wells, which may be more likely to seep chemicals into the groundwater. Fracking companies have not always been forthcoming about what chemicals are used in their wells, but some states are passing laws that require full public disclosure (McGraw, 2016).

The issue of earthquakes or earth tremors is a very recent concern. There has been a notable rise in earthquakes in the central United States, particularly in Oklahoma, where fracking has increased in recent years. The U.S. Geological Survey states that fracking wells are not causing most of these tremors; rather, other wastewater disposal wells are to blame (USGS, 2018).

Other complex issues are related to fracking. It has drastic effects on the landscape, as the drilling pads and wells are large and obvious. The traffic near a well is significantly increased as over 200 semi-truck loads must be transported to the site. In a quiet, rural region this is a significant change. Furthermore, industry representatives claim that fracking emits half the CO_2 of coal and thus is beneficial in the fight to slow climate change. But this is offset by the fact that methane escapes during the drilling process and when the fuel is transported via pipelines. Methane is approximately 30 times more potent than CO_2 as a heat-trapping gas, so fracking does not necessarily reduce greenhouse gas emissions.

Finally, there is the question of what to do with the wastewater that comes back to the surface after the fracking process is complete. In some locations, it can be reinjected into impermeable rock, but not if the underlying rock is not porous enough. In this case, off-site wastewater treatment facilities must be built. Fracking is an example of a complex environmental management issue, because America's energy demands continue to rise, and the technology has allowed this new form of extraction. The Trump administration has stated that it seeks to increase fracking in the future. It is unclear, however, what the long-term social and environmental impacts will be.

See also: Section 1: Energy, Mining; *Section 3:* Renewable Energy

FURTHER READING

BBC. 2015. "What Is Fracking and Why Is It Controversial?" December 16. http://www.bbc.com/news/uk-14432401.

EPA. 2017. "Hydraulic Fracturing for Oil and Gas and Its Potential Impact on Drinking Water Resources." U.S. Environmental Protection Agency. https://www.epa.gov/hfstudy.

McGraw, Seamus. 2016. "Is Fracking Safe? The 10 Most Controversial Claims about Natural Gas Drilling." *Popular Mechanics.* May 1. http://www.popularmechanics.com/science/energy/g161/top-10-myths-about-natural-gas-drilling-6386593/.

USGS. 2018. "Induced Earthquakes." U.S. Geological Survey. https://earthquake.usgs.gov/research/induced/myths.php.

Genetically Modified Crops

Genetically modified (GM) or genetically engineered (GE) crops are very controversial. For every article stating the safety and benefits of GE crops, there is an opposing article describing the problems. In short, the scientific community mostly believes the crops are safe, but the environmental community is not convinced because there are no long-term studies of the impact of genetic engineering on nature and humans. In addition, GE crops are driven by profit, and there are ethical issues related to corporations "owning" the genetic imprints of living things.

GE seeds are distinct from past seed improvement efforts where farmers cross-pollinate or create hybrids. Cross-pollination is simply when a farmer transfers the pollen from one plant to another within the same species, but the resulting plants are not consistent. Hybrids try to produce seeds with more consistent improved characteristics, so it often involves scientists using pure strains of plants that are then cross-pollinated to create more predictable results, but the seeds from these plants do not produce offspring. GE seeds are completely different because the genetic material from one organism can be inserted with great precision into the genes of a completely different organism, even a nonplant species (DeJohn, 2018).

According to the U.S. Department of Agriculture, over 90% of the corn, soybeans, and cotton grown in the United States is genetically engineered. All of this has happened fairly recently. In the late 1990s, only about 10% of crops were GE (USDA-ERS, 2017). Now though, every person who eats processed food like chips or packaged cookies in the United States has eaten GE food, because GE corn and soy products are found in all such items.

The rest of the world has not accepted the GE crops so readily. Indeed, just four countries (the United States, Canada, Brazil, and Argentina) grow 90% of all GE crops. Many countries have banned them, and many others have strict controls on importing GE foods (Freedman, 2013).

Environmental activist and anti-globalization author Vandana Shiva speaks at the ReclaimRealFood Food Seminar and Workshop at AXE on March 24, 2013, in Venice, California. (Amanda Edwards/WireImage)

In Europe, the fight against GE crops has been strong and successful. Europeans argue that if a GE crop became invasive, it would be catastrophic. For example, what if genetic modification caused a "super pest" to evolve that could no longer be controlled? Europeans believe that planting GE crops should be halted until the technology is proved absolutely safe. This follows the precautionary principle, which is part of EU environmental law (EU, 2018). To be clear, GE is not the same as selective crop breeding because it enables scientists to insert genes from other species of plants, bacteria, viruses, or animals (Freedman, 2013).

There are two main types of GE crops: herbicide tolerant (HT) and insect resistant (IR). HT crops have been genetically modified to survive the application of specific herbicides (weed-killing chemicals) that previously would have destroyed both the weed and the crop. IR crops contain a specific gene that is toxic to some insects so the plant is protected from these pests (USDA-ERS, 2017). These IT crops are registered as a pesticide, because this toxin is within the plant itself.

Possible benefits of GE crops, as noted by their manufacturers and some scientists, include increased crop production, lower food costs, pest- and climate-resistant crops, and the ability to lower malnutrition (Freedman, 2013). These benefits have not yet been realized. Food prices have not decreased, and the number of people facing famine and malnutrition around the world has not decreased. But these crops have not been accepted in much of the world, including most of Europe, much of Africa, Asia, and some of Latin America (Freedman, 2013).

Mexico has banned the use of GMO corn seeds due to concerns about contamination. Mexico is the center of the origin of corn, meaning that in Mexico, there are numerous varieties of corn—59 indigenous strains. Some very early forms of the crop are still found, and even some wild plants related to corn are found in the country. Thus, the genetic diversity of corn is extremely rich in Mexico. There are both environmental and social concerns related to protecting this genetic diversity. First, scientists and farmers do not want to contaminate and wipe out the ancient genetic basis of one of the world's most important crops. It is feared that GMO seeds are like pollution that could harm all other corn seeds. The old varieties could be helpful in the future when trying, for example, to breed plants that are resilient to climate change. Second, the culture is so closely linked to corn that the sacred Mayan creation story describes Mexicans as "people of corn." Now many people, particularly indigenous populations, fear that GMO pollution will forever change the genetic composition of corn, which is equivalent to destroying the heart of their culture (Villagran, 2013).

Another critic of GE crops is Vandana Shiva, an environmental activist and author of more than 20 books. She claims GE seeds take away the livelihood of small-scale farmers in the developing world. According to her and many others, large multinational agricultural corporations have engineered, patented, and transformed seeds into costly "inputs" that threaten to control all farmers. To Shiva, agricultural biotechnology is like a global war of a few giant seed companies seeking victory over billions of farmers who depend on seeds they save themselves to plant year after year. "For me, the idea of owning intellectual-property rights for seeds is a bad, pathetic attempt at seed dictatorship," Shiva said recently (Specter, 2014).

As noted in *Scientific American*, most research on GE crops indicates that they are safe to eat and have potential to increase food production, but pro-GE scientists "are often dismissive" of any questions or evidence opposing them. In fact, scientists who point out the possible problems with GE crops have had their credibility attacked, which makes scientists keep quiet. There is a lot of money from GE agricultural corporations seeking to promote research that advances the use of GE crops in agriculture. Whereas much of the world has rejected GE crops, North America has fully embraced them. In the future, more long-term independent testing of GE crops is necessary (Freedman, 2013).

See also: Section 1: Agriculture, Case Study 2; *Section 3:* Alternative Agriculture

FURTHER READING

DeJohn, Suzanne. 2018. "How Genetic Engineering Differs from Traditional Plant Breeding." Gardeners.com. http://www.gardeners.com/how-to/genetic-engineering-traditional-plant-breeding/7926.html.

EU. 2018. "Facts on GMOs in the EU." European Commission. July 13. http://europa.eu/rapid/press-release_MEMO-00-43_en.htm.

Freedman, David H. 2013. "The Truth about Genetically Modified Food." *Scientific American*. https://www.scientificamerican.com/article/the-truth-about-genetically-modified-food/.

Specter, Michael. 2014. "Seeds of Doubt: An Activist's Controversial Crusade against Genetically Modified Crops." *The New Yorker*. August 25. http://www.newyorker.com/magazine/2014/08/25/seeds-of-doubt.

USDA-ERS. 2017. "Recent Trends in GE Adoption." U.S. Department of Agriculture-Economic Research Service. https://www.ers.usda.gov/data-products/adoption-of-genetically-engineered-crops-in-the-us/recent-trends-in-ge-adoption.aspx.

Villagran, Lauren. 2013. "'People of Corn' Protest GMO Strain in Mexico." *Christian Science Monitor*. May 16. https://www.csmonitor.com/World/Americas/Latin-America-Monitor/2013/0516/People-of-corn-protest-GMO-strain-in-Mexico.

Hazardous Waste

Hazardous waste, which can be in liquid, solid, gas, or sludge form, is waste that is "dangerous or capable of having a harmful effect on human health or the environment" (EPA, 2017). There are many sources of hazardous waste, including manufacturing, agriculture, and household items; this includes things like batteries, cleaners, electronic waste, medical waste, oil, paints, pesticides, or the many by-products of manufacturing.

Geographically, there are distinct contrasts between how hazardous waste is managed across the globe. For example in the United States and Europe, there are strict regulations on toxic waste, whereas in other parts of the world, it is imported because a country is dependent on income generated by accepting these hazardous materials. Thus the Global North has environmental regulations and agencies specifically designated to take on the task of hazardous waste management, whereas people in the Global South may be exposed to dangerous substances with little

U.S. Environmental Protection Agency (EPA) hazardous waste collection sites, like this one in Cameron, Louisiana, separate and store various toxic materials for proper disposal. (Robert Kaufmann/Federal Emergency Management Agency)

government oversight. The United States and Bangladesh provide two clear examples of these significant variations.

The U.S. Environmental Protection Agency (EPA) regulates hazardous waste under Subtitle C of the Resource Conservation and Recovery Act (RCRA) of 1976. There are specific definitions and a process through which hazardous waste is identified, and the EPA works with local and state agencies to oversee management. Even with all these national policies, environmentalists and some communities complain that enforcement by local, state, and federal governments is lacking. At the same time many industries and corporations claim that the EPA regulations are too strict and harm the economy. These industries spend millions of dollars a year to lobby Congress and state legislatures to weaken toxic waste rules.

The EPA defines specific categories of hazardous solid waste, each of which has specific rules for disposal and management, within the two main categories of household hazardous waste and industrial hazardous waste.

Household hazardous waste includes products that have toxic, corrosive, or reactive materials, as well as items that contain potentially hazardous ingredients. Examples include batteries, cleaners, light bulbs, medical waste, oil, and paints. Safe management of household hazardous wastes involves these stages: collection, reuse, recycling, and disposal. Much of this management is facilitated by the local governments, as specified by the EPA in household hazardous waste regulations.

Industrial hazardous wastes are from industrial facilities, manufacturing and processing sites, workshops, nuclear facilities, and chemical units. There are four

Brownfield Sites

A brownfield site is a property with "the presence or potential presence of a hazardous substance, pollutant, or contaminant" that may complicate its future use (EPA, 2018). A key aspect of a brownfield site is that it is targeted for redevelopment (OSHA, 2018). By cleaning up and reinvesting in these properties, a community can increase its tax base, stimulate job growth, use existing infrastructure, decrease urban sprawl, and help protect the environment (EPA, 2018).

It is estimated that there are more than 450,000 brownfields in the United States. These sites are assumed to have some pollution due to their previous industrial use, but cannot be heavily contaminated or in the midst of remediation and cannot be on the Superfund National Priority List (OSHA, 2018).

Sometimes a large brownfield site dominates a whole neighborhood where industrial buildings are abandoned or commercial buildings are closed. In other cases, it can be quite small, like an abandoned gas station or dry cleaners that contaminated the soil when it was in operation.

State and federal programs encourage brownfield redevelopment for economic revitalization in surrounding neighborhoods. The U.S. EPA Brownfields Program began in 1995 and expanded under the Small Business Liability Relief and Brownfields Revitalization Act of 2002, which is an amendment to the Comprehensive Environmental Response, Compensation, and Liability Act (CERCLA—also known as Superfund) of 1980.

CERCLA was the federal law to oversee the cleanup of many hazardous-waste sites. It also made the purchaser of any property liable for any contaminants on the property. The new buyer obviously does not want to be responsible for paying for the cleanup of any toxins remaining on a brownfield site. Because of this liability rule, it is necessary to undertake a comprehensive environmental site assessment to determine the actual presence of contaminants on the property before redevelopment.

Since its inception in 1995, the EPA's Brownfields Program assisted communities and states to work together to assess, safely clean up, and sustainably reuse brownfields. Between federal and state funding, 144,000 cleanups have been completed and over 1 million acres were made ready for reuse. This has created benefits throughout the communities, including decreased storm water runoff, new jobs created, and significant increases in property values nearby. There are many examples of brownfield sites that have been redeveloped into community gardens. This is possible if the soils are not contaminated or if the soil is removed and new soils are managed to promote healthy urban agriculture projects (EPA, 2018).

Further Reading

EPA. 2018. "Overview of the Brownfields Program." Environmental Protection Agency. https://www.epa.gov/brownfields/brownfields-program-accomplishments-and-benefits.

OSHA. 2018. "Brownfields Health & Safety." Occupational Safety and Health Administration. https://www.osha.gov/SLTC/brownfields/brnfld_qna.html.

---

types of industrial hazardous wastes, as outlined by the EPA: listed waste, universal waste, characteristic waste, and mixed waste. Listed waste is further divided into four lists related to specific manufacturing processes such as petroleum refining, wood treatment, and discarded commercial chemical products and solvents. Universal waste is mostly related to batteries, pesticides, and mercury-containing equipment. Characteristic wastes have specific qualities of being corrosive, reactive, ignitable, and toxic and must be carefully managed separate from other wastes. Mixed wastes contain both radioactive and hazardous waste components, which is complicated to regulate and manage. Overall, the four categories of industrial hazardous waste each require specific, detailed methods of transportation and disposal (EPA, 2017).

These regulations, which are similar to those found in Canada, Australia, and European countries, are in sharp contrast to countries in the Global South, which often have lax or nonexistent rules. In some countries, the economic situation is so dire, they are forced to accept toxic waste from wealthy countries just to receive monetary payments.

"More than 200 million people around the world are at risk of exposure to toxic waste" according to the BBC (Parameswaran, 2013). The World Health Organization agrees that over 12 million deaths each year, or one in four, are caused by environmental problems (WHO, 2016). Toxic waste in the Global South is a significant and growing concern. An annual report identifies the following industries as causing the greatest burdens of illness and disease: used lead, acid battery recycling, mining and ore processing, tanneries, dumpsites, industrial estates, smelting, artisanal small-scale gold mining, product manufacturing, chemical manufacturing, and the dye industry. These industries have significant negative health impacts on people in low- and middle-income countries, according to the 2016 report that was based on research carried out at over 3,000 sites in more than 49 countries (Green Cross, 2016).

One example is Hazaribagh, Bangladesh, which is known as one of the most polluted cities in the world due to the leather industry, which employs nearly 12,000 workers. The leather tannery companies supply no aprons, gloves, or protective gear, so workers wear no hand protection, even when handling dangerous toxic chemicals. The water pollution is extreme, as each day the leather industry tanneries dump 22,000 cubic liters (5,800 gallons) of toxic waste into the local Buriganga River. This river is the main water supply for the city of Dhaka, with a population of over 14 million. One of the main toxic chemicals from the leather industry is hexavalent chromium, which is a known carcinogen. The air is also heavily polluted due to the industry (Green Cross, 2016). The region has no other economic options, so people must stay and are thus continuously exposed to these toxins.

See also: Section 1: E-waste, Solid Waste, Superfund

FURTHER READING

EPA. 1976. "Summary of the Resource Conservation and Recovery Act. 42 U.S.C. §6901 et seq." U.S. Environmental Protection Agency. https://www.epa.gov/laws-regulations/summary-resource-conservation-and-recovery-act.

EPA. 2017. "Learn the Basics of Hazardous Waste." U.S. Environmental Protection Agency. August 16. www.epa.gov/hw/learn-basics-hazardous-waste#hwid.

Green Cross. 2016. "The World's Worst Pollution Problems 2016: The Toxics Beneath Our Feet." http://www.worstpolluted.org/.

Parameswaran, Siva. 2013. "Toxic Waste 'Major Global Threat.'" BBC. November 20. http://www.bbc.com/news/science-environment-24994209.

WHO. 2016. "An Estimated 12.6 Million Deaths Each Year Are Attributable to Unhealthy Environments." March 15. http://www.who.int/mediacentre/news/releases/2016/deaths-attributable-to-unhealthy-environments/en/

Mining

Mining is the process of locating and extracting valuable minerals from the Earth. There are three main phases in mining: exploration, extraction, and mine closure, each of which has both environmental and social impacts, and are thus regulated in most countries. There are four main types of mining: underground, surface (open pit), placer, and in situ mining. Mining companies decide which method to use based on the type of mineral to be extracted, its location, and the value of the mineral compared to the cost of the method. There are pros and cons of each method. Underground mining is more expensive, but can reach large, valuable deposits. Surface mines are cheaper and are used for less valuable mineral deposits. Placer mining conjures up that image of the old frontiersman trying to sift gold out of a stream, but it is now more sophisticated and includes sifting out valuable minerals from beach sand, rivers, and other locations. In situ ("in place") mining is mostly used for uranium, an extremely valuable mineral, whereby it is dissolved and processed on the surface without moving rock from the ground (AGI, 2018).

There are potential environmental problems associated with all types of mining, including acid drainage, inadequate reclamation, tailing waste (mining waste rock mixed with extraction chemicals), truck noise and dust in the region, waste rock disposal, and water contamination. Underground mines can cause problems of land subsidence from tunnels and mineral removal, loud noises and vibrations from blasting under the Earth, and water contamination with high sediment loading (Canada, 2017).

Surface mines cause large land disturbance with vast, visible impacts on the landscape. For example, one of the largest open pit mines in the world is southwest of Salt Lake City, Utah. It is over 900 hectares (3.5 square miles), 1.2 kilometers (0.75 miles) deep, and over 4 kilometers (1.9 miles) wide and extracts vast amounts of copper, gold, silver, and molybdenum (Mining.com, 2016). Another example of surface mining is called mountaintop removal, and as this name suggests, is "any method of surface coal mining that destroys a mountaintop or ridgeline" (Appalachian Voices, 2017). This has significant environmental impacts and social implications, so there are significant local community efforts to halt this type of mining.

The World Mining Congress notes that the 61 different minerals that are mined globally are typically arranged in five groups: iron and ferro-alloy metals (11 types, including iron, cobalt, titanium); nonferrous metals (18 types, including aluminum, copper, REE, tin); precious metals (3 types, including gold, silver, platinum); industrial minerals (21 types, including asbestos, gypsum, salt, sulfur); and mineral fuels (8 types, including coal, natural gas, petroleum). A total of 17,269,688,784 metric tons (mt) (1 metric ton = 1,000 kg) of minerals were mined globally in 2015, the vast majority of which (14.7 billion mt) was mineral fuels followed by iron metals at 1.6 billion mt, nonferrous metals at 96 million mt, industrial minerals at 791 million mt, and precious metals at 31,000 mt (Reichl et al., 2017).

Considered a nonferrous metal, rare earth elements (REEs) were almost unheard of 50 years ago, but that changed with advances in electronics technology and availability. REEs are used in consumer and business items such as mobile phones, computers, flat screen televisions, LED lighting, and more. They are also important components to defense technology. Countries such as the United States, Japan, and many in Europe are increasingly dependent on REEs, but the vast majority of these minerals (85% to –95) are mined only in China (Van Gosen et al., 2014). There is concern about China's dominance of the production and supply of REEs, as the world has become dependent on these critical elements. Exploration into REE deposits in other parts of the world has increased in recent years.

In addition to REE, there is high global demand for precious metals, which stimulates illegal mining activities. Artisanal and small-scale mining (ASM) is informal, often illegal, mining activity using no technology or with minimal machinery. It is estimated that 13 million people are engaged in ASM, primarily in developing countries, where they are driven by economic desperation (Hentschel et al., 2002). ASM causes environmental degradation because it occurs with no water and air pollution regulation. In addition, many people are poisoned or killed through illegal mining. ASM miners are often harmed by lack of safety equipment, improper use of chemicals, no ventilation, and lack of training.

Mining is a dangerous job, even in regulated mining operations. It is estimated that over 12,000 miners are killed in accidents each year, which is a high rate. Mining employs approximately 1% of the global workforce, but generates 8% of fatal accidents (BBC, 2010). In the United States, the National Institute for Occupational Safety and Health (NIOSH), as part of the U.S. Centers for Disease Control and Prevention, keeps track of safety issues related to mining. The NIOSH website provides informative maps and data regarding mines (number and location), employees (number and hours), fatalities (number, rate, cause), and injuries (number, rate, location, type).

These data indicate that in the United States as of 2015, there were 13,294 mining operations (12,637 surface and 657 active underground operations). These mines break down by material as follows: 1,460 coal, 315 metal, 924 nonmetal, 4,303 stone, and 6,292 sand and gravel mines (CDC, 2017). The maps show interesting geographic patterns; for example, underground mines are concentrated in Appalachia, whereas surface mines show higher density near the coasts as well as eastern states. In 2015, there were 237,812 mine employees in the United States. That year, the total number of nonfatal lost-time injuries was 4,517 and there were 26 mining fatalities (coal = 11, metal = 5, nonmetal = 2, stone = 2, sand and gravel = 6) (CDC, 2017).

See also: Section 1: Case Study 5, Hazardous Waste, Water Pollution

FURTHER READING

AGI. 2018. "Mining." American Geosciences Institute. https://www.americangeosciences.org/critical-issues/mining.

Appalachian Voices. 2017. "Mountaintop Removal 101." http://appvoices.org/end-mountaintop-removal/mtr101/

BBC. 2010. "The Dangers of Mining Around the World." BBC News. October 14. http://www.bbc.com/news/world-latin-america-11533349.

Canada. 2017. "Environmental Code of Practice for Metal Mines." Government of Canada. https://www.ec.gc.ca/lcpe-cepa/default.asp?lang=En&n=CBE3CD59-1&offset=5.

CDC. 2017. "Statistics: All Mining." The National Institute for Occupational Safety and Health. https://www.cdc.gov/niosh/mining/statistics/allmining.html.

Hentschel, Thomas, Felix Hruschka, and Michael Priester. 2002. "Global Report on Artisanal and Small-Scale Mining." International Institute for Environment and Development. http://www.ddiglobal.org/login/resources/g00723.pdf.

Mining.com. 2016. "How Big Is Big?" Matter Solutions. May 25. http://www.mining.com/web/how-big-is-big/.

Reichl, C., M. Schatz, and G. Zsak. 2017. "Minerals Production." International Organizing Committee for the World Mining Congresses. http://www.wmc.org.pl/sites/default/files/WMD2017.pdf.

Van Gosen, B. S., P. L. Verplanck, K. R. Long, Joseph Gambogi, and R. R. Seal, R. R., II. 2014. "The Rare-Earth Elements—Vital to Modern Technologies and Lifestyles." U.S. Geological Survey Fact Sheet. https://pubs.usgs.gov/fs/2014/3078/.

Overfishing

Overfishing is when more fish are caught than can be naturally replaced through natural reproduction. Today, over 90% of the earth's fisheries are overexploited, fully exploited, or have collapsed (MBA, 2018). This has both environmental and social consequences. Marine ecology can be permanently disrupted and species lost forever. Economic and social well-being of coastal communities can be destroyed.

The top 10 countries that dominate fishing in international waters are Japan, South Korea, Taiwan, Spain, United States, Chile, China, Indonesia, Philippines, and France. These 10 nations take about $8.5 billion of high seas catch each year, which is about 70% of the value of all fish sold in the world. Most (90%) of the world's fish catch is from the "exclusive economic zones" (EEZs), which are areas of the sea that each country claims going out as far as 370 kilometers (200 miles) from their coast (Economist, 2017). As overfishing makes it increasingly difficult to locate and extract fish, international arguments occur regarding fishing rights, the allowable distances from shorelines, and how much a country can claim. For example, fishing rights in the South China Sea and the new access to fish in the melting arctic waters are already in dispute.

Overfishing has accelerated rapidly since the mid-1900s, when the fishing industry became increasingly industrialized and large profit-driven commercial fleets replaced the local fishermen's boats. These industrial fleets had technology to locate, extract, and process huge numbers of target-species fish. Customers came to expect wide availability of specific varieties of fish at affordable prices. In 1989, 90 million tons of ocean catch was extracted, and since then the yields have stagnated or declined. The stock of the most popular species, such as orange roughy, Chilean sea bass, and bluefin tuna, has collapsed. By the early 2000s, studies showed that only 10% of large ocean fish remained, compared to before industrial fishing times (NGS, 2010).

One well-known example of a fishery's collapse is the cod along the east coast of North America. Near Grand Banks, Newfoundland, Canada, cod were once so abundant a fisherman only needed to dip a bucket into the ocean and it would quickly fill up with fish. But this changed with the industrialization of fishing in the 1950s. Small local fishing boats were outcompeted by the large factory-like trawlers raking in tons of fish all day, every day. In an hour they could haul up 200 tons of fish, more than older boats would catch in a week. By 1968, the peak catch was 800,000 tons, but by 1975, this had dropped more than 60%. As the cod catches declined, the industrial ships used special sonar to target the remaining fish. The government did not step in to set regulations, despite warnings from marine scientists. By 1992, the cod catch was the lowest ever, and the government closed the entire area to cod fishing. This moratorium caused job losses of 40,000 people whose jobs relied on the cod industry (equipment, processing, transportation, etc.) (Greenpeace, 2009). Although initially assumed to be temporary, the moratorium is still in place because the cod numbers never recovered. In fact in 2003, two cod species were placed on the Canadian list of endangered species. Along the U.S. coast of New England, cod catches are at an all-time low in recent years and limits on fishing are in place.

Overfishing is intensified by illegal fishing, bycatch, and habitat damage. Illegal fishing is occurring because the fisheries are collapsing and people are desperate to catch fish. They break the law by ignoring rules and limits for fishing. It is estimated that about 20% of the global fish catch is illegal. Bycatch is all the dead species that are "thrown away" or dumped back in the ocean after the target species is removed from the nets. For example, baited longlines can extend for over 50 miles and attract anything that swims by. Bottom trawls actually drag nets across the seabed and catch everything in their path. When this fishing gear is hauled up, it brings thousands of "extra" dead animals along with the desired fish. Nontarget species die from being caught for so long—various other fish species and also sea turtles, sharks, and dolphins, to name a few. Shrimp fisheries are particularly violent, as for every pound of shrimp caught, up to six pounds of other species are discarded bycatch (MBA, 2018). Habitat damage is occurring due to plastic garbage in the oceans, acidification due to climate change, and fishing gear destroying the natural marine ecosystems. For example, seabed trawling can be compared to clear-cutting a forest.

Overfishing is bad for fish and also bad for people. The goal of sustainable fisheries management, then, is to have a steady number of fish so that the number extracted is equal to the number born each year. If the fishing industry takes more than this sustainable yield, as is occurring today, then in the long term there will be less of the fisheries resource in the future. Each country can do as they wish within their own EEZ, and some are choosing to protect their fisheries.

Management actions can include catch limits, using safer gear, stopping illegal fishing, and helping rebuild marine habitats. Catch limits or quotas on the number of fish allowed to be caught can be successful, but this also has some drawbacks. For example, fishermen want to keep the largest fish they can find, so they throw back the smaller specimens, which die anyway. And quotas are very political, so species that get attention (and funding) from lobbying groups tend to have higher quotas, despite scientists warning about threats to specific species.

Other key aspects of sustainable management include the use of less damaging fishing gear that can reduce bycatch and habitat damage. Some countries have set up regulations that require the use of devices that allow sea turtles to escape from the nets, and other rules state that trawls must reduce their impact on the ocean floor.

Management will also need to include more protected areas both in international waters and within countries' EEZs. These Marine Protected Areas would be particularly helpful if there were "no-take" zones where fishing was completely banned. This would help marine ecosystems recover by providing breeding spaces so fish populations could increase and stabilize. Overall, sustainable fisheries management should take an ecosystems approach, looking at the marine habitat as a whole, rather than the traditional regulations that just address individual species in isolation.

See also: Section 1: Case Study 3, Case Study 4, *Section 3:* Sustainable Diet

FURTHER READING

Economist. 2017. "Getting Serious about Overfishing." May 27. https://www.economist .com/news/briefing/21722629-oceans-face-dire-threats-better-regulated-fisheries -would-help-getting-serious-about.
Greenpeace. 2009. "The Collapse of the Canadian Newfoundland Cod Fishery." May 8. http://www.greenpeace.org/international/en/campaigns/oceans/seafood/under standing-the-problem/overfishing-history/cod-fishery-canadian/.
MBA. 2018. "Wild Seafood." Monterey Bay Aquarium Seafood Watch. https://www.seafood watch.org/ocean-issues/wild-seafood.
NGS. 2010. "Overfishing." *National Geographic.* April 27. http://www.nationalgeographic .com/environment/oceans/critical-issues-overfishing/.

Solid Waste (Garbage!)

Garbage, trash, or municipal solid waste (MSW) is all the stuff people commonly throw away, including packaging, consumer goods, food, yard clippings, clothing, personal items, and electronics. Sources of MSW are residential, business, and commercial.

Globally, about 11.2 billion tons of solid waste are collected annually and the decay of the organic materials within it comprise 5% of total greenhouse gas (GHG) emissions (UNEP, 2018). In addition, the waste stream is increasingly composed of electronics that contain hazardous materials and present extra challenges for safe management. There is also illegal dumping or burning of trash that is outside the MSW waste stream, and this particularly affects poor people and people in the Global South.

The World Bank is actively working in developing countries to improve solid waste management through loans and technical advising. Indeed, their objectives exemplify the many complex issues that must be addressed in waste management. Infrastructure must be built or upgraded in order to sort and treat waste, and this includes trucks, landfills, and bins. Legal institutions are needed in order to develop strong policies and coordinated waste management programs.

Finances must be stable, but solid waste management is very expensive, often representing a high percentage of a poorer city's budget. Citizens must be engaged and participate in waste reduction and recycling efforts. Social issues include the labor sector because waste management can provide jobs, but there must be regulations for worker safety and rules against child labor. In addition, in many poverty-stricken places, people's livelihoods are actually based on waste picking. So although implementing a formal MSW collection system is necessary for improving the urban environment, it could harm poor people's economic situation.

Many environmental benefits can be gained from good waste management strategies, including the ability to address climate change by reducing organic waste that generates GHG, especially if disposal technologies that capture landfill gas could be adopted. Waste disposal in waterways could be reduced. Solid waste management promotes human health by reducing open burning, reducing disease vectors, and even reducing crime. Indeed, large waste items can be used as road blocks and glass bottles as weapons in situations of civil unrest (World Bank, 2017).

In wealthier countries, the issues are quite different. Waste management systems are well established, but policies and new technology can create shifts in the waste stream. Data on waste management in the 27 European Union nations show marked change from 1995 to 2015. There was a significant decrease in the percentage of waste that was landfilled (58% decrease), which was balanced by increases in incineration (99%), recycling (176%), and composting (184%). This is partly due to an EU directive stating that member nations had to reduce the amount of biodegradable municipal waste going into landfills. Overall, Europeans produce 477 kilograms (1,051 pounds) of trash per person each year (EU, 2018).

Americans each create 729 kilograms (1,606 pounds) of solid waste per year. In 2014, about 258 million tons of MSW were generated in the United States, of which 53% was landfilled, 35% was recycled and composted, and 12% was burned for energy, a concept known as "waste to energy." Indeed, MSW is officially listed as a "renewable energy resource" by the U.S. Department of Energy (EIA, 2018). This underscores society's continuous and reliable waste production—seemingly infinite and always available. The good news is that in some places, it can be put to a beneficial use.

Energy recovery from waste is the conversion of nonrecyclable waste materials into usable heat, electricity, or fuel through a variety of processes, including combustion (confined and controlled burning), pyrolization (high-temperature burning in the absence of oxygen), and landfill gas recovery.

Typically, MSW is burned at special facilities that use the heat from the fire to make steam for generating electricity. In 2015, 71 "waste to energy" power plants in the United States burned about 29 million tons of MSW in 2015 and generated nearly 14 billion kilowatt hours of electricity. In addition to electricity generation, the second benefit of these plants is that burning MSW reduces the volume of waste going to the landfill by about 87%. And now many large landfills also generate electricity by using the methane gas that is produced from decomposing biomass in landfills (EIA, 2018).

In the United States, modern landfills are located, designed, operated, and monitored to promote environmental safely. But it is estimated that there are over 10,000 old dump sites spread across the country that have been abandoned and may not meet current regulations. Concerns include groundwater contamination and landfill gas, both of which could occur as the massive piles of waste break down and seep into the ground and air. Current landfills are regulated under the U.S. Resource Conservation and Recovery Act (RCRA) of 1976, but are managed by each state, so rules vary in terms of items that can be accepted and site management requirements (EPA, 2017).

See also: Section 1: Case Study 4, E-waste, Hazardous Waste; *Section 3:* Sustainable Cities

FURTHER READING

EIA. 2018. "Energy from Municipal Solid Waste." U.S. Energy Information Administration. https://www.eia.gov/energyexplained/?page=biomass_waste_to_energy.
EPA. 2016. "Advancing Sustainable Materials Management: 2014 Fact Sheet." U.S. Environmental Protection Agency. https://www.epa.gov/sites/production/files/2016-11/documents/2014_smmfactsheet_508.pdf.
EPA. 2017. "Municipal Solid Waste Landfills." U.S. Environmental Protection Agency. https://www.epa.gov/landfills/municipal-solid-waste-landfills.
EU. 2018. "Municipal Waste Statistics." Eurostat. http://ec.europa.eu/eurostat/statistics-explained/index.php/Municipal_waste_statistics.
UNEP. 2018. "Solid Waste Management." United Nations Environmental Programme. http://drustage.unep.org/resourceefficiency/what-we-do/policy-strategy/resource-efficient-cities/focus-areas-cities/solid-waste-management.
World Bank. 2017. "Solid Waste Management." http://www.worldbank.org/en/topic/urbandevelopment/brief/solid-waste-management.

Superfund

In 1980, the U.S. Congress passed the Comprehensive Environmental Response, Compensation, and Liability Act (CERCLA) that established Superfund to clean up toxic waste sites. This law was passed because of public outrage and citizens demanding government action to regulate hazardous waste sites, such as the well-known case of the Love Canal neighborhood in Niagara Falls, New York.

In the area of Love Canal, soil and water was contaminated because Hooker Chemicals & Plastics Corporation dumped over 20,000 tons of hazardous chemicals there in the 1940s and 1950s. This company was sold to a different company, and the land was sold to other people. Then the ground was covered up, and the land was developed with a school and residential housing. By the 1970s, local people saw and smelled chemicals and had increased cancer rates and birth defects. It took the local people many years of protesting to get enough attention that the government took action. Lois Gibbs was a mother with no prior experience in environmental activism, but when she saw the health problems in her neighborhood and learned that her son's school was built on a toxic waste site, she took action.

In 1978, Lois Gibbs discovered that her 5-year-old son's school was built on a toxic dump site. She went door-to-door gathering signatures and organizing a fight against local, state, and national officials, finally forcing them to clean up her neighborhood in Love Canal, New York. (Kimberly Butler/The LIFE Images Collection/Getty Images)

Organizing a neighborhood group, she led a petition drive that went from the local, to the state, to the national government. Eventually the Love Canal site was declared a national emergency area and nearly 1,000 families were evacuated from about 10 square blocks with a bigger region divided into seven areas of various safety and habitability levels (EPA, 2017).

There were significant cover-ups and secrecy because nobody wanted to be liable and held responsible for paying for the cleanup around these types of sites, but eventually the public outcry made politicians listen. The Superfund program was created within the EPA to clean up the thousands of active and abandoned toxic waste sites across the United States. In many cases, the company that originally caused the pollution no longer exists and the property has been sold several times, so no one is responsible for the cost of the cleanup. This places a high burden on local communities attempting to find out what toxins may have been dumped decades earlier—and a financial burden of paying for the costly cleanup.

Since 1982, the EPA has put more than 1,700 sites on the Superfund National Priorities List (NGS, 2014). The law was set up to help communities that are left to struggle with toxic sites. So when polluters cannot be forced to pay for the cleanup, the Superfund would pay.

Superfund was initially financed through a new tax on oil and chemical companies. But when that tax expired in 1995, the Republican-led Congress refused to renew it, so the funding decreased significantly and the U.S. taxpayers became the source of Superfund payments. Thus, the original Superfund was a "polluter pays" approach by taxing the industry, but it is now another program that must compete for general tax dollars. Annually, the U.S. government payments for Superfund cleanups went from $2 billion down to $1.1 billion between 1999 and 2013. In real dollars, the level of Superfund money had declined 35% and about half as many sites are cleaned up annually now compared with the 1990s (King, 2017). The number of sites needing clean up continues to increase, even if the money allocated to do Superfund cleanup work is dwindling.

As of mid-2017, there are over 1,300 Superfund sites, located in every state, and "about 53 million Americans—about one of every six—lives within three miles of a Superfund site." Concerned citizens will want to track future U.S. federal budget negotiations, as the budget proposals should address the EPA Superfund program (King, 2017). The geographic location of the Superfund sites can be found on the EPA website at https://www.epa.gov/superfund/search-superfund-sites-where-you-live. The EPA Superfund website also provides extensive information on community involvement, cleanup support and training, specific types of contaminants, and redevelopment initiatives (EPA, 2018).

See also: Section 1: Hazardous Waste, Solid Waste; *Section 3:* Environmental Justice, Sustainable Cities

FURTHER READING

EPA. 2017. "Love Canal: Press Releases and Articles." U.S. Environmental Protection Agency. https://www.epa.gov/history/love-canal.
EPA. 2018. "Superfund." U.S. Environmental Protection Agency. https://www.epa.gov/superfund.
King, Ledyard. 2017. "President Trump's Budget Would Cut Superfund Toxic Cleanup Program by 30%." *USA Today.* April 7. https://www.usatoday.com/story/news/politics/2017/04/07/president-trumps-budget-would-cut-superfund-toxic-cleanup-program-30/100168436/.
NGS. 2014. "How Close Are You to a Superfund Site?" *National Geographic.* http://www.nationalgeographic.com/superfund/#charts.
Vaughn, Jacqueline. 2017. *Environmental Activism: A Reference Handbook.* Santa Barbara, CA: ABC-CLIO.

Technology: Innovation and Consequences

Technology is "the practical application of knowledge so that something entirely new can be done, or so that something can be done in a completely new way" (European Space Agency, 2012). New technology can help make life simpler, easier, and more fun. But technology is not without problems, and it is important that informed citizens understand both the benefits and the possible drawbacks brought by innovations.

The introduction of new technology can cause unintended consequences or unplanned outcomes. These consequences can be beneficial, detrimental, or controversial. Because impacts can occur much later and much farther away, sometimes people overlook or do not link unintended consequences with the new technology. It is difficult to prove that an activity that occurred long ago and far away

Assembly Line Manufacturing

In 1908, the Ford Motor Company introduced the Model T, which cost $825. The car was no-frills, and the price was reasonable for a highly innovative product such as a "horseless carriage"—now called a "car"! Interestingly, this one innovation drastically changed America and the world in terms of dependence on fossil fuels and thus the production of greenhouse gas (GHG) emissions. But in addition to these direct changes, there was the indirect influence of the Ford assembly line (UM-Dearborn, 2010).

This new system of manufacturing became the norm over the next several decades, so that today nearly all consumer goods are produced in large factories with laborers repeating one task over and over, all day long. Think about this shift in society and people's relationship to natural resources and the environment (Nye, 2013).

Before assembly lines, every consumer product was handmade. There were skilled craftsmen who sought out wood from specific trees to build their perfectly balanced and sanded chair, with a thoughtful design in the cut-out backrest and hand-polished stain. Each craftsman had specific skills and earned a living wage by selling their special items, often to people they knew personally. Every consumer bought just what they needed, because each item was expensive.

Assembly line manufacturing changed the system forever. Now raw materials were shipped in based on lowest cost, and each worker used a specialized machine to make one small component. Productivity, in terms of items produced in the least possible amount of time, increased significantly.

On the other hand, this type of production changed people's use of environmental resources and views of nature in production. Natural resources in a factory were used based on their cheap prices and ready availability. Laborers were no longer experts and did not particularly care about what they produced (UM-Dearborn, 2010).

Natural resources became just like any other "input" into a system (Nye, 2013). They were removed from the consumer item and from the consumer. Now, in the modern day, this is the only system most people know. Very few items are handcrafted any longer, and this influences how people view natural resources. These "inputs" for manufacturing are not seen as part of nature. They are simply seen as a monetary item.

It would be useful if a consumer went to a superstore to buy a new chair and the label stated: "This wood is mahogany from the Amazon rainforest and it was not sustainably harvested—tropical rainforests provide habitat for 80% of the world's terrestrial biodiversity and the current global extinction rate is 50,000 species per year" (Rainforest Alliance, 2018). But consumer products do not come with detailed information about environmental impacts. Concerned consumers must educate themselves about the natural resources used in their purchases (Urquhart, 2002).

Further Reading

Nye, David E. 2013. *America's Assembly Line*. Cambridge, MA: MIT Press.
Rainforest Alliance. 2018. "Forests: Our Very Lives Depend on Them." https://www.rainforest-alliance.org/issues/forests.
UM-Dearborn. 2010. "Automobile in American Life and Society." University of Michigan-Dearborn and Benson Ford Research Center. http://www.autolife.umd.umich.edu/.
Urquhart, Gerald R. 2002. "Saving the Rainforest: Purchasing Wood Products." https://msu.edu/user/urquhart/tour/tropical_woods.html.

might be causing harm today in a new location—like toxic air pollution that causes health problems downwind years later.

There are four types of unintended consequences: positive, negative, perverse, and controversial. Positive consequences are when a technology actually had beneficial side effects in addition to the intended effect: preserving land for a hiking trail might also lead to increased bird populations. Negative consequences are when a harmful side effect occurs in addition to the intended effect of the technology: compact fluorescent light bulbs use 70% less energy, but contain mercury that can pollute water if not discarded properly. Perverse consequences occur when technology actually produces the exact opposite of the intended result: pesticides are strong chemicals that kill weeds, but over time hardy weeds evolve that are resistant to the chemicals, so stronger pesticides must be sprayed. Finally, controversial consequences to technology are when people have varying views; some people say that genetically modified crops are beneficial and can increase global food supplies, but others say they cause environmental and social problems.

There are a few theories about the effects of new technology on the environment. The rebound effect is the term used to describe when a new technology that is supposed to help the environment actually causes unintended environmental consequences. This often occurs as the new technology is more environmentally efficient, so it encourages people to do more of a specific behavior, which actually leads to worse environmental impacts in the long run. Take the example of email; certainly sending one email is more energy efficient and produces fewer GHG than typing up a letter, printing it out on paper, addressing a stamped envelope, mailing the letter, sorting the mail, and transporting and delivering the envelope. But people now send zillions of emails, far more than the letters they used to send. So computer servers have been increasing, more laptops and desktops have been produced, and entire businesses now deal with email software and hardware. All of this demands a lot of energy and has a big environmental impact, which is the rebound effect of the new technology. But the rebound varies by geographic location.

People adopt new technology at different rates; some wait in line for hours to buy the latest electronic device as soon as it is available, whereas others are happy to wait and buy it much later. This relates not just to cell phones, but to bigger ideas and actions that affect the environment and society. So the Adoption of Innovation Model was developed to describe this process.

Innovators make up 2% of the population and are adventurously willing to be one of the first to try new technologies. Early adopters (14% of people) are next, followed by early majority (34% of people), who accept a new technology as it becomes more proven and more information is available. Late majority adopters are the skeptical 34% of people who only accept innovation once it is well known and approved by most people in a place. Finally, the laggards are the 16% of people who are very traditional and not willing to try new technology until it is actually not innovative anymore but is generally accepted within the mainstream before these folks will try it. All of these stages are greatly affected by influential people in society, opinion leaders, who act as agents of change, whether at the local level up to the national or global media (Rogers, 2003). Advertising money can also affect individual decision making regarding the acceptance and adoption of new technology.

Even green technology can lead to consequences that are negative, perverse, or least controversial. Solar panels allow people to produce clean renewable energy, but the panels contain heavy metals that can contaminate the environment if not recycled and disposed of properly. Wind farms produce energy from a constant renewable source, but residents complain about how they look and sound, and often do not want them placed near residential areas; on the other hand, siting wind farms in remote areas can lead to new roads that allow more resource extraction in sensitive areas. Another interesting example of unintended consequences is that green technology can change people's behaviors, which can lead to negative results. For example, consumers who buy energy-efficient washing machines tend to do more loads of laundry! And some consumers who save money buying more energy-efficient products may then spend that money to buy more consumer items—things that they may not really need—thus unintentionally causing more environmental degradation due to increased consumption (Zehner, 2012).

Because of the many consequences of new technology, it is important to consider possible impacts early in the research and development stage and to assess the actual impacts once an innovation has occurred. Technology assessment seeks to evaluate the impacts of new technologies in the most neutral way possible, but often political and economic forces also have an influence. Sometimes innovative technology can cause unforeseen impacts on society and the environment that have not previously been considered—think about the ethical issues of cloning animals or even humans. Overall, then, it is important that all impacts of technology be assessed with full involvement of all stakeholders.

See also: Section 1: Case Study 1, Case Study 2, Fracking, Genetically Modified Crops

FURTHER READING

European Space Agency. 2012. "What Is Technology?" http://www.esa.int/Our_Activities/Space_Engineering_Technology/What_is_technology.

Mulder, Karel F. 2013. "Impact of New Technologies: How to Assess the Intended and Unintended Effects of New Technologies" in *Handbook of Sustainable Engineering.* Kauffman, Joanne and Kun-Mo Lee, eds. Dordrecht, Netherlands: Springer.

Nowak, Peter. 2012. "The Unintended Environmental Impact of New Technology." Canadian Business. August 1. http://www.canadianbusiness.com/blogs-and-comment/the-unintended-environmental-impact-of-new-technology/.

Rogers, Everett M. 2003. *Diffusion of Innovation,* 5th ed. New York: Free Press, Simon & Schuster.

Zehner, Ozzie. 2012. *Green Illusions: The Dirty Secrets of Clean Energy and the Future of Environmentalism (Our Sustainable Future).* Lincoln: University of Nebraska Press.

Urbanization

The basic definition of urbanization is "the process by which towns and cities are formed and become larger as more and more people begin living and working in central areas," and the term was first used in 1888 (Merriam-Webster, 2018). For environmental geographers working in the real world, however, "there is no common global definition of what constitutes an urban settlement" (UN, 2014).

The definition of "urban" that is used by governmental statistical offices varies by country and can even change over time. Typical variables for defining urban areas are population density, a threshold population number, amount of paved roads and infrastructure, electricity grids, piped water, and sewers or percentage of nonagricultural employment. When making international comparisons, the United Nations currently uses the definitions that are employed by each nation, but this causes concern about cross-case comparisons. There are efforts to develop a consistent definition to assist in global research and planning. These approaches tend to be very geographic: use of remote sensing, analysis of satellite images of land-use/land-cover change, or viewing light signatures from urban areas at night. Although new methods for determining and measuring the extent of urbanization will be useful in the future, they will not provide an historical framework that allows researchers to understand urban change over time. Thus, all global comparisons available today are based on estimates and projections built on data from each individual country's definitions.

Taking a geographic view of urbanization, there are three themes. First, regional trends show contrasts over time. Second, there are also sharp contrasts between wealthy and poor living in urban areas. Finally, urbanization can cause significant environmental degradation.

Urban areas have grown rapidly, from 746 million in 1950 to 3.9 billion in 2014. Although 54% of the world's population lives in urban areas, this rate is lowest in Africa (40%) and Asia (48%), which remain mostly rural. By contrast, about 80% of people in the Global North and Latin America and the Caribbean are urban. These geographic variations are quickly changing, however, as there is rapid urbanization in Africa and Asia, and by 2050 these regions are expected to be about 60% urbanized. About one in eight urban dwellers live in one of the world's 28 megacities with more than 10 million inhabitants each. But by 2040, there are projected to be 41 megacities, and most of these will be in the Global South (UN, 2014). Such a fast shift in population often means that the urban areas cannot keep up with the rapid influx of people (UN, 2018).

Urban areas can provide opportunities, particularly jobs, upward mobility, more gender equity and women's rights, and better health care. Because women tend to have more opportunities in urban areas that are outside the traditional rural patriarchal social rules, they often obtain an education, get a job, and have fewer children. Thus, urbanization is linked to lower population growth rates, as seen in Latin America and sub-Saharan Africa. In contrast, economic inequalities in cities are increasing. Cities demonstrate the harsh inequalities between rich and poor, where wealthy neighborhoods exist separately from slums and informal settlements. Poor people are often marginalized to the edge of cities, often in dense slums where there is no health care, no clean water or sewers, and a high level of pollution. The number of residents in slums is increasing rapidly, with an estimated 860 million people now in urban slums worldwide. Sub-Saharan Africa has the highest rates, with almost 62% of urban people living in slum conditions (UNFPA, 2018).

The environmental impacts of urbanization present both opportunities and significant challenges. Certainly, the cities provide an opportunity for increased sustainability, if policies are implemented and enforced. Dense urban areas can use resources more efficiently, plan sustainable land-use, and leave natural areas intact to promote biodiversity. On the other hand, there is significant environmental degradation in poorly planned urban areas.

Urban people require food, energy, water, and land. They produce large amounts of waste concentrated in one area. Indeed, urban populations tend to be more affluent than their rural counterparts, so they actually consume more food and more goods than rural populations. Energy use in urban areas, combined with paved surfaces, leads to the urban heat island effect, where temperatures are higher in the city than in the surrounding rural areas. This leads to additional energy consumption, with people using fans or air conditioners to cool their homes.

In addition, urban industries produce pollution and urban heat islands trap these air pollutants, causing poor air quality. Infiltration of precipitation is reduced in urban areas with concrete impermeable surfaces, causing rapid flooding and water pollution downstream. Clearly, urban environmental problems like water and air pollution are linked to human health concerns, which are higher in rapidly growing cities with lax pollution regulations and lack of money.

"Governance," which means the policy framework and authority to make a place function, is a key issue in urban environmental sustainability (Torrey, 2004). Rapidly growing cities and overlapping authorities for roads, housing, industry, utilities, and pollution are very inefficient and do not provide a comprehensive geographic overview for city leaders. A lack of complete data and unified authority to act can make the implementation of urban environmental management policies very difficult.

See also: Section 1: Case Study 1; *Section 2:* Human Population; *Section 3:* Sustainable Cities, Sustainable Development

FURTHER READING

Merriam-Webster. 2018. "Urbanization." https://www.merriam-webster.com/dictionary/urbanization.

Torrey, Barbara Boyle. 2004. "Urbanization: An Environmental Force to Be Reckoned With." Population Reference Bureau. http://www.prb.org/Publications/Articles/2004/UrbanizationAnEnvironmentalForcetoBeReckonedWith.aspx.

UN. 2014. "World Urbanization Prospects: The 2014 Revision, Highlights." United Nations Department of Economic and Social Affairs. https://esa.un.org/unpd/wup/publications/files/wup2014-highlights.Pdf.

UN. 2018. "Urbanization." United Nations Department of Economic and Social Affairs. http://www.un.org/en/development/desa/population/theme/urbanization/index.shtml.

UNFPA. 2018. "Urbanization." United Nations Population Fund. http://www.unfpa.org/urbanization.

Water Pollution

"Day after day, we pour millions of tons of untreated sewage and industrial and agricultural wastes into the world's water systems. Clean water has become scarce and will become even scarcer with the onset of climate change. And the poor continue to suffer first and most from pollution, water shortages and the lack of adequate sanitation" (UN, 2014). This statement from then–United Nations Secretary General Ban Ki-moon provides a useful introduction to the issue of water pollution.

Water pollution occurs when substances that are harmful to humans or the environment contaminate the water. This may include sewage, chemicals from hazardous waste sites, fertilizers from agriculture, or heavy metals from manufacturing, to name a few examples. Worldwide, nearly 2 billion people drink contaminated water that could be detrimental to their health (NIH, 2017).

The topic of water quality covers the provision of clean drinking water to every person on Earth while at the same time addressing the infrastructure to clean up the industrial pollutants and sewage waste created by human activities. The natural environment also depends on clean water. Overall, water quality issues are of global concern, but regulation is typically addressed at the national level.

Oil pollutes the water off Failaka, Island, Kuwait on April 4, 1991. Warfare destroyed wells, causing oil to leak and spread throughout the Persian Gulf. (Department of Defense)

The "Guidelines for Canadian Drinking Water Quality" are established by the Federal-Provincial-Territorial Committee on Drinking Water and published by Health Canada. The Canada Water Act of 1970 established a framework for collaboration among the federal and provincial (state) governments in water resource management. The two main sections relate to comprehensive water resource management and water quality management. It states that "no person shall deposit or permit the deposit of waste of any type in any waters composing a water quality management area" (Part II. R.S., c. 5(1st Supp.), s. 8.) (Government of Canada, 2018). This exemplifies the need, even within one nation, for multiple geographic scales and political agencies to work together when managing water resources.

For European countries, the European Union Drinking Water Directive of 1998 is the main law that regulates water quality for human consumption. Its objective is to "protect human health from adverse effects of any contamination of water intended for human consumption." Following the World Health Organization's guidelines for drinking water, a total of 48 chemical, microbiological, and indicator parameters must be tested regularly (EU, 2015). Member states of the EU are each responsible for their own national water legislation, which could have higher (but not lower) standards than the EU rules.

In the United States, the Clean Water Act (CWA) of 1972 set up the structure for the Environmental Protection Agency to establish and enforce water quality standards and to regulate water pollutants that can be discharged into navigable streams and rivers. CWA water regulations require permits in order for industry, municipalities, animal feeding operations, and other facilities to release water pollutants from their facility. This is called point-source pollution because it is from a specific pipe that comes from the industry and can thus be readily monitored. Three points to note here: 1) the EPA has, through CWA, drastically improved the quality of the nation's water; 2) U.S. water law is closely associated with navigable waters, because these are seen to have economic importance for the nation; and 3) point-source pollution has been the focus of regulation.

Most water pollution regulations focus on point-source pollution, which the CWA defines as: "discernible, confined and discrete conveyance, such as a pipe, ditch, channel, tunnel, conduit, discrete fissure, or container . . . from which pollutants are or may be discharged." By law, the term point-source also includes concentrated animal feeding operations, which are places where large numbers of animals are confined and fed. On the other hand, return flows from irrigated agriculture are not point sources (EPA, 2017). Instead, this nonpoint-source water pollution is diffuse, like water from a farm field, golf course, or natural ditches. These areas are harder to regulate because there is no simple way to monitor and test water as it flows overland.

Although regulations are determined by countries, water pollution is an important geographic issue for all people and ecosystems. "Water is life. It is a precondition for human, animal and plant life as well as an indispensable resource for the economy" (EU, 2018).

See also: Section 1: Case Study 2, Case Study 3

FURTHER READING

EPA. 2017. "Summary of the Clean Water Act 33 U.S.C. §1251 et seq. (1972)." U.S. Environmental Protection Agency. https://www.epa.gov/laws-regulations/summary-clean-water-act.

EU. 2015. "Drinking Water Legislation: The Directive Overview." European Commission. http://ec.europa.eu/environment/water/water-drink/legislation_en.html.

EU. 2018. "Environment: Water." European Commission. http://ec.europa.eu/environment/water/.

Government of Canada. 2018. "Canada Water Act. R.S.C., 1985, c. C-11." http://laws-lois.justice.gc.ca/eng/acts/C-11/FullText.html.

NIH. 2017. "Water Pollution." National Institutes of Health. https://www.niehs.nih.gov/health/topics/agents/water-poll/index.cfm.

UN. 2014. "Water Quality." United Nations Department of Economic and Social Affairs. http://www.un.org/waterforlifedecade/quality.shtml.

Water Scarcity

Water scarcity affects people on every continent around the world and can be in the form of physical scarcity or economic scarcity. Physical water scarcity occurs when there is simply not enough fresh water for the given population. Economic water scarcity is where countries lack the infrastructure needed to take water from rivers and aquifers. Over 1 billion people (almost 20% of the world's population) live in areas of physical scarcity. Another 1.6 billion people (almost 25% of humans) face economic water shortage (UN, 2014). Nearly 2.5 billion people live with no access to improved sanitation, meaning that over 30% of people in the world do not have access to a toilet. In fact, this lack of water infrastructure means that "more people have a mobile phone than a toilet" in some countries (Water.org. 2018).

Geographers understand that water scarcity has both environmental and social causes. Indeed, there is enough fresh water for all the people on Earth, but it is unevenly distributed and much of it is polluted or wasted. Many of the Earth's water resources are not managed sustainably. And water use has grown at more than twice the rate of population increase in the past 100 years and so some regions have chronic water shortages. "Water scarcity is a relative concept and can occur at any level of supply or demand" (UN, 2014). In other words, water scarcity varies from place to place, because different cultures have different expectations and behaviors related to water use. People in wealthy countries expect lush green golf courses in the desert. But in other cultures, people easily live with less water. In addition to social variations, water scarcity can be caused by an absolute constraint, such as reduced water supplies due to climate change.

The United Nations notes that water scarcity is linked to social and economic development in many countries. Whereas people in the Global North typically have clean, safe drinking water flowing into their homes with the turn of a faucet, many people in the Global South do not have access to clean, safe water for drinking, cooking, showering, or indoor plumbing. This affects development goals in several ways. It causes poverty and food scarcity. It disrupts schools and educational achievements. It affects gender equality, because females spend many hours a day walking and hauling water from a community well, rather than attending schools or jobs. Unreliable clean drinking water and lack of wastewater treatment leads to transmission of numerous diseases (UN, 2014). For example, water scarcity causes people to store water near their homes, which increases the number of mosquitos that can carry dengue fever, malaria, and other diseases (Reuters, 2009). Dirty water increases risks of cholera, typhoid fever, dysentery, and other infections. Two million people, mostly children, die each year from diarrheal diseases caused by drinking dirty water (WWF, 2018).

The human population has been able to grow and thrive by controlling the natural water system through digging water wells, building dams, and setting up vast irrigation systems. But these natural systems are increasingly stressed, due to drainage, agriculture, pollution, and climate change. More than half the world's wetlands have already disappeared, which reduces nature's ability to filter water and provide fish habitat (WWF, 2018).

Agriculture uses 70% of the world's fresh water, but much of this is wasted due to inefficient irrigation or trying to grow high-water crops in arid regions (WWF, 2018). Likewise dietary choices affect water uses, as it takes 2,400 liters (630 gallons) of water to produce one hamburger (NGS, 2018). Indeed, producing a calorie of meat requires 10 times as much water as a calorie of food crops (Mosher and Calderone, 2016). Water pollution in agriculture includes pesticide runoff from farm fields. Likewise human waste and industry both create pollution problems. Climate change from high levels of greenhouse gas emissions is changing weather patterns around the world, with droughts likely in some places and floods elsewhere; glaciers are already disappearing in some areas. These climate variables affect downstream freshwater supplies.

The terms water stress and water scarcity have specific definitions that are based on population numbers and the amount of water in a given area. Water stress occurs when annual water supplies drop below 1.7 million liters (450 million gallons) per person. Water scarcity is when annual water supplies drop below 1 million liters (264,000 gallons) per person; below 500,000 liters (132,000 gallons) is "absolute scarcity" and requires immediate action (UN, 2014).

According to the United Nations, with the current climate change scenario, half the world's population will be living in areas of high water stress by 2030, including up to 250 million people in Africa. In addition, water scarcity in some arid and semi-arid places will displace up to 700 million people. Sub-Saharan Africa has the most water-stressed countries of any region (UN, 2014).

Although 70% of Earth is covered by water, it is mostly in salty oceans. Only 2.5% of the planet's water is fresh water and just 1% is accessible, because much of it is frozen in glaciers and snowfields. So, in reality, only 0.007% of the Earth's water is available for human use, and much of this is being degraded by human activities (NGS, 2018).

See also: Section 1: Case Study 2; *Section 3:* Case Study 15, Water Conservation

FURTHER READING

Mosher, Dave and Julia Calderone. 2016. "Secret Government Report Warns of 'Potentially Catastrophic' Water Crisis." *Business Insider.* April 23. http://www.businessinsider.com/nestle-water-scarcity-meat-consumption-cable-2016-4.

NGS, 2018. "Freshwater Crisis." *National Geographic.* https://www.nationalgeographic.com/environment/freshwater/freshwater-crisis/.

Reuters. 2009. "Facts on World's Water Shortage." June 16. http://www.reuters.com/article/us-water-beverages-factbox-idUSTRE55F6WW20090616.

UN. 2014. "Water Scarcity." United Nations Department of Economic and Social Affairs. November 24. http://www.un.org/waterforlifedecade/scarcity.shtml.

Water.org. 2018. "The Water Crisis." www.Water.org.

WWF. 2018. "Water Scarcity: Overview." World Wildlife Fund. https://www.worldwildlife.org/threats/water-scarcity.

Section 2
Introduction: How the Environment Affects Humans

The natural environment has had a significant impact on humans for most of their history on Earth. From climate extremes, to harsh landscapes, to pest infestations and natural hazards, humans have often been at the will of nature. People's close relationship with nature also meant that it played a key role in the development of their cultures, languages, and religions.

Early humans worshiped gods that represented nature, and archeologists have found carved figurines that they believe to represent an Earth goddess from 10,000 BCE (12,000 years ago). The Greek Earth goddess, Gaea, was the mother of all life for that society in about 3,000 BCE, and the Roman goddess, Terra, also represented humans' worship of the Earth in about 2,000 BCE (Neumann, 1955). Similar Earth gods were present across the planet: from Nigeria with the Ibo goddess named Ane, to North America with the Hopi goddess named Tuuwaqatsi, to Mongolia's goddess named Etugen Eke (Baring and Cashford, 1993). From early times to fairly recent times, we see the emotional, cultural, and religious influence of nature on our lives.

The first section of this book gave examples of how humans have affected the environment, Section 2 will describe how nature has affected humans. Think about this from both a positive and negative perspective—natural beauty provides emotional calm and happiness, but natural hazards, like an earthquake, are destructive and life threatening. In this section, both these angles will be explored, because nature provides both opportunities and constraints. For example, parks and green space give people wonderful recreational opportunities, but floods force them to adapt to natural forces.

HUMANS' NEED FOR NATURE

In the late 1800s, at a time when most humans only viewed nature as resources to be taken from the Earth and used for their own benefit, John Muir stood out by saying that people need to preserve nature. He said that nature is necessary for human health and emotional well-being. "Thousands of tired, nerve-shaken, over-civilized people are beginning to find out that going to the mountains is going home; that wildness is a necessity" wrote John Muir in 1901 in his book called *Our National Parks* (Muir, 1901: 1). Indeed it is because of Muir's dedication to fighting for preservation that the U.S. National Parks exist today. Administered through the U.S. Department of Interior, the National Parks Service now protects 59 parks, 87 national monuments, 19 national preserves, 51 national historic parks, 78 national historic sites, and other important sites (NPS, 2018). Millions of people visit national parks or historic sites each year, and thus benefit from the early work by Muir and others.

Aldo Leopold was a forester and conservationist, but he realized that a healthy environment was more than just an issue of logical management. By 1949, he was writing about the need for a land ethic—just as people have an ethical code for our treatment of other people in our communities, there needs to be an ethical relationship with the land. He wrote: "The land ethic simply enlarges the boundaries of the community to include soils, waters, plants, and animals, or collectively: the land." (Leopold, 1949: 203). Thus, nature does not belong to humans; rather, we must develop an ethical relationship with it in order to preserve and maintain ecological balance. His book was the first to describe an ecocentric approach for people's relationship to the environment (DesJardins, 2013). Leopold popularized the idea of nature as a whole system and of ecology as a complex system.

The Gaia hypothesis was developed in the 1960s by James Lovelock, who was a scientist, inventor, and consultant for the U.S. National Aeronautics and Space Administration (NASA). The Gaia hypothesis states that the Earth interacts with all life on it and these systems have evolved together as a single living system. In other words, the total sum of the Earth is greater than its individual parts of air, water, soil, carbon, etc. Instead, the Earth is a living, self-regulating system. To be honest, scientists have disproven this hypothesis, due in part to the fact that humans clearly do have the ability to greatly disrupt or even destroy our planet. But the value of Gaia is the concept that the Earth is composed of many complex, interrelated ecological systems that must be kept in balance (Lovelock, 1972; Lovelock and Margulis, 1974; New Scientist, 2018).

ENVIRONMENTAL SYSTEMS AND SCIENCE

Studying and understanding Earth's environmental systems has been the goal of scientists for centuries. The scientific method is used to gather information and create theories to help people understand natural systems. Science provides the basis of knowledge that is created from observing, correlating, and experimenting in a systematic way. It assumes that objective reality is based on certain principles and laws that occur consistently, so causes and effects are explainable. The scientific method is used to question, hypothesize, test (experiment), and draw conclusions. Over time, if numerous scientists repeat similar research projects and consistently obtain the same results, their conclusions are broadly accepted and might be known as a theory. This only refers to really significant, proven principles like the theory of gravity (Universal Law of Gravitation).

What is really cool about this approach is that every person can follow the same stated methods and come up with the same results, or likewise, develop their own methods that other people can follow. In other words, science is something every person can learn about, follow, and contribute to. It is not just a sudden random thought; rather, it is a very logical process, which provides confidence that the results are meaningful.

In fact, to get scientific research published, it must go through a very strict process of review (Spier, 2002). It is called "peer review" and it means that once a scientist submits her research findings to a journal, the editor of that journal must send the draft article out to several (usually three) "peer" reviewers, meaning that these people are experts in that area of research. The reviewers each read the paper and make sure everything checks out right. Then they provide comments back to the editor. Sometimes reviewers say: "This is junk and it shouldn't be published!" Or they make suggestions for clarifying the research process. All this is done anonymously so the author does not know who reviewed their paper and the reviewers do not know who wrote the draft paper. Once all the comments are given back to the editor, she decides what to do. She can tell the author to edit the paper based on the reviewers' comments. Or she can reject the paper and say "Nope! This is not good science."

So, for a research article to actually come into print, it has followed a strict scientific process that has been checked by other expert scientists and reviewed by several editors. This peer-review process means that there is integrity, or truth, to what is published in professional scientific journals. But keep in mind that social media does not have a peer-review process. Anybody can post anything, especially opinions, with no review of facts or scientific methods. Scientifically verified facts are not opinions, and vice versa—opinions are not facts.

Remember that scientists must be very specific when they pose research questions to be sure that they ask specific questions that can be answered by following their given methods. For example, "pesticides hurt the environment" is not a testable hypothesis because we do not have a measurement for what "hurting the environment" means. In contrast, a testable hypothesis could be stated as: "The pesticide atrazine causes deformities in the frog species *Xenopus laevis*." This is testable using the scientific method of observation, question, and experiment. And yes, it does, according to numerous peer-reviewed research articles, including a paper in *Proceedings of the National Academy of Science* (Hayes et al., 2010).

From this scientific process, scientists have learned a lot about how the Earth's environmental systems work. The hydrologic cycle is a closed system through which water (H_2O) moves from precipitation, to surface runoff into rivers, flows into the ocean, and evaporates back into the atmosphere to again fall as precipitation. There are also cycles for nitrogen, carbon, and other key elements. Each of these cycles affects humans by narrowing their options for resource use or misuse. Humans must operate within these natural systems or over time they will be damaged beyond repair.

SILENT SPRING AND AWARENESS

People did not always realize that they could permanently alter nature, but in 1962, the book *Silent Spring* by Rachel Carson brought this fact to the forefront. This book tells the story of a springtime when we heard no birds singing—because humans' use of the pesticide DDT had killed them all.

The author was an established biologist when she wrote the book, yet the pesticide industry still attacked her and said she was a silly woman and a communist for writing about the need to reduce chemical use. But Carson wrote the book to reach the general public and tell them, in a most accessible way, of her scientifically based concerns about chemicals in the environment.

Indeed at this time, in the early 1960s, there were no rules or regulations about chemical use. So people sprayed anything, anytime, with little concern for environmental or human health outcomes. Most people had no idea that a chemical could be harmful to humans; they just assumed that the chemicals only killed the targeted species (bugs that ate crops, for example). But this is not the case; innocent species are also affected by pesticides. *Silent Spring* was initially controversial, then popular, and then it helped spur the environmental movement.

Because of the book's popularity, President Kennedy initiated a committee to investigate the use of synthetic chemicals in our environment. Rachel Carson was invited to speak to a congressional subcommittee where she explained: "Our heedless and destructive acts enter into the vast cycles of the earth and in time return to bring hazard to ourselves" (Griswold, 2012). Indeed, short-term economic gain cannot win over the long-term laws of nature. People must realize this and control their actions so that environmental systems survive and humans can live.

Not everyone acts out of inherent concern for nature, as Muir or Leopold would suggest. Instead people often make better ecological choices simply because they selfishly realize that nature will no longer provide for them if it is severely degraded. This may not be morally upstanding or ethically sound, but it does still spur people to protect the environment. So is this a bad thing? As long as a person makes a choice that keeps the environment healthy, does it matter what their motivation behind it is?

Indeed, part of the problem is lack of awareness. Do people realize how much they depend on nature? Probably not. Today, there is a huge disconnect between people's reliance on environmental systems and the overarching belief that humans control nature. Many people live in highly urbanized areas where skyscrapers obscure their view of the sky, they see no green gardens or parks, and they have water piped into their faucets with little knowledge of where it comes from. This disconnect is dangerous, because it makes people ignorant to our ultimate dependence on the environment. People need to understand how much they depend on the environment and how much their existence is influenced by physical geography.

LANDSCAPES AND GEOGRAPHY

Landscapes influence humans. Think of how early settlers to the U.S. West had such difficulty crossing the Rocky Mountains until the railroad managed to cut through the rugged terrain. Clearly the mountains were a physical barrier to settlement. Before the railroad, it took almost six months to travel between New York and California, but after the transcontinental railroad was completed, it only took five days (History.com, 2018).

Likewise reaching back farther into history the oceans were a significant barrier to human migration until the compass made seafaring somewhat possible, although still dangerous. The trip from Europe to North America, for example, took five months in 1750 (Mittelberger, 1750). Now it is only a six-hour airplane flight.

Overall, landforms and physical geography provided both opportunities and constraints for humans throughout history. Rich soils provided the opportunity for agriculture, but deserts constrained humans, as food and water are necessities for settlement.

The Fertile Crescent is a region of watersheds starting with the Nile River in North Africa and arcing up the east coast of the Mediterranean Sea to link into the Euphrates River and Tigris River in present-day Iraq, Syria, and Turkey. This region had the right hydrology, soils, and landscape to promote the opportunity for nomadic people to settle and harvest wild grains, which they then managed to domesticate and plant as crops (ICARDA, 2017). This was a big shift in human lifestyle between roaming the landscape as hunters and gatherers to settling in one region and specializing in certain jobs. Some people became farmers, others dealt more in trade, or learning—and even developed a written form of language—some 3,000 years ago (Diamond, 1997; Mark, 2011). All this and more humans owe to the perfect environmental conditions in that region. With different weather patterns or terrain, these achievements would not have been possible.

But rivers provide more than just fertile land; they are also a key transportation route. Without the rivers of Europe, the Vikings would not have been able to raid deep into Europe. Without the Mississippi River, the Europeans would not have been able to explore the central part of the United States so easily. Rivers provided seafaring people with a highway for exploration that would not have been possible otherwise.

Another way to think of the important influence that the environment has on humans is to consider mountainous regions. Mountainous regions have fewer people because these areas have less agricultural land and less development of transportation routes and industries. If you layer data on both elevation and population density, there is a strong correlation between high mountains and lower numbers of people. Japan has a mountainous interior and so its population is concentrated more along the coastlines rather than in its rugged interior. Another example can be noted in the topography along the western regions of South America, where the Andes Mountains provide high elevation, and there is a lower population in these regions of Bolivia and Peru. Also obvious is the low population density seen in Tibet and Nepal where the Himalaya Mountains provide rugged terrain.

NATURAL HAZARDS

In addition to steadfast landscapes influencing humans throughout history, sudden environmental events can have a major impact. Environmental geographers have studied natural hazards extensively. An accepted definition comes from well-known geographers who define natural hazards as "those elements of the physical environment" that are harmful to humans and are caused by "forces extraneous" to us (Burton et al., 1978). Keep in mind that natural processes, like a volcano, are only a hazard to humans if they are nearby; for millions of years, these natural processes simply occurred but were not a "hazard" because people were not affected. Humans are a bit selfish this way—they only consider an event to be a hazard if it is harmful to them!

Natural hazards have had significant impacts on humans throughout history (Coppola, 2007). A few examples include the famous Roman city of Pompeii, which was suddenly covered in lava when the volcano Mount Vesuvius erupted in 79 CE, killing at least 1,500 people. There are many lesser known events that had high death tolls. In 1556, a huge earthquake, centered in Shaanxi, China, killed approximately 830,000 people. And in 1839, an enormous tropical cyclone made landfall at Coringa, India, killing approximately 300,000 people (Leroy and Gracheva, 2013). These are just a few of the hundreds of examples in which the forces of nature had significant impact on humans.

There are many types of natural hazards, which can be categorized as 1) hydrologic (flood, avalanche); 2) climatological (extreme temperatures, drought, wildfires); 3) geophysical (earthquake, tsunami, volcano, landslide, subsidence); 4) biological (disease epidemics and insect/animal plagues); and 4) meteorological (hurricane, tornado, storm surge). Note that the terms hurricane, typhoon, and tropical cyclone are used to denote the same phenomenon, only the name varies by region (IFRC, 2018; Montz et al., 2017; OAS, 1990).

Overall, then, there is a wide variety in the types, causes, and geographical extent of these numerous types of natural hazards. Still, they all have two things in common. First, they are naturally occurring physical phenomena. Second, they can be caused either by rapid- or slow-onset events (IFRC, 2018).

The U.S. Geological Survey (USGS) Natural Hazards Mission Area includes six science programs: Coastal and Marine Geology, Earthquake Hazards, Geomagnetism, Global Seismographic Network, Landslide Hazards, and Volcano Hazards. They work with other agencies, such as the U.S. National Oceanic and Atmospheric Administration (NOAA) to study and make warnings for storms and natural hazards (USGS, 2017). They note that "scientific research—founded on detailed observations and improved understanding of the responsible physical processes—can help to understand and reduce natural hazard risks and to make and effectively communicate reliable statements about hazard characteristics, such as frequency, magnitude, extent, onset, consequences, and where possible, the time of future events" (Holmes, et al., 2013: 1).

NATURAL HAZARDS TO NATURAL DISASTERS

Natural hazards are a reality on Earth. Natural disasters are created by human actions. In fact, the United Nations Office of Disaster Risk Reduction states: "There is no such thing as a 'natural' disaster, only natural hazards" (UNISDR, 2018). As a result, scientists work to understand the natural processes that cause natural hazards, but it is human actions that are responsible for the negative outcomes that people feel after a hazard event. Poverty, education, location, and land-use choices can greatly intensify the damage to humans' lives and livelihoods. Prevention goes a long way.

At the global level, the United Nations Environmental Programme (UNEP) has a specific office—the Disasters and Conflicts Programme—that works to reduce the threat of hazards to humans. This group works to build partnerships across national boundaries to provide expertise and efficient solutions to disasters and conflicts in three areas: risk reduction, preparedness and response, and recovery (UNEP, 2018). Further, the United Nations' Environmental Emergencies Centre (UNEEC) and the UN Office for the Coordination of Humanitarian Affairs (OCHA) have online training modules. These help educate people on environmental emergency preparedness and response. These online courses are free and open to everybody (UNEEC, 2018).

The United Nations Office for Disaster Risk Reduction was established by the United Nations in 1999 to implement the United Nations International Strategy for Disaster Reduction (UNISDR). The overall goal is to "ensure synergies among the disaster reduction activities of the United Nations system and regional organizations and activities in socio-economic and humanitarian fields" (UNISDR, 2018). Thus this group calls itself "a system of partnerships," and they provide information to assist many disaster relief groups (UNISDR-Partners, 2018). Governments, United Nations programs, regional groups, international financial institutions, and non-governmental organizations are all partners in this effort to share information and knowledge. Partners include groups that many people are familiar with: the UN Environmental Programme (UNEP), UN Development Programme (UNDP), UN Education, Scientific and Cultural Organization (UNESCO), International Union for Conservation of Nature, World Wildlife Fund, The Nature Conservancy, European Union, Council of Europe, Stockholm Environment Institute, and many others (UNISDR-Partners, 2018).

As a subgroup within the UNISDR, the Partnership for Environment and Disaster Risk Reduction (PEDRR) promotes an ecosystem-based disaster risk reduction approach, which they want to make mainstream within development planning at global, national, and local levels. They have a vision of "[r]esilient communities as a result of improved ecosystem management for disaster risk reduction (DRR) and climate change adaptation (CCA)." The scientists work together through PEDRR to combine their expertise and suggest policy changes to promote beneficial inter-relationships among ecosystems, livelihoods, and disasters. Specifically, PEDRR notes that:

> Ecosystem management is an integral part of disaster risk reduction.
> Disasters due to natural hazards, such as tropical cyclones, avalanches and wildfires, can have adverse environmental consequences. On the other hand, degraded environments can cause or exacerbate the negative impacts of disasters.
> Healthy and well-managed ecosystems—such as coral reefs, mangroves, forests and wetlands—reduce disaster risk by acting as natural buffers or protective barriers, for instance through flood and landslide mitigation and water filtration and absorption. At the same time, fully-functioning ecosystems build local resilience

against disasters by sustaining livelihoods and providing important products to local populations. (PEDRR, 2018: 1)

Additional key policies regarding disasters include the Hyogo Framework for Action (HFA) 2005–2015 and the current Sendai Framework for Disaster Risk Reduction 2015–2030. These are international agreements that most nations in the world have signed. They outline the goals of coordination among multiple stakeholders through a network of global, national, and local partners (UNISDR-Partners, 2018).

This integration is necessary because specific conditions act to particularly aggravate an already hazardous situation. For example, poverty makes people more vulnerable to the risks of disasters. Climate change may intensify some storms in the future. These factors can lead to increased frequency and severity of disasters and more complexity in the cleanup necessary in the aftermath of the disasters (IFRC, 2018).

Climate change, also aggravated by human causes, is typically considered a natural hazard, because it is linked to natural environmental phenomena and outcomes to the natural environment. Climate change provides a meaningful example of two key concepts in natural hazards: mitigation and adaption. Mitigation is what people can try to do beforehand to reduce or change the hazard. This example—reducing climate change—involves reducing the flow of heat-trapping greenhouse gases into the atmosphere. Adaptation, on the other hand, is what people can do after a disaster or in the midst of a recurring disaster to reduce their vulnerability. People must adapt to life in a changing climate, and this will mean adjusting to actual or expected future climate challenges (NASA, 2018). For example, as sea levels rise, people will be forced to move inland from coastal areas.

ENVIRONMENT AND VULNERABILITY

There are significant variations in the effects of natural disasters in a more developed country and a less developed country. One recent example is Hurricane Matthew that struck the Caribbean and eastern United States in October 2016. The hurricane hit Haiti, then Cuba, and traveled a long distance up the Atlantic coast of Florida, Georgia, South Carolina, North Carolina, and Virginia. Although the storm covered a much larger area of the United States, many more lives were lost in Haiti. Due to lack of funding for evacuation, fewer warning systems, inadequate housing, and lack of options due to poverty, the death toll in Haiti was approximately 1,000, but in the United States fewer than 50 lives were lost. This disaster occurred within a few years of the devastating 2010 earthquake in Haiti, which caused an estimated 300,000 deaths and left the population particularly vulnerable (Ahmedoct, 2016).

In terms of human life, time and again, poorer regions are hardest hit by natural disasters. Whereas in wealthier areas, property damage is the most significant outcome after a natural disaster.

CASE STUDIES 6-10

The Case Studies and Key Concepts presented in this section describe the many ways the environment affects humans, despite the fact that modern society tends to emphasize that humans are in control. Environmental geographers realize that nature still has a significant influence on people's lives.

Case Studies 6 and 7 describe current problems related to climate change. Yes, climate change is driven by human action, but this topic still exemplifies how the environment affects humans. Therefore, the reason climate change is relevant in this section of the book is that it fully demonstrates how humans have been temporarily fooled. Indeed, modern society thought it was in control of the environment and that dumping gigatons of greenhouse gases into the atmosphere year after year would be okay. Wrong! Although the Earth's atmospheric system is truly amazing, it has limits. Humans have pushed it too far and are causing major damage to the Earth's climate.

Case Study 6 describes the facts of climate change and what is causing it. This environmental information is then linked to policy aspects of the issue. Overall, the title describes the central theme: climate change is occurring, regardless of politics.

Case Study 7 builds on this notion, but provides a real-world example of current impacts of climate change. In this case, the topic of climate refugees is introduced, with a specific focus on rising sea levels that are inundating many low-lying island nations. Entire countries are literally being overtaken by saltwater. What options do these citizens have? Other nations will not accept them because "refugee" status is defined as fleeing from political persecution, not climate alteration.

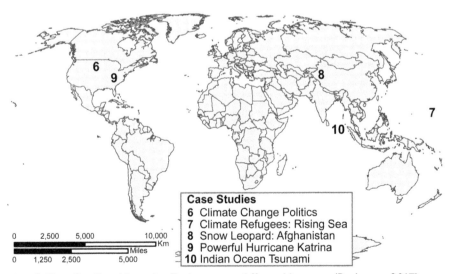

Section 2 Case Studies: How the Environment Affects Humans. (Bathgate, 2017)

Case Study 8 illuminates people's intricate relationship with nature. The geographic setting is remote, mountainous Afghanistan where people eke out a living by herding sheep and goats. They have few modern conveniences, and their livelihoods were threatened by the snow leopard, a species pushed to the edge of extinction from human activities. Recently, with outside help from conservation groups, this has become a fascinating example of potential ecotourism. People are now working with their natural environment for mutual benefit and preservation of this rare leopard.

Case Study 9 and Case Study 10 depict the true power of nature over humans. Luckily scientists have the technical know-how to predict the most devastating storms, but sadly the warning systems cost money, which is not always available. Hurricane Katrina was a massive storm that hit the Gulf of Mexico in 2005, killing 1,800 people and displacing hundreds of thousands. Poor people were particularly hit hard, as they had fewer options for escape or rebuilding in the aftermath. Still, the devastation of this storm contrasts with the high loss of life caused by the 2004 South Asian Tsunami, in which 230,000 people died. In this case, there was no warning system and little education about potential tsunamis, which meant more people were extremely vulnerable.

Overall, these five Case Studies describe the many ways that the environment influences people's lives both on a daily basis and in extreme events. The Key Concepts carry on this theme, providing details about natural hazards, emergency management, and resource limits. In addition, several approaches for interacting with nature are described: ecotourism, environmental ethics, adaptation, and mitigation, among others.

FURTHER READING

Ahmedoct, Azam. 2016. "After Hurricane, Haiti Confronts Scars from 2010 Earthquake Recovery." *New York Times*. October 21. https://www.nytimes.com/2016/10/22/world/americas/hurricane-matthew-haiti-earthquake.html?_r=0.

Baring, Anne and Jules Cashford. 1993. *The Myth of the Goddess: Evolution of an Image*. London: Penguin.

Burton, Ian, Robert W. Kates, and Gilbert F. White. 1978. *The Environment as Hazard*. New York: Oxford University Press.

Carson, Rachel. 1962. *Silent Spring*. Boston: Houghton-Mifflin.

Coppola, Damon. 2007. *Introduction to International Disaster Management*. Elsevier: Oxford. https://www.elsevier.com/books/introduction-to-international-disaster-management/coppola/978-0-12-801477-6.

DesJardins, Joseph R. 2013. *Environmental Ethics: An Introduction to Environmental Philosophy*, 5th ed. Boston: Wadsworth.

Diamond, Jared. 1997. *Guns, Germs, and Steel*. New York: Norton.

Griswold, Eliza. 2012. "How 'Silent Spring' Ignited the Environmental Movement." *New York Times Magazine*. September 21. http://www.nytimes.com/2012/09/23/magazine/how-silent-spring-ignited-the-environmental-movement.html?_r=0).

Hayes, Tyrone B., V. Khoury, A. Narayan, M. Nazir, A. Park, T. Brown, . . . S. Gallipeau. 2010. "Atrazine Induces Complete Feminization and Chemical Castration in Male African Clawed Frog (*Xenopus laevis*)." *Proceedings of the National Academy of Science of the U.S.A.* 107(10): 4612–4617.

History.com. 2018. "Express Train Crosses the Nation in 83 Hours." This Day in History: June 4, 1876. http://www.history.com/this-day-in-history/express-train-crosses-the-nation-in-83-hours.

Holmes, Robert R., Jr., Lucile M. Jones, Jeffery C. Eidenshink, Jonathan W. Godt, Stephen H. Kirby, Jeffrey J. Love, . . . Suzanne C. Perry. 2013. "U.S. Geological Survey Natural Hazards Science Strategy—Promoting the Safety, Security, and Economic Well-Being of the Nation." U.S. Geological Survey Circular 1383-F. https://pubs.usgs.gov/circ/1383f.

ICARDA. 2017. "The Fertile Crescent." International Center for Agricultural Research in the Dry Areas. Consortium of International Agricultural Research Centers. https://www.icarda.org/fertile-crescent.

IFRC. 2018. "Types of Disasters: Definition of Hazard." International Federation of Red Cross and Red Crescent Societies. http://www.ifrc.org/en/what-we-do/disaster-management/about-disasters/definition-of-hazard.

Leopold, Aldo. 1949. *A Sand County Almanac and Sketches Here and There*. New York: Oxford University Press.

Leroy, Suzanne and Raisa Gracheva. 2013. "Historical Events." *Encyclopedia of Natural Hazards*. Dordrecht, The Netherlands: Springer Science.

Lovelock, James E. 1972. "Gaia as Seen through the Atmosphere." *Atmospheric Environment* 6 (8): 579–580.

Lovelock, James E. and Lynn Margulis. 1974. "Atmospheric Homeostasis by and for the Biosphere: The Gaia Hypothesis." *Tellus*. Series A. 26(1–2): 2–10.

Mark, Joshua. 2011. "Writing." *Ancient History Encyclopedia*. April 28. http://www.ancient.eu/writing.

Mittelberger, Gottlieb. 1750. "Journey to Pennsylvania in the Year 1750 and Return to Germany in the Year 1754." German Society of Pennsylvania. http://www.swarthmore.edu/SocSci/bdorsey1/41docs/40-mit.html.

Montz, Burrell E., Graham A. Tobin, and Ronald R. Hagelman III. 2017. *Natural Hazards: Explanation and Integration*, 2nd ed. New York: Guilford.

Muir, John. 1901. *Our National Parks*. Boston: Houghton, Mifflin and Company. http://vault.sierraclub.org/john_muir_exhibit/writings/our_national_parks.

NASA. 2018. "Responding to Climate Change." U.S. National Aeronautics and Space Administration. https://climate.nasa.gov/solutions/adaptation-mitigation.

Neumann, Erich. 1955. *The Great Mother: An Analysis of the Archetype*. Princeton, NJ: Princeton University Press.

New Scientist. 2018. "James Lovelock and the Gaia Hypothesis." *Newscientist.com*. https://www.newscientist.com/round-up/gaia.

NPS. 2018. "Frequently Asked Questions." U.S. National Park Service. https://www.nps.gov/aboutus/faqs.htm.

OAS. 1990. "Disaster, Planning and Development: Managing Natural Hazards to Reduce Loss." Organization of American States. https://www.oas.org/dsd/publications/Unit/oea54e/ch05.htm.

PEDRR. 2018. "Understanding the Link between Environment, Livelihoods and Disaster." Partnership for Environment and Disaster Risk Reduction. http://pedrr.org/about-us.

Spier, Ray. 2002. "The History of the Peer-Review Process." *Trends in Biotechnology* 20(8): 357–358.

UNEEC. 2018. "About the EEC." United Nations Environmental Emergencies Centre. http://www.eecentre.org/about-eec.

UNEP. 2018. "Disasters and Conflicts." United Nations Environmental Programme. http://web.unep.org/disastersandconflicts/who-we-are/overview.

UNISDR. 2018. "Who We Are." United Nations Office for Disaster Risk Reduction. https://www.unisdr.org/who-we-are.
UNISDR-Partners. 2018. "ISDR Partnership." United Nations Office for Disaster Risk Reduction. http://www.preventionweb.net/english/hyogo/isdr.
USGS. 2017. "Natural Hazards." U.S. Geological Survey. https://www2.usgs.gov/natural_hazards.

Case Studies

Case Study 6: Climate Change Is Occurring—Regardless of Politics

In 2014, then–United Nations Secretary Ban Ki-moon noted that "science has spoken" and the world must phase out the use of fossil fuels by the year 2100 in order to avoid irreversible climate change. He called for all world leaders to act quickly (Feeney, 2014). Indeed hundreds of climate experts, working together to assess the current data, agree that "most of the observed increase in global average temperatures since the mid-twentieth century is very likely due to the observed increase in anthropogenic GHG concentrations" (UN Chronicle, 2007).

Climate change is the most dangerous environmental challenge for our planet. It also represents the complex and integrated approach that environmental geographers use to understand and solve environmental problems. There are human components (the economy, policies, and beliefs, for example), and there are environmental factors (atmospheric science, greenhouse gas emissions, and rising seas, for example). These multiple human and environmental factors all influence one another too. This makes a shift in policy both very difficult, as it requires changes in economic priorities, and absolutely necessary in order to address the scientifically proven environmental change that is occurring.

THE CONTROVERSY

This is a global phenomenon, but the United States is particularly notable for its skepticism in accepting the facts about climate change and people's contribution to it (Foran, 2016). This is at least partially due to a misinformation campaign paid for by corporations that profit from fossil fuels (Hall, 2015; Nuccitelli, 2015). U.S. politicians, often under the influence of fossil fuel industry lobbyists, have been slow to act in combating climate change. But the scientific data are clear: climate change is occurring and human activity is responsible.

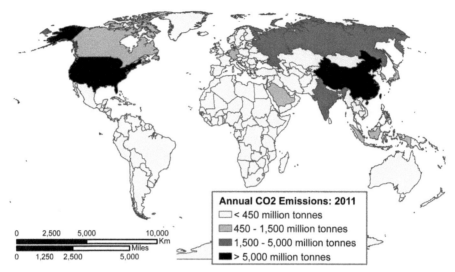

Annual Carbon Dioxide (CO_2) Emissions, 2011. (Bathgate, 2017)

While some politicians, industries, and media outlets still try to stir doubt, the fact is that we have known about climate change for a long time. By 1977 scientists were fairly certain that Earth was warming and that we would experience environmental consequences in the next century. President Carter put solar panels on the White House and began the conversation about conserving energy. In 1978, there was a weak attempt to stimulate climate research under the National Climate Program Act, but this was quickly shut down when President Reagan took office. He also took the solar panels off the White House. The real shift came in the late 1980s when the media began to cover record heat waves and droughts. In 1988, several climate scientists, including NASA scientist Dr. James E. Hansen, testified before the U.S. Senate Energy and Natural Resources Committee (AIP, 2018). Hansen said that "evidence is pretty strong that the greenhouse effect is here." He and the other scientists recommended a rapid reduction in burning fossil fuels and an effort to reforest large areas, because trees absorb carbon dioxide and store carbon (Shabecoff, 1988). The popular press picked up on this topic, and the July 11, 1988, *Newsweek* cover warned about the danger of the "Greenhouse Effect."

Interestingly, this scientific consensus has been clearly documented for many years; already in 2004 we see articles that state the overwhelming endorsement among climate scientists (Oreskes, 2004). It continues right up to the present (NASA Consensus, 2018). In 2016, another study showed that "the number of papers rejecting AGW [Anthropogenic, or human-caused, Global Warming] is a miniscule proportion of the published research, with the percentage slightly decreasing over time. Among papers expressing a position on AGW, an overwhelming percentage (97.2% based on self-ratings, 97.1% based on abstract ratings) endorses the scientific consensus on AGW" (Cook et al., 2016: 6).

Back in 2009, a statement was issued from 18 eminent scientific associations, including the American Association for the Advancement of Science, American Geophysical Union, and the American Meteorological Society. This letter stated: "Observations throughout the world make it clear that climate change is occurring, and rigorous scientific research demonstrates that the greenhouse gases emitted by human activities are the primary driver" (American Scientific Societies, 2009). And this consensus continues to grow. Hundreds more American and international organizations have noted that climate change is caused by human actions, including the American Medical Association, the U.S. National Academy of Sciences, the European Geosciences Union, the European Science Foundation, and the World Health Organization, to name a few (NASA Consensus, 2018; Worldwide Scientific Organizations, 2018). Recently, concern about human influence on the Earth's climate comes from a wide variety of sources. The U.S. military notes that climate change is an issue of national security. Even the Pope has urged the global community to take action (Bolstad, 2016; Dennis, 2016).

HUMAN INFLUENCE ON CLIMATE CHANGE

Let's review the basics. The greenhouse effect is a good thing, in general, as it helps moderate the Earth's climate. So our atmosphere is like a cozy quilt that keeps us from getting too cold. Mars, our relatively nearby neighbor, has a very thin atmosphere, about 100 times thinner than ours here on Earth. So Mars is really, really cold: its average temperature is −60°C (−80° F). In the winter, on the poles, it can drop to −125°C (−195°F). Even in the summer, if the temperature rises to 20°C (70°F) at noon at the equator—that night it would drop to −73°C (−100°F) (Space.com, 2012). Earth is lucky to have more moderate temperatures that allow long growing seasons for crops; biodiversity of species; and oceans, fresh water, and ice, which give us our wonderful hydrologic cycle, on which life on Earth depends.

Second, although we owe our very existence to our atmosphere, we humans have not been treating it well. Since our Industrial Revolution, beginning in the 1880s, we have spewed tons of chemicals into the atmosphere, primarily due to burning fossil fuels. The pre-industrial concentration of carbon dioxide in the atmosphere was 280 parts per million (ppm), but it has risen to over 407 ppm in 2018 (ESRL 2018; Kahn, 2017; National Wildlife Foundation, 2018). This is the highest amount of CO_2 in our atmosphere in the last several million years (NASA, 2013) and it is causing global warming.

"Warming of the climate system is unequivocal, and since the 1950s, many of the observed changes are unprecedented over decades to millennia. The atmosphere and ocean have warmed, the amounts of snow and ice have diminished, the sea level has risen, and the concentrations of greenhouse gases have increased" (IPCC, 2013: 2). So the fact is that climate change is occurring.

What makes this an immediate problem, and one people must address is that "[h]uman influence on the climate system is clear, and recent anthropogenic emissions of greenhouse gases are the highest in history. Recent climate changes have had widespread impacts on human and natural systems" (IPCC Synthesis, 2014: 2).

Human actions, particularly burning fossil fuels and deforestation, have caused atmospheric CO_2 to increase significantly since the Industrial Revolution (NASA, 2018). Informative geographic information system (GIS) maps can be found from the U.S. Energy Information Administration (EIA), which analyzes and provides energy information (EIA, 2018). Global temperature increases are expected to be within the range of 0.5°F to 8.6°F by 2100, with a likely increase of at least 2.7°F, in the 21st century (EPA, 2017; IPCC Projections, 2007). As noted by NASA, evidence of rapid climate change is seen in sea level rise, global temperature rise, warming oceans, shrinking ice sheets, declining arctic sea ice, glacial retreat, extreme weather events, ocean acidification, and decreased snow cover (NASA, 2018).

THE INTERGOVERNMENTAL PANEL ON CLIMATE CHANGE

The international group that assesses the scientific data and information related to climate change is the Intergovernmental Panel on Climate Change (IPCC). In 1988, the IPCC was established by the World Meteorological Organization (WMO) and United Nations Environment Programme (UNEP) to help government leaders gain information on the science, impacts, and risks of climate change (AIP, 2018). There are now 195 nations that are members of the IPCC. Their work helps policy makers determine the best options for mitigating the causes of the changing climate and adapting to changes that occur. In addition to providing consistent data, the IPCC reports help inform policies that are negotiated at the UN Climate Conference (United Nations Framework Convention on Climate Change) each year. The IPCC commonly presents models or scenarios that project future climate changes based on detailed scientific data; this helps people understand how current actions could actually affect future generations (IPCC, 2018).

To be clear, the IPCC is a group of hundreds of expert climate scientists who volunteer their time to write the reports in teams. There is a structure of expertise: coordinating lead authors work with many lead authors, who gain knowledge from hundreds more expert contributing authors. Each IPCC report goes through many drafts and reviews by thousands of additional scientists before publication. This ensures that the information is objective, comprehensive, and developed in an open way. The IPCC scientists are organized in four broad topics:

- Working Group I: Physical Science Basis
- Working Group II: Impacts, Adaptation, and Vulnerability
- Working Group III: Mitigation of Climate Change
- Task Force on National Greenhouse Gas Inventories (TFI)

There is also a group that works to merge the findings of these groups in order to write a single synthesis document. Another offshoot of these groups is a Task

Group on Data and Scenario Support for Impact and Climate Analysis (TGICA), which helps with distributing and applying change-related data and scenarios (IPCC, 2018).

The IPCC published global assessment reports in 1990, 1995, 2001, 2007, and 2014. This fifth assessment report was written by over 800 scientists from 80 countries and assessed more than 30,000 scientific papers (United Nations, 2014). The key findings of the *Synthesis Report* are as follows: "1) Human influence on the climate system is clear; 2) The more we disrupt our climate, the more we risk severe, pervasive, and irreversible impacts; and 3) We have the means to limit climate change and build a more prosperous, sustainable future" (United Nations, 2014).

LOOKING AHEAD

So with all these data, what is the world doing about climate change? The Paris Agreement was negotiated by representatives from 195 countries at the 21st Conference of the Parties (COP) of the United Nations Framework Convention on Climate Change (UNFCCC) held in Paris, France, in December 2015. It focuses on limiting and mitigating greenhouse gas emissions, reducing fossil fuel usage, and funding these changes. The biggest goal of the Paris Agreement is to limit global warming to 1.5° to 2°C (2.7° to 3.6°F). Even though all these nations agreed to these goals, the Paris Agreement could only go into effect if it was ratified by the governments of 55 countries representing 55% of global greenhouse gas emissions. Indeed, all 55 countries ratified it, and the Paris Agreement went into effect on November 4, 2016. Now all countries must work toward the stated goals.

Continued fossil fuel dependence will have disastrous results for the climate. Changes made now, however, can have a real effect on the magnitude of future warming. If GHG emissions are drastically cut, there will be fewer devastating changes in the future. If emissions remain at current levels, the negative effects of climate change will continue to harm Earth and its people.

See also: Section 1: Climate Change, Climate Change Policies; *Section 2:* Anthropocene, Human Modification of Ecosystems; *Section 3:* Environmental Policy

FURTHER READING

AIP. 2018. "The Discovery of Global Warming: Timeline." American Institute of Physics. http://history.aip.org/climate/timeline.htm#T011.

American Scientific Societies. 2009. "Statement on Climate Change." http://www.aaas.org/sites/default/files/migrate/uploads/1021climate_letter1.pdf.

Bolstad, Erika. 2016. "Military Leaders Urge Trump to See Climate as a Security Threat." *Scientific American.* November 15. https://www.scientificamerican.com/article/military-leaders-urge-trump-to-see-climate-as-a-security-threat/.

Cook, J. Dana Nuccitelli, Sarah A. Green, Mark Richardson, Barbel Winkler, Rob Painting, . . . Andrew Skuce, 2016. "Consensus on Consensus: A Synthesis of Consensus Estimates on Human-Caused Global Warming." *Environmental Research Letters* 8(2): 7 pp.). http://iopscience.iop.org/article/10.1088/1748-9326/11/4/048002.

Dennis, Brady. 2016. "Pope Francis: 'Never Been Such a Clear Need for Science' to Protect the Planet." *Washington Post.* November 29. https://www.washingtonpost.com/news/energy-environment/wp/2016/11/29/pope-francis-urges-world-leaders-not-to-delay-climate-change-efforts/?utm_term=.477c72b19e24.

EIA. 2018. "U.S. Energy Mapping System." U.S. Energy Information Administration. www.eia.gov.

EPA. 2017. "Climate Change Science: Future of Climate Change." U.S. Environmental Protection Agency. https://19january2017snapshot.epa.gov/climatechange_.html.

ESRL. 2018. "Trends in Atmospheric Carbon Dioxide." Earth System Research Laboratory. https://www.esrl.noaa.gov/gmd/ccgg/trends/.

Feeney, Nolan. 2014. "Phase Out Fossil Fuels by 2100 or Face 'Irreversible' Climate Impact." *Time.* November 2. http://time.com/3553269/un-climate-change-report/.

Foran, Clare. 2016. "Donald Trump and the Triumph of Climate-Change Denial." *The Atlantic.* December 25. https://www.theatlantic.com/politics/archive/2016/12/donald-trump-climate-change-skeptic-denial/510359/.

Hall, Shannon. 2015. "Exxon Knew about Climate Change Almost 40 Years Ago." *Scientific American.* https://www.scientificamerican.com/article/exxon-knew-about-climate-change-almost-40-years-ago/.

IPCC. 2013. "Climate Change 2013: The Physical Science Basis, Summary for Policymakers." Intergovernmental Panel on Climate Change. www.climatechange2013.org.

IPCC. 2018. "Home." Intergovernmental Panel on Climate Change. http://www.ipcc.ch/.

IPCC Projections. 2007. "IPCC Fourth Assessment Report: Climate Change 2007." Intergovernmental Panel on Climate Change. www.ipcc.ch/publications_and_data/ar4/wg1/en/spmsspm-projections-of.html.

IPCC Synthesis. 2007. "Climate Change 2007: Synthesis Report." Intergovernmental Panel on Climate Change. http://www.ipcc.ch/report/ar5/syr/.

IPCC Synthesis. 2014. "Climate Change 2014: Synthesis Report: Summary for Policymakers." Intergovernmental Panel on Climate Change. http://www.ipcc.ch/report/ar5/syr/.

Kahn, Brian. 2017. "We Just Breached the 410 PPM Threshold for CO2." *Scientific American.* April 21. www.scientificamerican.com/article/we-just-breached-the-410-ppm-threshold-for-co2/.

NASA. 2013. "NASA Scientists React to 400 Ppm Carbon Milestone." National Aeronautics and Space Administration. May 21. https://climate.nasa.gov/400ppmquotes/.

NASA. 2018. "Climate Change: How Do We Know?" National Aeronautics and Space Administration. https://climate.nasa.gov/evidence/.

NASA Consensus. 2018. "Scientific Consensus: Earth's Climate Is Warming." National Aeronautics and Space Administration. February 16. http://climate.nasa.gov/scientific-consensus/.

National Wildlife Foundation. 2018. "Fast Facts about Climate Change." https://www.nwf.org/Eco-Schools-USA/Become-an-Eco-School/Pathways/Climate-Change/Facts.aspx.

Nuccitelli, Dana. 2014. "Time Is Running Out on Climate Denial." *The Guardian.* December 30. https://www.theguardian.com/environment/climate-consensus-97-per-cent/2014/dec/30/all.

Nuccitelli, Dana. 2015. "Two-Faced Exxon: The Misinformation Campaign against Its Own Scientists." *The Guardian.* November 25. https://www.theguardian.com/environment/climate-consensus-97-per-cent/2015/nov/25/all.

Oreskes, Naomi. 2004. "Beyond the Ivory Tower: The Scientific Consensus on Climate Change." *Science* 306(5702): 1686.

Shabecoff, Philip. 1988. "Global Warming Has Begun, Expert Tells Senate." *New York Times.* June 24. http://www.nytimes.com/1988/06/24/us/global-warming-has-begun-expert-tells-senate.html.

Space.com. 2012. "Temperature on Mars." http://www.space.com/16907-what-is-the-temperature-of-mars.html.

UN Chronicle. 2007. "Highlights of the Fourth IPCC Assessment Report." *The Magazine of the United Nations* June. 44(2). https://unchronicle.un.org/article/warming-climate-system-unequivocal-highlights-fourth-ipcc-assessment-report.

United Nations. 2014. "IPCC Launches Complete Synthesis Report." http://www.un.org/climatechange/blog/2015/03/ipcc-launches-complete-synthesis-report/.

Worldwide Scientific Organizations. 2018. "List of Worldwide Scientific Organizations." Governor's Office of Planning and Research. http://www.opr.ca.gov/facts/list-of-scientific-organizations.html.

Case Study 7: Climate Refugees—Island Nations Disappear Because of Rising Seas

Across the planet, the effects of climate change are causing a "flow of climate refugees around the globe" (Goode, 2016). Droughts and poor harvests have caused thousands of desperate climate migrants to flee their homes. At the same time, sea levels are rising and inundating low-lying countries. Environmental geographers work to assess both the environmental patterns and social outcomes of these global changes.

There is an uneven distribution between who is causing climate change (industrialized nations with high greenhouse gas emissions) and who is bearing the brunt of its impacts (people in nonindustrialized countries). This case study focuses on Kiribati, a small island nation in the south Pacific Ocean. Forced to move as sea levels rise, citizens of this country exemplify a new type of human migration that has recently emerged: climate refugees.

RISING SEA LEVELS

Climate change, specifically global warming, is causing sea levels to rise. Three factors are driving this: 1) melting land ice, especially from Greenland and western Antarctica; 2) melting sea ice, glaciers, and polar ice; and 3) expansion of the warming sea water (National Geographic Reference, 2018). Melting of sea and land ice is caused, of course, by increased temperatures, like an ice cube in a warm cola or a snowman melting in your front yard. What causes ocean water to expand? Oceans have absorbed the vast majority of increased atmospheric heat caused by human greenhouse gas (GHG) emissions, and heating this seawater makes it expand, further driving sea level rise (Lewin, 2015; USGCRP, 2014).

As environmental geographers, we are interested in the social impacts of these changes and in the ecological variables, most notably how we can best measure and study this issue. Of course, geographers are experts in remote sensing, and their analysis of the satellite images from 1993 to the present show the rate of change to be 3.4 millimeters (0.134 inches) per year (NASA Climate, 2018). Evidence shows that global sea level has risen by about 200 millimeters (8 inches) since reliable record keeping with ocean buoys began in 1880 (USGCRP, 2014).

Projecting future sea level rise is extremely difficult because of complex feedback loops—that's when one thing affects another, which causes something else, which then affects the first thing (Hortona et al., 2014). For example, ice is white, so it reflects sunlight and stays cooler, but when sea ice melts, the underlying ocean is darker and that surface absorbs sunlight and heats up more quickly, resulting in further melting. Here is another feedback example: melting land ice adds fresh water to the ocean, lowering the salinity of the sea, which alters its density and changes the pattern of currents and can stimulate more climate change. Adding to the complexity of this issue, sea level rise is not geographically uniform across the globe due to the ocean's circulation (CSE, 2018; DeConto and Pollard, 2016).

Further, we do not know when and to what extent humans will reduce their GHG emissions, which would drastically affect the level of warming and related sea level increase. Thus, experts make models based on assuming low, medium, or high emissions in the future. In the case of a high warming scenario of 4.5°C (8.1°F) by the year 2100 and 8°C (14.4°F) in 2300, the median likely ranges are 0.7 to 1.2 meters (2.3 to 3.9 feet) by 2100 and 2.0 to 3.0 meters (6.6 to 9.8 feet) by 2300, "calling into question the future survival of some coastal cities and low-lying island nations" (Hortona et al., 2014). Other models even show that oceans could rise by close to 2 meters in total (more than 6 feet) by the end of this century (Dennis and Mooney, 2016).

THE VICTIMS

Several excellent websites show maps of estimated future sea level rise, displayed as landmasses becoming inundated by the ocean (Climate Central, 2018; Geology.com, 2018; Lewin, 2015). On these maps, one can actually see how Florida would shrink as the sea overtakes much of the coastline and widens the rivers where they flow into the sea. New York, San Francisco, and even Washington, D.C., will be very different places if sea levels rise as predicted.

In 2016, the U.S. government began offering grants to help communities adapt to climate change; $1 billion was available to towns in 13 affected states. For example, Isle de Jean Charles is a small community on the coast of Louisiana that has lost over 90% of its land area since the 1950s, and they have received $48 million to resettle on drier ground (Davenport and Robertson, 2016). In Alaska, over 30 coastal villages are assessing their options and may also decide to move (Goode, 2016). Elsewhere across the globe, the Netherlands has spent billions of Euros (dollars) constructing barriers to the sea—and plans are underway for even more engineering solutions to address rising seas due to climate change (Folger, 2013).

These examples contrast sharply to poorer countries that have no money for adaptation. Citizens here have few options when faced with ecological disaster.

KIRIBATI

Kiribati exemplifies the complex realities of climate change in one small country. It is an island country in the south Pacific Ocean with a population of 106,925. It is located about 4,000 km (2,400 miles) south of Hawaii and about 8,000 km (5,000 miles) northeast of Australia (Google Earth, 2018). This nation is composed of 21 tiny inhabited islands (of 33 total islands), spread out over an area of 3.5 million square kilometers (1.35 million square miles, or about double the size of Alaska) with 1,143 kilometers of coastline (CIA, 2018). This tropical nation is made of low-lying coral atolls surrounded by reefs; thus Kiribati is severely affected by rising seas (Kiribati, 2018).

Some scientists say that Kiribati may become uninhabitable within decades because of an onslaught of environmental problems linked to climate change (Mathiesen, 2015). Government reports describe the complete devastation that rising seas would bring, from severe erosion to declining freshwater supplies. The former president, Anote Tong, actually tried to buy land in the country of Fiji and develop a program that encouraged citizens to move there. Fiji is an island nation that sits higher above sea level, but it is located some 600 km (1,000 miles) away. Among Kiribati citizens, opposition was strong because they thought that President Tong was trying to gain international fame rather than truly helping his people with immediate problems like unemployment. Indeed, he was forced out of office before his relocation plan was implemented (Ives, 2016).

In addition, starting in 2003 the World Bank funded a $17.7 million climate mitigation effort in Kiribati that built coastal seawalls, planted mangroves, and

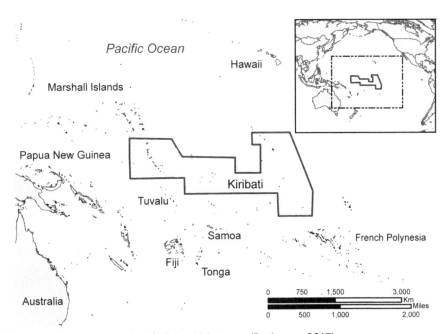

Kiribati and Neighboring Pacific Island Nations. (Bathgate, 2017)

developed water management plans. But the geographer Simon Donner, who led a 2014 review, concluded that the efforts were not successful and in some places caused more erosion. Experts now say that it is an issue of money. To live on Kiribati in the near future will cost a lot of money as the salty sea is infiltrating the drinking water systems. Desalinization is pricey and not a viable option for this nation (Ives, 2016).

Kiribati citizens who want to migrate elsewhere have few options; no countries are willing to take climate refugees (Mathiesen, 2015). Ioane Teitiota, a migrant worker from Kiribati, tried to seek asylum in New Zealand, stating that he had no potable water to grow crops in his homeland due to saltwater intrusion from climate change. The courts rejected this argument because there is no legal standing for climate refugees; however, Teitiota is appealing to the UNHCR. This may be the first case to push the legal definition of "refugee" to include those fleeing climate change (Ives, 2016).

Overall, many in Kiribati feel that migration to developed countries like Australia and New Zealand are the only real choice, as numerous villages will likely be underwater by 2050. These lovely tropical islands, losing ground to the sea, will no longer support people due to lack of drinking water (Mathiesen, 2015). Individuals have little choice but to stay and see how long it takes for their nation to be underwater.

Climate refugees, like those from Kiribati, are not considered political asylum seekers, so they do not have legal refugee standing (Sengupta, 2016). Therefore, other countries, like Australia and New Zealand, do not have to let them in. Where can people go when their livelihood is disrupted by climate change? The global community must now address a new problem: how to assist climate refugees.

LEGAL STATUS OF CLIMATE REFUGEES

In 1950, the United Nations High Commissioner for Refugees (UNHCR) was created to assist the millions of Europeans who were forced to flee and were homeless due to the destruction caused by World War II. By 2015, the UNHCR had helped over 50 million refugees to survive and rebuild their lives (UNHCR History, 2018).

The 1951 Convention Relating to the Status of Refugees is an international treaty that notes that a refugee cannot be returned to a country if their life or freedom is threatened. Specifically, in Article 1.A.2, a "refugee" is defined as a person who has a "well-founded fear of being persecuted for reasons of race, religion, nationality, membership of a particular social group or political opinion" (UNHCR Convention, 2018). While the 1951 Convention is now customary international law, the definition of a refugee does not include people fleeing climate change and related disasters. This could include rising sea levels, droughts, severe weather events, excessive rainfall and related landslides, etc.

Indeed, climate refugees are a new concept and no country will allow them in, because they do not meet the traditional definition of a refugee. This makes the work of the UNHCR even more important, as this international organization must increasingly assist the many vulnerable people who are forced to leave their homes due to changing temperature and precipitation patterns. Climate change particularly affects food and water availability (UNHCR Climate Change, 2018).

The UNHCR emphasizes that this is not some future possible scenario; climate refugees are a reality today. Each year millions of people are forced from their homes due to hazards related to climate change. In addition, climate may not always be the sole driver in a refugee movement, but rather may intensify conflict and make displacement so much worse when it occurs—think, for example, of people fleeing war, only to be trapped in a region with dwindling drinking water supplies (UNHCR FAQ, 2016). Estimates vary widely, but media commonly refer to the concept that 200 million people will be displaced by climate change by 2050; that is equal to about two-thirds of the entire population of the United States, so this represents a huge number of people being forced to flee their homelands (Basu, 2009).

In terms of climate refugees, Africa has seen weather-related crises, but South and East Asia have been hardest hit overall and will likely experience the most severe impacts of climate change (UNHCR FAQ, 2016). This is due to the geographic factors of landforms and the low-lying countries in Asia.

Overall, there is both an immediate danger of higher ocean waters flooding our coastal cities and the longer-term impact of nations disappearing completely (Pilkey and Young, 2009). According to a recent report by the World Bank, Pacific Island nations are truly on the "front line" of the battle with climate change and natural disasters (World Bank, 2013).

LOOKING AHEAD

President of the Seychelles, James Michel, eloquently called for all small island nations to unite and demand action against climate change: "We cannot accept that climate change be treated as an inevitability. We cannot accept that any island be lost to sea level rise. We cannot accept that our islands be submerged by the rising oceans" (ABC, 2014).

As noted by ocean experts writing several years ago: "It's a cruel irony that many of these societies, for the most part non-industrial, have played almost no role in the global warming that lies behind so much of the current sea level rise; in that sense, they are truly innocent victims of the industrialized world" (Pilkey and Young, 2009: 6).

Does the industrial world know or care? It remains to be seen whether the countries that produce the overwhelming majority of GHG will do anything to help the countries that are drowning in saltwater.

See also: Section 1: Climate Change, Climate Change Policies; *Section 2:* Adaptation; *Section 3:* Environmental Justice, Environmental Policy, Renewable Energy, Sustainable Development

FURTHER READING

ABC. 2014. "Climate Change Meeting Told Small Island Nations Must Unite or Drown in Rising Seas." ABC News Australia. November 11. http://www.abc.net.au/news

/2014-11-12/unite-or-drown,-small-island-nations-told-at-climate-meeting/588 4116.

Basu, Moni. 2009. "Climate Change May Displace Up to 200 Million." CNN. June 10. www.cnn.com/2009/TECH/06/10/climate.change.refugees/index.html.

CIA. 2018. "Kiribati." World Fact Book. https://www.cia.gov/library/publications /resources/the-world-factbook/geos/kr.html.

CSE. 2018. "Rising Sea Level." Center for Scientific Education. https://scied.ucar.edu /longcontent/rising-sea-level.

Climate Central. 2018. "Maps and Tools." http://sealevel.climatecentral.org/maps.

Davenport, Coral and Campbell Robertson. 2016. "Resettling the First American Climate Refugees." *New York Times*. https://www.nytimes.com/2016/05/03/us/resettling-the -first-american-climate-refugees.html.

DeConto, Robert M. and David Pollard. 2016. "Contribution of Antarctica to Past and Future Sea-Level Rise." *Nature* 531: 591–597.

Dennis, Brady and Chris Mooney. 2016. "Scientists Nearly Double Sea Level Rise Projections for 2100, Because of Antarctica." *Washington Post*. March 30. https://www .washingtonpost.com/news/energy-environment/wp/2016/03/30/antarctic-loss -could-double-expected-sea-level-rise-by-2100-scientists-say/.

Folger, Tim. 2013. "Rising Seas." *National Geographic Magazine*. September. http://www .nationalgeographic.com/magazine/2013/09/rising-seas-coastal-impact-climate -change/.

Geology.com. 2018. "Global Sea Level Rise Map." http://geology.com/sea-level-rise/.

Goode, Erica. 2016. "A Wrenching Choice for Alaska Towns in the Path of Climate Change." *New York Times*. November 29. https://www.nytimes.com/interactive/2016/11/29 /science/alaska-global-warming.html.

Google Earth. 2018. "Kiribati." https://earth.google.com.

Hortona, Benjamin P., Stefan Rahmstorfc, Simon E. Engelhartd, and Andrew C. Kempe. 2014. "Expert Assessment of Sea-Level Rise by AD 2100 and AD 2300." *Quaternary Science Reviews* 84: 1–6.

Ives, Mike. 2016. "A Remote Pacific Nation, Threatened by Rising Seas." *New York Times*. July 2. https://www.nytimes.com/2016/07/03/world/asia/climate-change-kiribati .html?_r=0.

Kiribati. 2018. "Climate Change Effects." Office of the President, Republic of Kiribati. http://www.climate.gov.ki/.

Lewin, Sarah. 2015. "New NASA Model Maps Sea Level Rise Like Never Before (Video)." Space.com. August 26. http://www.space.com/30379-nasa-sea-level-rise -model-video.html.

Mathiesen, Karl. 2015. "Losing Paradise: The People Displaced by Atomic Bombs, and Now Climate Change." *The Guardian*. March 9.

NASA Climate. 2018. "Sea Level Change: Observations from Space." National Aeronautics and Space Administration. https://sealevel.nasa.gov/.

NASA Facts. 2018. "Global Climate Change: Vital Signs of the Planet: Facts." National Aeronautics and Space Administration. http://climate.nasa.gov/vital-signs/sea-level/.

National Geographic Reference. 2018. "Sea Level Rise." http://www.nationalgeographic .com/environment/global-warming/sea-level-rise/.

Pilkey, Orrin H. and Rob Young. 2009. *The Rising Sea*. Washington, D.C.: Island Press.

Sengupta, Somini. 2016. "Heat, Hunger and War Force Africans onto a 'Road on Fire,'" *New York Times*. December 15. https://www.nytimes.com/interactive/2016/12/15 /world/africa/agadez-climate-change.html.

UNHCR Climate Change. 2018. "Climate Change and Disasters." United Nations High Commissioner for Refugees. http://www.unhcr.org/en-us/climate-change-and- disasters.html.

UNHCR Convention. 2018. "The 1951 Refugee Convention." United Nations High Commissioner for Refugees. www.unhcr.org/en-us/1951-refugee-convention.html.
UNHCR FAQ. 2016. "Frequently Asked Questions: Climate Change and Disaster Displacement." United Nations High Commissioner for Refugees. http://www.unhcr.org/en-us/news/latest/2016/ 11/581f52dc4/frequently-asked-questions-climate-change-disaster-displacement.html.
UNHCR History. 2018. "United Nations High Commissioner for Refugees." http://www.unhcr.org/en-us/history-of-unhcr.html.
USGCRP. 2014. "Sea Level Rise." U.S. Global Change Research Program. http://nca2014.globalchange.gov/report/our-changing-climate/sea-level-rise.
World Bank. 2013. "Acting on Climate Change and Disaster Risk for the Pacific." http://documents.worldbank.org/curated/en/354821468098054153/Acting-on-climate-change-and-disaster-risk-for-the-Pacific.

Case Study 8: Endangered Snow Leopard in Afghanistan— Local Efforts to Promote Conservation

This case study provides an example of how humans have both negative and positive impacts on the environment. Environmental geographers often study these types of issues because they are able to analyze the complex influences and interactions that often occur between human and ecological systems.

For example, human actions such as land-use change, farming, and illegal hunting have drastically reduced the population of some large species of wildlife. Elephants, lions, tigers, and other animals are on the brink of extinction due to human activities. The snow leopard of Afghanistan is another endangered animal with global numbers in the low thousands. Specifically, this case study will focus on efforts to protect this endangered species in the Wakhan Corridor of the Badakhshan Province of Afghanistan. Effective collaboration among people and organizations at the local, national, and international levels can lead to positive outcomes.

AFGHANISTAN AND THE WAKHAN CORRIDOR

Afghanistan is a country in Southern Asia located north and west of Pakistan and east of Iran. It is landlocked with no coast, and its climate is arid to semi-arid, with cold winters and hot summers. It is almost the size of Texas (CIA, 2018).

Due to its central location, Afghanistan has been under relentless conquest from multiple outside powers: the Persians, Alexander the Great, the Huns, Genghis Khan, and Mongols, to name a few. Finally in the late 1700s an Afghan empire was developed, only to be taken over by the British, who then granted the country its independence in 1921. After World War II, Afghanistan had increasingly friendly ties with the Soviet Union and by the late 1970s it had a communist government. Then a conservative Islamic rebel group, with funding from the United States, battled the communist Soviet-backed government. There has essentially been war in Afghanistan since this time. The Russians pulled out, but then the fundamentalist Taliban and Al Qaeda exerted more power. The U.S.-backed forces claimed victory, and in 2004 President Hamid Karzai was elected president. The

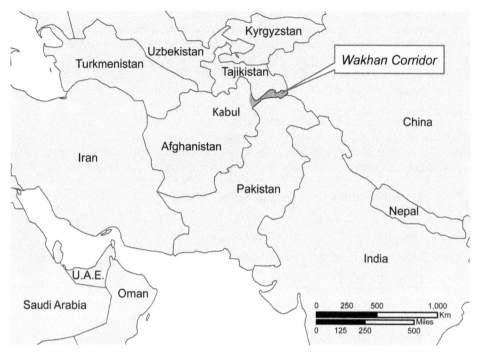

Wakhan Corridor, Northeastern Afghanistan. (Bathgate, 2017)

Afghan government's power is mostly within the region of the capital, Kabul. Outside of that area, there are still power struggles and violence from many armed factions. This country has experienced and continues to experience violence on a daily basis (Jenkins, 2005; PBS, 2011.)

Droughts and famine exacerbated the desperate situation of the Afghan people. The main environmental problems are lack of drinking water, soil degradation, overgrazing, deforestation, and desertification. After decades of war, the Afghani population of 33 million faces widespread lack of basic needs. Over 35% of the country lives in poverty, nearly 40% are unemployed, and only 38% can read. The birthrate is high, with Afghan women having, on average, five children. Tragically, infant mortality rates are the highest in the world with 112.8 deaths per 1,000 live births (CIA, 2018).

Given this complicated geopolitical history and current economic adversity, the Afghan people face many challenges. Different parts of the country have distinct issues. For example, in the far northeastern part of the country, remoteness is both beneficial and problematic.

In the Badakhshan Province of Afghanistan, the Wakhan Corridor is a narrow stretch of land that is about 350 km (220 miles) long and its width varies from 13 to 65 km (8 to 40 miles). Its main neighboring countries are Tajikistan to the north and Pakistan to the south, and there is a short 56-mile common border with China to the east (Finkel, 2013; Google Earth, 2018).

This remote location means that the region has avoided some of Afghanistan's decades of violent war. In addition, the high elevations, rugged terrain, and almost barren landscapes host relatively few people. In this remote Wakhan Corridor region, there are fewer 20,000 people, mostly sheep herders in small villages (Skowronska, 2018). The Kyrgyz nomads are a group of several hundred that clings to the harsh frigid landscape in only semi-permanent structures (Finkel, 2013). This region is rich with local culture, much of it nomadic and steeped in a close relationship between people and their harsh, mountainous environment (Jenkins, 2005.)

THE SNOW LEOPARD

People share the rugged Wakhan landscape with the snow leopard, which is called "wāwrīn ppwrī" in the local Pashto language (written: واورین پرانگ). The species' scientific name is *Panthera uncial* and it is a stocky, thick-furred, wild cat weighing 27 to 55 kg (60 to 121 lb.) and standing about 60 cm (24 inches) at the shoulder (Snow Leopard Trust, 2017).

This animal is on the International Union for the Conservation of Nature (IUCN) Red List of Threatened Species. The IUCN is the global authority on the status of our animal and plant species. Their Red List indicates the risk of extinction of various species across regions and countries. The IUCN works with hundreds of local conservation groups to gather information on a given species, determine if sufficient data are available, and to then assess its ranking. This list is presented in categories from least concern, threatened, vulnerable, endangered, to extinct (IUCN, 2018).

The entire population of snow leopards has decreased by over 20% in the last 16 years, due to loss of habitat, loss of prey, and illegal poaching and trade (McCarthy et al., 2018). Although prohibited, the spotted leopard hides still can be found in markets in the capital city of Kabul fetching high prices, often over $2,000, which is almost triple the average annual income of most Afghanis (O'Donnell, 2016; World Bank, 2018).

The snow leopard's habitat is high-elevation grasslands that are vulnerable to the impacts of livestock (sheep, yaks, cows, and goats), which people have brought into the Wakhan region in large numbers. Habitat destruction caused by people's livestock has led to decreases in native prey for the snow leopards, which forces them to take these farm animals for their food source (McCarthy et al., 2018). Herders are then motivated to kill the big cats to protect their livestock herds, which leads to a vicious circle.

BRINGING BACK THE SNOW LEOPARD

Recently, there have been some positive activities indicating that there is hope for bringing the snow leopard back from the brink of extinction. Initially, in 2005, the national Afghani government implemented a law stating that killing any snow leopard is illegal, but in remote regions like the Wakhan Corridor there is little policing. So the law was mostly ignored, and the leopards were often poached and sold at high prices.

More effective are the actions taken to help local people remain economically viable by protecting their livelihoods. Starting in 2009, conservation groups began building enclosed pens with mesh roofs for local livestock because the local economy depends on herding (O'Donnell, 2016).

In addition, conservation education programs have been implemented by environmental groups, notably the Wildlife Conservation Society (WCS), which is a New York–based organization. The WCS is supported by $3 million in funding from the United Nations Development Programme (UNEP) to oversee snow leopard research in the Wakhan area for the next few years. These efforts are already providing a great deal of data about the little-known big cats. By capturing, GPS collaring, and tracking several animals, scientists are beginning to learn more about the elusive snow leopard. For example, they are excellent swimmers and often cross large, frigid rivers in search of prey (O'Donnell, 2016).

Community-based natural resource management (CBNRM) has proven to be the best approach for obtaining a long-term, sustainable outcome for the wildlife. Experts say that community involvement is crucial. The WCS is working with all 14 schools in the Wakhan Corridor to bring an environmental education program to children. This includes things like training of teachers and involving families in Parents' Day events with wildlife themes. Now school children are taught about the importance of preserving the snow leopard and its habitat (WCS, 2018).

Conservation groups are employing locals in the Wakhan Corridor to work as wildlife rangers. Some of them help scientists shoot the leopards with tranquilizer darts so they can collar the animals with GPS trackers to learn about their geographic range. This is a high-tech approach for a region that otherwise has a very low-tech standard of living. Keep in mind that many of these new local rangers had previously used actual guns to shoot, kill, and sell the animals' pelts to help eke out a living (Sieff, 2013). Now the conservation salary provides a good wage for some local workers.

Because of all this interest, the Wakhan National Park was established in 2014 to encompass 1 million square hectares (4,200 square miles), but unfortunately, the Afghani government has very little money to support it. Still, an estimated 140 big cats live in the park, and scientists say that its main prey (Siberian ibex and golden marmot) are also thriving.

Although people in the Wakhan Corridor have few modern conveniences in daily life, they are learning important scientific information about an endangered species and how they may best interact with it. This is a remote region, but with careful planning, wildlife organizations, scientists, and locals hope that ecotourism may eventually help in maintaining habitats for this solitary animal.

ECOTOURISM

In recent years, strong and effective collaboration has developed among international environmental organizations and local communities to protect the snow leopard in the Wakhan Corridor. Local people are learning that the snow leopard might bring tourists who wish to catch a glimpse of this rare animal. Ecotourism is an important new concept in the Wakhan Corridor.

Tourism, as typically carried out, is usually harmful to the natural environment. Building tourism infrastructure often has negative impacts. For example, just consider how building a new hotel and parking lot would destroy natural habitats. If this is a chain hotel with corporate ownership from outside the area, most of the profits would leave the area and not help the local economy.

Luckily, many people are interested in trying to create beneficial ecotourism sites. Ecotourism is commonly defined as "responsible travel to natural areas that conserves the environment, sustains the well-being of the local people, and involves interpretation and education" (TIES, 2015). Indeed, The International Ecotourism Society explains that beneficial ecotourism is travel to natural areas that conserves the environment, educates tourists and residents, and helps sustain local people. When ecotourism is done correctly, it can be beneficial to both people and environment, making this of particular interest to environmental geographers.

The reason that ecotourism might work in the Wakhan Corridor is that the region hosts stunning mountainous landscapes that are virtually untouched by humans, as well as the endangered snow leopard—and a willingness by local people to participate in the process of species protection. In addition, environmental groups have put money and leadership into the region.

LOOKING AHEAD

Adding to the scientific knowledge base is crucial, both in terms of ecological data and social factors. Zoologists must understand the habitat needs of the snow leopards, but equally important are the human influences. People in this region are extremely poor, and traditionally they have seen the leopards as a threat to both their livelihoods and even their very existence.

People now understand that protecting the snow leopard habitat might bring foreign tourists and their money. Children in school may not have Internet access, but they are taught about wildlife conservation, and it is linked to the long-term health of their communities. This, combined with real action to help protect the community's livestock, is beginning to make a difference.

Despite the fact that environmental policies are relatively few and still fairly weak in Afghanistan due to its decades-long wars, some people are optimistic (Sieff, 2013). Mostapha Zaher, director of the Afghanistan National Environment Protection Agency, has a hopeful view of the region and its future. "When peace returns to Afghanistan—and it will, as no war lasts forever—Wakhan has great potential for ecotourism" (O'Donnell, 2016).

With the majestic snow leopard as an indicator species in a strikingly barren and beautiful mountain landscape, the Wakhan Corridor may provide opportunities for travelers willing to venture far off the beaten path. If done properly, ecotourism could aid the local economy and further help protect the endangered leopard. In the meantime, local people are providing a crucial service to our global environment by taking steps to preserve this endangered species and its habitat.

See also: Section 1: Biodiversity Loss, Endangered Species; *Section 2:* Habitat and Wildlife, Protected Areas and National Parks; *Section 3:* Environmental Nongovernmental Organization, Sustainable Development

FURTHER READING

CIA. 2018. "Afghanistan." World Fact Book. https://www.cia.gov/library/publications/the-world-factbook/geos/af.html.
Finkel, Michael. 2013. "Afghanistan's Wakhan Corridor and Kyrgyz Nomads." *National Geographic.* http://ngm.nationalgeographic.com/2013/02/wakhan-corridor/finkel-text.
Google Earth. 2018. "Wakhan, Afghanistan." https://earth.google.com.
IUCN. 2018. "The IUCN Red List of Threatened Species." International Union for Conservation of Nature and Natural Resources. http://www.iucnredlist.org/.
Jenkins, Mark. 2005. "A Short Walk in the Wakhan Corridor." *Outside Magazine.* https://www.outsideonline.com/1885016/short-walk-wakhan-corridor.
McCarthy, T., D. Mallon, R. Jackson, P. Zahler, and K. McCarthy. 2018. "Panthera uncia." The IUCN Red List of Threatened Species. http://dx.doi.org/10.2305/IUCN.UK.2008.RLTS.T22732A9381126.en.
O'Donnell, Lynne. 2016. "Snow Leopards' Return Brings Hope to Remote Afghan Region." Associated Press. September 12. http://bigstory.ap.org/eaa6f3d7fcf3467185fe6a0112974bc0&utm.
PBS. 2011. "A Historical Timeline of Afghanistan." *PBS News Hour.* May 4. http://www.pbs.org/newshour/updates/asia-jan-june11-timeline-afghanistan/.
Sieff, Kevin. 2013. "In Afghanistan, a Quest to Save the Snow Leopard." *Washington Post.* July 22. https://www.washingtonpost.com/world/asia_pacific/in-afghanistan-a-quest-to-save-the-snow-leopard/2013/07/21/098bd5c4-ed7c-11e2-bb32-725c8351a69e_story.html.
Skowronska, Malgorzata. 2018. "Women in Red: Afghanistan's Forgotten Corner." CBS News. https://www.cbsnews.com/pictures/wakhan-corridor-afghanistan-forgotten-corner/.
Snow Leopard Trust. 2017. "Key Snow Leopard Facts." https://www.snowleopard.org/.
TIES. 2015. "Ecotourism Principles." The International Ecotourism Society. http://www.ecotourism.org/what-is-ecotourism.
WCS. 2018. "Badakhshan Region." Wildlife Conservation Society. https://afghanistan.wcs.org/Additional-info/Badakhshan-2.
World Bank. 2018. "Afghanistan." http://data.worldbank.org/country/afghanistan.

Case Study 9: The Power of Hurricane Katrina (2005)—Evacuation and the Aftermath

The ominous warning issued by the National Weather Service National Hurricane Center in Miami, Florida, at 8 a.m. on August 29, 2005, stated: "HURRICANE KATRINA INTERMEDIATE ADVISORY ... LARGE AND EXTREMELY DANGEROUS CATEGORY FOUR HURRICANE KATRINA POUNDING SOUTHEASTERN LOUISIANA AND SOUTHERN MISSISSIPPI" (National Hurricane Center, 2005).

Hurricane Katrina began as a tropical storm over the Bahamas. It gained energy as it moved west and became a hurricane before hitting landfall near Aventura, Florida, on August 25. It crossed over land and then strengthened considerably over the open waters of the Gulf of Mexico. The storm doubled in size, as the unusually warm waters of the Gulf provided additional energy. When Hurricane Katrina made landfall east of New Orleans, Louisiana, on August 29, 2005, it was a Category 3 hurricane on the Saffir–Simpson Hurricane Wind Scale (SSHWS) with one-minute sustained winds of 240 km/h (150 mph). Hurricane Katrina was also huge, stretching some 640 kilometers (400 miles) across. It finally lost hurricane strength about six hours later after travelling 240 kilometers (150 miles) inland (History.com, 2009; NWS, 2018).

It was the largest storm ever seen in the Gulf of Mexico, and it affected some 230,000 square kilometers (90,000 square miles) of the United States (History.com, 2009). Although the immediate danger was extreme, the days, months, and years afterward continued to bring loss to the region. Over 1,800 people died, and damages totaled over $100 billion (Worland, 2015). Poor and minority residents of Louisiana, Mississippi, and Alabama were particularly harmed by the storm and its aftermath (History.com, 2009).

THE HURRICANE AND ITS AFTERMATH

Relying on excellent meteorological predictions and warnings, Louisiana Governor Kathleen Blanco declared a state of emergency on August 26, before the storm hit. She called President George W. Bush during his vacation to plead for help from the U.S. Federal Emergency Management Agency (FEMA) (Frontline, 2005). In the city of New Orleans, Mayor Ray Nagin announced the first-ever mandatory evacuation order the day before the storm (Frontline, 2005). In about one day, 80% of the city's 500,000 population had evacuated. But about 100,000 people did not have access to a car. The Superdome professional football stadium

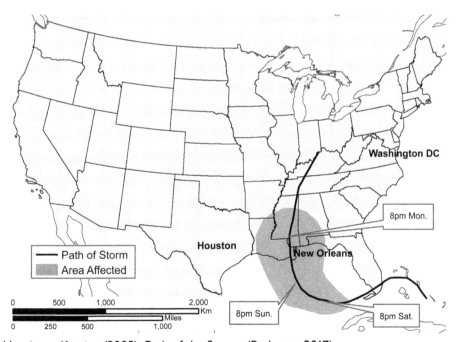

Hurricane Katrina (2005): Path of the Storm. (Bathgate, 2017)

was designated an emergency shelter, and indeed thousands of people sought shelter there, but even more tried to survive the storm in their homes (History.com, 2009).

First came hours of rain, making the ground waterlogged and weakening levees and storm-wall footings. Canals and streams were flooded. Then came the winds and the storm surge, which was as high as nine meters (30 feet) in some places (History.com, 2009). Levees began to fail as water seeped underneath through soaked soil. Other levees were overtopped and then broke and washed away. People in their homes scrambled to rooftops, hoping to be rescued (Frontline, 2005).

In the end, more than 1 million people were forced to leave Louisiana and Mississippi. About 250,000 of these refugees ended up in Houston, Texas, some of whom were in a 500-bus caravan from the area (Turner, 2015). Eighty percent of New Orleans was under water and 70% of occupied housing was damaged. Poor and minority neighborhoods were particularly hard-hit. Hurricane Katrina destroyed parts of Louisiana, Mississippi, and Alabama, but the worst destruction was concentrated in New Orleans where the population was about 67% black and 30% in poverty (History.com, 2009).

For those who were able to leave, it was difficult. For those forced to stay, it was nearly impossible. The Superdome had limited supplies, with water and electricity unreliable during the storm. Eventually 15,000 refugees ended up living here for weeks. The city and FEMA had no plans for all the additional refugees. Again, for poor people without a car, there were no options in the flooded city. In fact, when some refugees tried to walk over a bridge to the nearby suburb of Gretna, police with loaded shotguns forced them to turn back (History.com, 2009).

Hurricane Katrina is the most expensive disaster in the history of the insurance industry. For people who could afford property insurance, their companies paid over $40 billion in claims for homes, vehicles, and businesses that were damaged. The U.S. National Flood Insurance Program paid out about $16 billion in claims. In addition, FEMA has given over $15 billion to states to repair infrastructure such as roads and schools. FEMA also paid almost $7 billion to more than 1 million people to help in their recovery (FEMA, 2015).

But for those living in poverty, without the ability to pay for property insurance, the storm took everything. More than 1 million people in the Gulf region were displaced by the storm. Over 275,000 homes were destroyed (NWS, 2018).

Before Hurricane Katrina, the population of New Orleans was approximately 500,000 (excluding the suburbs). The population fell by one-half when tallied one year after the storm. It recovered somewhat, but is now only about three-quarters of what it was before the storm. This is because the inner city had more poverty, so once people left, they could not afford to return and rebuild. They had fewer options than people who had higher incomes.

FEMA

This crisis made it clear the U.S. government was not prepared to handle such a disaster. FEMA took days to set up operations in New Orleans and even then did not appear to have a plan. High officials, including President Bush and FEMA

Director Michael Brown, did not realize the extent of the destruction and desperate need for aid (Frontline, 2005). They were later criticized for a slow and inadequate response to the crisis. Indeed, a report written one year later after investigation by the U.S. Department of Homeland Security, which oversees FEMA, found there had been significant problems. In fact, "criticism is warranted" for the "slow and ineffective response to Hurricane Katrina." Overall, the report described a lack of attention focused on natural disaster planning, as terrorist concerns had become the focus of FEMA (Ahlers, 2006).

Since the hurricane, FEMA has developed a National Disaster Recovery Framework that is a strategy to guide recovery efforts after a major emergency. FEMA also established Incident Management Assistance Teams that are rapid responders, able to deploy within two hours and arrive onsite within 12 hours of the disaster. They also improved their search and rescue capability and set up regional emergency communications groups for coordination among federal, state, and local emergency responders (FEMA, 2015).

THE LEVEES

The geography of New Orleans makes it particularly vulnerable to flooding and natural disasters. Water surrounds New Orleans, with the Mississippi River, Lake Pontchartrain, and the Gulf of Mexico all nearby. This city is actually sited an average of two meters (six feet) below sea level, so the city relies on a 560-kilometer (350-mile) levee system to keep the water out (Burnett, 2015). The physical geography of the city is like a bowl, with the city sitting down inside. This is a real problem, as it prevents water from draining away after a storm. Once a levee is broken, water will continue to flow into city streets.

The Army Corps of Engineers has been building levees along the Mississippi River since the late 1800s. Initially they used soil to make embankments to protect against floods. Technology helped advance the possible efficiency of levees, but there is a complicated balance between what protection is needed versus what the U.S. Congress and taxpayers are willing to pay the Corps to build and maintain this protection. The level of risk that people are willing to accept must be balanced with the scientifically modeled likelihood of flood events and the possible harm that such an event could cause (USACE, 2018). After Hurricane Katrina, the Army Corps of Engineers was blamed for faulty levee construction and cutting corners to save money on flood protections.

Since Hurricane Katrina, the U.S. Army Corps of Engineers has spent $14.5 billion to build a sophisticated hurricane protection system surrounding New Orleans. The Corps has built huge new flood gates, installed advanced computer storm modeling, strengthened 350 miles of barriers, and updated pumping stations. This new protection is supposed to protect the city in a 100-year storm, meaning that a storm the size of which only has a 1% chance of happening in any given year. But Katrina is considered a 400-year storm, so although it would be a rare occurrence, another storm of this size would still destroy the city (Burnett, 2015).

LOOKING AHEAD

With such devastation, people wonder if this could happen again. Scientists are studying the link between climate change, hurricanes, and certain meteorological trends. Rising sea levels caused by climate change have increased the devastation that hurricanes can cause. Storm surges happen when coastal waters escalate above their normal levels during a storm due to wind and weather that pushes more water to shore. These surges increase with higher seas (NOS, 2015). In addition, the frequency and intensity of hurricanes may increase because climate change has caused higher ocean temperatures (Worland, 2015).

So planning for emergency response after a hurricane is extremely important. Planning and preparation beforehand is equally important, because this can reduce the overall impact of the disaster. As the lead climate prediction and scientific weather agency of the United States, NOAA works with states and local communities to help them prepare for risk. For the next decade, NOAA has identified several Coastal Community Preparedness Goals: 1) identify risks and vulnerability, because these vary by development, geography, ocean depth variations, and people's capabilities; 2) understand options such as building codes and redevelopment options, such as green design; 3) plan and collaborate with politicians, businesses, scientists, and others to be prepared before disaster hits; and 4) remember that both people and technology are part of the solution (NOS, 2015).

NOAA also provides technical assistance for coastal resilience. The website https://coast.noaa.gov/digitalcoast/tools/ shows numerous helpful tools such as the Sea Level Rise Viewer and the Coastal Flood Exposure Mapper, which can help communities gather facts and make informed natural hazard plans. In addition, NOAA provides coastal resilience grant money for coastal communities to build green infrastructure and increase coastal habitats that will lessen the severity of future floods (Office for Coastal Management, 2017).

The bigger issue highlighted by Hurricane Katrina and its aftermath is the great disparity in who is most affected by natural disasters. The poor, elderly, and minority residents are most vulnerable because they have few options. Think about people with no car or money for transportation; they cannot get out of harm's way. People who do not have enough money to buy insurance coverage are left with nothing if their home is destroyed by a storm. People without an education or job skills may have a more difficult time rebuilding their lives if they are forced to flee a region.

The death toll in Hurricane Katrina was low, given the magnitude of the storm. Certainly, comparing this hurricane to other natural disasters in developing countries, the United States was very lucky to have scientific information to provide some lead time, an excellent warning system, alerts that were mostly followed, and an infrastructure that mostly helped keep the number of fatalities low.

Yet Hurricane Katrina destroyed many neighborhoods in New Orleans, some of which will never be rebuilt. The cost of cleanup and rebuilding was massive—over $100 billion (NWS, 2018). This disaster starkly displayed that the risk of natural disasters is not equitable. In other words, it is not fair that poor people must bear more risk than wealthy people. Community planning for risk mitigation should address these issues of social sustainability in the future.

See also: Section 2: Adaptation, Flood, Hurricane, Mitigation, Natural Hazards; *Section 3:* Environmental Justice

FURTHER READING

Ahlers, Mike M. 2006. "Report: Criticism of FEMA's Katrina Response Deserved Inspector General: 'Much of the Criticism Is Warranted.'" CNN. April 14. http://www.cnn.com/2006/POLITICS/04/14/fema.ig/.

Burnett, John. 2015. "Billions Spent on Flood Barriers, But New Orleans Still a 'Fishbowl.'" *All Things Considered*, National Public Radio. August 28. http://www.npr.org/2015/08/28/432059261/billions-spent-on-flood-barriers-but-new-orleans-still-a-fishbowl.

FEMA. 2015."FEMA Outlines a Decade of Progress after Hurricane Katrina." Federal Emergency Management Agency. July 30. https://www.fema.gov/news-release/2015/07/30/fema-outlines-decade-progress-after-hurricane-katrina.

Frontline. 2005. "The Storm." Public Broadcasting Service. November 22. http://www.pbs.org/wgbh/pages/frontline/storm/etc/cron.html.

History.com. 2009. "Hurricane Katrina." http://www.history.com/topics/hurricane-katrina.

National Hurricane Center. 2005. "Hurricane KATRINA." National Oceanic and Atmospheric Administration. http://www.nhc.noaa.gov/archive/2005/pub/al122005.public_b.026.shtml.

NOS. 2015. "Hurricane Katrina: Ten Years Later." National Ocean Service. http://oceanservice.noaa.gov/news/aug15/katrina-ten-years-later.html.

NWS. 2018. "Hurricane Katrina." National Weather Service. https://www.weather.gov/jetstream/katrina.

Office for Coastal Management. 2017. "Tools." National Oceanic and Atmospheric Administration. https://coast.noaa.gov/digitalcoast/tools/.

Turner, Allan. 2015. "Ten Years Later, Katrina Evacuees Now Part of Houston Fabric." *Houston Chronicle*. August 21. http://www.houstonchronicle.com/news/houston-texas/houston/article/Ten-years-later-Katrina-evacuees-now-part-of-6458412.php.

USACE. 2018. "New Orleans District." U.S. Army Corps of Engineers. http://www.mvn.usace.army.mil/.

Worland, Justin. 2015. "Why Climate Change Could Make Hurricane Impact Worse." *Time*. August 27. http://time.com/4013637/climate-change-hurricanes-impact/.

Case Study 10: The 2004 South Asian Tsunami Natural Disaster

The Indian Ocean tsunami is the deadliest tsunami ever and one of the deadliest natural disasters in recorded world history. Waves were measured up to 30 meters in height, or about 108 feet (ABC Australia, 2014). Just imagine standing on a beach and seeing a wall of water that is as tall as a 10-floor building coming straight at you. This is, sadly, what happened to thousands of people, as there was no warning system in place (Kayal and Wald, 2004).

This tsunami was caused by one of the largest quakes ever recorded on Earth, which occurred on December 26, 2004, with an epicenter in the ocean near Sumatra, Indonesia. Called the Indian Ocean earthquake, it was over 9.0 on the

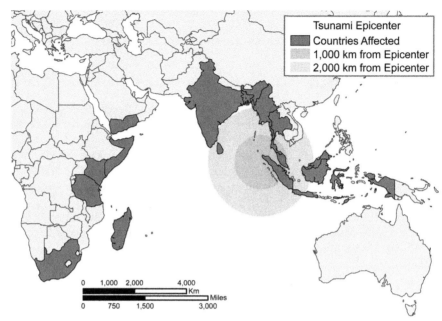

Areas Affected by the Indian Ocean Tsunami. (Bathgate, 2017)

moment magnitude scale (MMS) that seismologists use to measure quakes (ABC Australia, 2014; Kayal and Wald, 2004). The Richter scale, which is still used in the media, is somewhat similar in numbers. As a comparison, the famous 1906 earthquake that destroyed much of San Francisco, had an MMS of about 7.0. But since this is a logarithmic scale, the 9 MMS is 100 times greater than the 7 MMS (USGS, 2018).

Indeed, because of the magnitude of the Indian Ocean earthquake in 2004, it caused huge amounts of ocean water to move. This water then traveled hundreds of miles across the ocean and eventually surged onto land as a series of tsunamis that killed nearly 230,000 people (NWS, 2014). People were completely unprepared because they were located so far away that they never felt the tremors of the earthquake (Osborne, 2017).

This case study shows the environment continues to have a huge impact on humans despite advanced technology and the belief that people are somehow in control of nature. For an environmental geographer, this example of a natural disaster demonstrates the intersection between the Earth's complex physical environment, which is difficult to predict, and human factors like policy and economics that influence the ability to implement warning systems.

THE DISASTER

The Indian Ocean earthquake lasted nearly 10 minutes and moved hundreds of miles of ocean floor. It occurred approximately 160 kilometers (100 miles) west of Sumatra, 30 kilometers (18 miles) below the ocean floor. A 1,200-kilometers

(745-mile) long section of the Indian tectonic plate was thrust up 20 meters (66 feet) under the Burma tectonic plate, which caused the ocean floor to rise up. In fact, as the quake continued, it caused the ocean floor to move west by as much as 6 meters (20 feet) and lift up by 2 meters (6 feet) (Pickrell 2005).

As the ocean floor moved, it was like a sudden burp that shoved up the ocean water above it. This giant push of water caused a series of tsunami waves. The first wave hit Sumatra, Indonesia, about 25 minutes after the earthquake began (NWS, 2014). Then the waves grew and gathered energy as they spread across the ocean. Thailand was struck by tsunami waves about 2 hours later, and other countries along the Indian Ocean coast were hit a few hours later (ABC Australia, 2014). Even Madagascar and South Africa were affected, which are about 6,000 km (3,700 miles) away (Osborne, 2017).

WHERE?

These tsunamis caused widespread devastation along thousands of miles of Indian Ocean coastline throughout the region. Damage occurred in 14 countries (CNN, 2017). The estimate of people dead or missing and presumed dead is 227,898 according to official sources. Because no reliable censuses of population are taken in some of the affected countries, total counts will never be known. The distribution by country shows that Indonesia was hardest hit, with 167,540 deaths and $4.4 billion in damage. The other fatalities occurred in Sri Lanka (35,322), India (16,269), Thailand (8,212), Somalia (289), Maldives (108), Malaysia (75), Myanmar (61), Tanzania (13), Bangladesh (2), Seychelles (2), South Africa (2), Yemen (2), and Kenya (1). About 5,000 deaths were tourists from other countries (NWS, 2014; Osborne 2017; Pickrell, 2005).

Nearly 2 million people were left homeless, and 3 million more had no access to food or drinking water during the aftermath (UN, 2009). Long-term impacts include high numbers of orphans and lack of community structures in places where significant deaths occurred. Likewise, there are economic impacts. For example, the region relied on tourism for revenue, which ceased following the natural disaster. After the vast destruction, the world community came together to provide aid to the devastated nations; over $14 billion was pledged, although not all of this was actually received (UN, 2009).

WHY WAS THE TSUNAMI SO DEVASTATING?

Why were so many lives lost, given the modern forecasting and emergency management technologies that we have available? The answers to this question involve key geographic concepts related to both the physical environment and human geography.

In 2004, at the time of the devastating Indian Ocean tsunami, there was no warning system. This was due to a lower incidence of these disasters in this region, which means less overall risk. In addition, there was a lack of funding and absence of coordination among nations in the region. This contrasts sharply to the integrated warning system that has been in place for the Pacific Ocean region, which has much more tsunami activity. Although the Pacific Tsunami Center received some relevant information on the Indian Ocean earthquake, it was not enough to confirm that a tsunami was imminent along Indian Ocean coasts (PTWC, 2018). However, even if these data were clear, there was no emergency management system that could have informed citizens and announce evacuations (Rondonuwu, 2012).

Tsunami warning systems are very challenging because not all seafloor earthquakes cause tsunamis. In some cases, the tectonic plates may just "rub" and not shift in a way that moves a huge amount of water, so a tsunami is avoided. In addition, tsunamis are difficult to track. Sailors at sea often do not notice a tsunami wave that passed under them because on open water the height of the wave may be less than 30 centimeters (1 foot). Beneath the surface the shock wave of energy released by the earthquake travels extremely fast—up to 800 kilometers per hour (500 mph) in the case of the Indian Ocean tsunami (Pickrell, 2005). Once this energy gets near a coast with shallow water, it slows down, causing the top of the water to move faster than the bottom while the trailing part of the wave is still going very fast in deeper water. All this energy builds up and the height of the wave grows (Kayal and Wald, 2004; NGS, 2004).

Geographic features such as bays, reefs, and river mouths can reduce the energy of a tsunami. In some places the tsunami waves may only rise up a few feet, but in other places it can surge up to 30 meters (108 feet) in height. The damage caused by the tsunami was less severe in areas where coastal ecosystems had remained in good health. On the other hand, damage was more severe in cities, where protective coastal ecosystems had been destroyed by buildings, tourism, and overfishing.

People report that a tsunami acts like a giant, fast-moving river on top of the ocean and it sounds like a freight train (NGS, 2004). In other places, witnesses describe a rapid surge of the ocean, like a rapidly rising tide, rather than a wave. In this case, there is often extreme turbulence underwater, and this "undertow" can rapidly pull people under and out to sea. The energy in the tsunami can toss heavy objects like cars or buildings and strip away an entire beach within minutes. Likewise, this energy can affect inland areas with vast destruction. In the case of the Indian Ocean tsunami, for example, objects were destroyed up to 2 kilometers (1.3 miles) inland (Pickrell, 2005).

Tsunamis are also complex due to their series of waves and retreat. They sometimes have a series of waves, known as a wave train, so innocent people can assume that once a huge initial wave strikes, the surge is over. But in fact, there can be several huge waves, sometimes minutes or up to an hour apart. People think they are safe, so they go back toward the coast to survey damage, only to be pulled out to sea by the additional waves in the series. Likewise, people can be fooled by the force of the receding water. Just as the water surges in with rapid force, it also retreats very quickly. It was reported that some people were killed because they went to investigate the flooding, only to be drowned in the forceful receding of the Indian Ocean tsunami surge (NGS, 2004).

The environmental characteristics of a tsunami make it difficult to predict and track. Still, a warning system and education campaign to inform people about preparedness and response to possible tsunami warnings could greatly reduce the impact of tsunamis in the Indian Ocean in the future.

LOOKING AHEAD

Since the devastating earthquake and tsunami of 2004, a preliminary warning system has been put in place. Specifically, under United Nations guidance, the preparedness focuses on monitoring and communication (Rondonuwu, 2012). Only a small percentage of strong earthquakes actually produce a tsunami, so a warning system must be based on both seismic instruments and sea-based data to help scientists determine if a tsunami has actually been initiated. India, Thailand, Australia, and Indonesia (with help from Germany) have installed buoys to gather necessary data. Data centers in Hawaii and Japan are helping as the new system gets up and running (Casey, 2014).

In the meantime, countries have signed up to get the alerts. The key factor is how information is spread among the populations who could be in danger. Each of the 27 countries that border the Indian Ocean must develop their own warning system to provide alerts through sirens, radio announcements, and local governments. Upwards of $500 million overall has been spent by countries in the region on tsunami systems, according to Tony Elliott, who heads the Intergovernmental Coordination Group for the Indian Ocean Tsunami Warning and Mitigation System (Casey, 2014). Still, it is very difficult to prepare the appropriate infrastructure for hundreds of thousands of people to evacuate simultaneously from coastal areas. And it is challenging to educate people about how to get alerts, react quickly, and get to safety (BBC, 2005). These necessary stages in emergency management are further exacerbated by poverty and lack of transportation in some regions.

Overall, it is worth noting the complexity of preparing for a disaster of this magnitude, because of several factors, most notably national borders and poverty. Coordination in gathering and sharing scientific data is inhibited by national interests and costs. It is complicated to decide which country or countries will pay to coordinate all the monitoring efforts. Who will be in charge of disseminating the data? The countries would need to come together to determine which nation would be responsible for ultimately issuing a warning. At the same time each individual nation must determine, plan, and pay for an alert system and for the appropriate infrastructure to move many millions of people out of the path of a tsunami when one is detected (Casey, 2014). Finally, all of these things cost money, and many of the nations along the rim of the Indian Ocean are already trying to pay for education and basic infrastructure like clean water and waste removal for their populations. The additional burden of emergency preparedness is often an expense that these countries simply cannot afford.

The tragedy and loss of life from the 2004 South Asian Tsunami shows how the Earth's environment still affects humans on the planet. As much as technology tries to protect humans from nature, there are still natural disasters that can have devastating impacts.

See also: Section 2: Adaptation, Earthquake, Flood, Global North and Global South, Human Population, Mitigation, Natural Hazards; *Section 3:* Environmental Justice, Sustainable Development

FURTHER READING

ABC Australia. 2014. "Boxing Day Tsunami: How the Disaster Unfolded 10 Years Ago." ABC News Australia. December 23. http://www.abc.net.au/news/2014-12-24/boxing-day-tsunami-how-the-disaster-unfolded/5977568.

BBC. 2005. "Indian Ocean Tsunami Warning System." December 23. http://news.bbc.co.uk/2/hi/science/nature/4524642.stm.

Casey, Michael. 2014. "Tsunami 10 Years Later: Is the World Better Prepared for Disaster?" CBS News. December 24. http://www.cbsnews.com/news/tsunami-10-years-later-is-the-world-better-prepared-for-disaster/.

CNN. 2017. "Tsunami of 2004 Fast Facts." CNN Library. December 11. http://www.cnn.com/2013/08/23/world/tsunami-of-2004-fast-facts/.

Kayal, Michele and Matthew L. Wald. 2004. "At Warning Center, Alert for the Quake, None for a Tsunami." *New York Times*. December 28. http://www.nytimes.com/2004/12/28/science/at-warning-center-alert-for-the-quake-none-for-a-tsunami.

NGS. 2004. "The Deadliest Tsunami in History?" National Geographic News. December 27. http://news.nationalgeographic.com/news/2004/12/1227_041226_tsunami.html.

NWS. 2014. "December 26, 2004 Sumatra, Indonesia Earthquake and Tsunami—Tenth Anniversary Update." National Weather Service. ftp://ftp.ngdc.noaa.gov/hazards/publications/2004_1226.pdf.

Osborne, Hannah. 2017. "Boxing Day Tsunami: The Facts about the 2004 Indian Ocean Disaster: The Exact Death Toll Still Remains Unknown." *International Business Times*. March 24. http://www.ibtimes.co.uk/2004-indian-ocean-earthquake-tsunami-facts-1480629.

Pickrell, John. 2005. "Facts and Figures: Asian Tsunami Disaster." *New Scientist*. January 20. https://www.newscientist.com/article/dn9931-facts-and-figures-asian-tsunami-disaster/.

PTWC. 2018. "Tsunami Messages for All Regions." Pacific Tsunami Warning Center. http://ptwc.weather.gov.

Rondonuwu, Olivia. 2012. "Tsunami Alerts Pass Indonesia Quake Test, with Luck." Reuters. April 12. http://www.reuters.com/article/us-asia-quake-idUSBRE83B09G20120412.

UN. 2009. "Five Years after Indian Ocean Tsunami, Affected Nations Rebuilding Better." United Nations News. http://www.un.org/apps/news/story.asp?NewsID=33365#.WH-8AlMrIdU.

USGS. 2018. "Earthquake Hazards Program." U.S. Geological Survey. https://earthquake.usgs.gov/.

Key Concepts

Adaptation

Biologists use the word adaption to explain how animals are able to "live in environments that seem to place insurmountable obstacles in their way" (Schmidt-Nielsen, 1997: vii). In this context, an adaptation is a genetic change or mutation that helps an organism survive in its current environment. If this mutation is beneficial, it is handed down to future generations and eventually becomes a typical trait for the whole species. This process plays out as either structural or behavioral adaptations. Structural adaptation is a physical part of an organism, like protective coloring to camouflage it from predators. Behavioral adaptations are actions an animal takes to help it survive, like hibernation or migration. Adaptations commonly develop in response to a change in the animal's habitat.

For people in modern times, adaptation is closely linked to behaviors that modify the natural environment through the application of technology. Humans live in a wide variety of places, so throughout most of history, their behavioral adaptations have protected them from climatic stress. Building a fire kept people warm in a cave and later in a frontier cabin, for example, and more recently people invented electricity that blows heat throughout their homes. These adaptive actions allow people to live in colder places. Likewise, technologies like electric fans and air conditioning permit millions of people to adapt and live comfortably in hot regions. But these technological adaptations currently rely on fossil fuel energy, which is associated with numerous environmental problems. Human adaptions have serious effects on the natural environment, so societies often opt to manage these impacts through specific environmental protection policies.

Humans' inventions and use of technology have both intended and unintended consequences for the natural environment, which can have positive or negative ramifications. For example, technology related to energy exploration and production has overwhelmingly negative impacts on the environment, as offshore oil drilling, pipelines, and oil refineries have degraded water, land, and air. The geographic extent of these technologies has expanded from small-scale local production like an individual oil well, to global efforts to explore, discover, mine, and transport petroleum. At each scale, environmental and social impacts are present: local people might get new jobs which is a positive ramification, but increased pollution would negatively harm their community.

Energy sources changed from wood, to coal, and then to gas, as technological advances and discoveries changed people's expectations and ability to adapt to their environment. Centuries ago, nobody dreamed of living in a home that had a controlled temperature at a pleasant, moderate 21°C (70°F) year-round. Likewise, no one imagined having a society that was based on individual modes of transportation like private cars that operate by the simple turn of a key, rather than transportation from horses and carriages. Further, the widespread use of these motor vehicles led to extensive road systems and concrete that destroyed vast expanses of native forests and changed natural hydrological cycles. These fossil fuel–based technologies, which are now firmly part of industrialized society, are based on innovations that initially seemed positive and beneficial for human adaptation. But the related unintended consequences have led to significant negative environmental degradation, especially to the Earth's atmosphere.

Climate change is the most significant global environmental issue today. "From shifting weather patterns that threaten food production, to rising sea levels that increase the risk of catastrophic flooding, the impacts of climate change are global in scope and unprecedented in scale. Without drastic action today, adapting to these impacts in the future will be more difficult and costly" (UN, 2018).

People must now either mitigate or adapt to these vast changes in the natural environmental systems caused by humans. "Mitigation addresses the root causes, by reducing greenhouse gas emissions, while adaptation seeks to lower the risks posed by the consequences of climatic changes" (Clark, 2012). Mitigation actions are taken up-front so that negative environmental conditions do not occur. So if mitigation is done successfully, adaptation is not needed. But the reality is that, at this point, adaptation is necessary because people must take action to survive in an environment that is already experiencing the consequences of climate change.

Economic factors can influence humans' decisions to adapt or mitigate. Is it more expensive to shift away from fossil fuel use to slow climate change (mitigation) or to move cities to higher ground after rising seas from climate change causes massive flooding (adaptation)?

Related to a pure monetary cost and benefit analysis are indirect costs related to changing major economic drivers in an industrial society; thus, shifting away from fossil fuels has broader economic impacts, as some corporations and people will lose profits. This type of shift will require social and political change, which is difficult to launch. In the meantime, society can take small actions to mitigate human modifications of the environment, such as policies requiring the use of coal with lower sulfur content, demanding scrubbers to reduce emissions from smokestacks, or mandating higher fuel efficiency for vehicles.

Adaptation policies can be planned in advance or implemented quickly in response to an emergency. Adaptation strategies include building infrastructure, like seawalls to protect against rising seas or roadways that allow faster drainage during a huge rain event. In Europe, green infrastructure is seen as a good option for climate change adaptation because increasing "healthy ecosystems—serves the interests of both people and nature" where "the benefits from nature such as clean air and water, flood prevention, crop pollination, carbon storage or health and well-being are delivered to citizens" (EU, 2016).

People are more vulnerable to climate change impacts depending on three factors: exposure to hazards like droughts; sensitivity to those hazards such as dryland farming; and low capacity to adapt to those hazards due to lack of money for irrigation technology, for example. Of course, low-income people and countries are most vulnerable to the negative impacts of climate change and most unable to take adaptive steps (Clark, 2012).

As noted by the thousands of expert scientists who collaborated on the most recent Intergovernmental Panel on Climate Change report: "Continued emission of greenhouse gases will cause further warming and long-lasting changes in all impacts for people and ecosystems. Limiting climate change would require substantial and sustained reductions in greenhouse gas emissions which, together with adaptation, can limit climate change risks" (IPCC Synthesis, 2014: 8).

See also: Section 1: Case Study 3; *Section 2:* Case Study 6, Case Study 7, Case Study 9; *Section 3:* Case Study 15

FURTHER READING

Clark, Duncan. 2012. "What Is Climate Change Adaptation?" *The Guardian*. February 27. https://www.theguardian.com/environment/2012/feb/27/climate-change-adaptation.
EU. 2016. "Green Infrastructure." European Commission. June 20. http://ec.europa.eu/environment/nature/ecosystems/background.htm.
IPCC. 2018. "Sixth Assessment Report Cycle." Intergovernmental Panel on Climate Change. http://www.ipcc.ch/.
IPCC Synthesis. 2014. "Climate Change 2014: Synthesis Report: Summary for Policymakers." Intergovernmental Panel on Climate Change. http://www.ipcc.ch/report/ar5/syr/.
Schmidt-Nielsen, Knut. 1997. *Animal Physiology: Adaptation and Environment.* Cambridge: Cambridge University Press.
UN. 2018. "Climate Change." United Nations. http://www.un.org/en/sections/issues-depth/climate-change/index.html.

Anthropocene

Most textbooks say that the Earth is currently in the Holocene epoch, a geologic term meaning "recently new," but there is increasing evidence that the Anthropocene, or "human new," era is underway because people have polluted the oceans, caused mass extinctions, changed the atmosphere, and caused other long-term impacts to the planet.

The Earth and its solar system are about 4.54 billion years old, the Milky Way Galaxy is about 11 to 13 billion years old, and the universe is about 10 to 15 billion

Small-Scale Solar Power in Africa

Most people in North and South America and Europe have electricity. Asia is also catching up. Still, about 1.5 billion people in the world live without electricity, many of them in sub-Saharan Africa, where only about 14% of people in rural areas have electricity. Even for people who have electricity, power outages are common and electricity is fairly unreliable (McKibben, 2017). This affects all aspects of life—just imagine trying to do homework by a flickering gas lantern or trying to keep vaccines refrigerated, or farmers forced to sleep in their fields to turn on irrigation when intermittent electricity is available.

But building a traditional electric grid is very expensive. When the United States was electrified in the 1940s, the necessary resources were cheap: copper wire, timber for poles, and cheap coal for energy. Now all these natural resources are prohibitively expensive, and this makes the electric grid nearly impossible.

People are turning instead to small-scale solar energy to bring electricity to rural Africa. The price of solar energy is becoming inexpensive, as the price of solar panels has fallen and light bulbs have gotten more energy efficient. About five years ago, for example, a radio, a mobile-phone charger, and a solar electric system would have cost $1,000, and it is now $350 (McKibben, 2017).

Just as cell phones bypassed the traditional network of physical poles and wires, small-scale solar power will likely do the same thing. This is a geographic example of how green technology can be introduced as a new solution, avoiding the negative environmental impacts that historically occurred in industrialized countries. Indeed, there are many entrepreneurs who are making small-scale solar power a source of income in rural Africa.

For example, M-Kopa is a program in rural Kenya, Tanzania, and Uganda that helps poor people who live on less than $2 a day afford solar panels. "Kopa" is the Swahili word for "borrow," and it is at the core of many solar enterprises in sub-Saharan Africa. Customers pay about $35 up-front and about $0.50 per day payments via cell phone. They receive a solar panel and batteries to power three LED bulbs, a phone charger, and a radio. Using these items, people can start their own businesses, supply energy to neighbors for a fee, and even fund others. After a year, they own the system and the electricity is free. So far, M-Kopa has helped half a million homes gain electricity and the economic and educational opportunities that come with it (Del Bello, 2016).

Further Reading

Del Bello, Lou. 2016. "Six Bright Ideas Lighting Up Africa from the Grassroots." Climate Home. November 23. http://www.climatechangenews.com/2016/11/23/six-bright-ideas-lighting-up-africa-from-the-grassroots/.

McKibben, Bill. 2017. "The Race to Solar-Power Africa." *The New Yorker*. June 26. http://www.newyorker.com/magazine/2017/06/26/the-race-to-solar-power-africa.

years old (USGS, 2007). The various time periods of Earth's history have been given names by geologists who study rocks that were formed thousands, millions, or billions of years ago. Specifically, for the last 10,000 years, Earth has been in the Holocene epoch, which is a very stable time within the Quaternary Period that experienced several ice ages (*The Economist*, 2011).

This is now under question, however, because humans have had such impact on the Earth in modern times.

For geologists to determine that a new epoch has begun, they need to see definitive changes to the Earth, caused by movement of continents, huge volcanic eruptions, or a massive meteor strike. These global-scale changes "must be recorded in geological stratigraphic material, such as rock, glacier ice or marine sediments" (Lewis and Maslin, 2015). Many geologists have a hard time accepting that human activities over a relatively short time can cause such significant changes to Earth when these massive changes typically take thousands of years.

On the other hand, it is likely that millions of years from now, human impacts will be observable in the fossil record. It might not seem important what geologists decide to name the eras, but in fact, the idea that humans have fundamentally changed the Earth's systems is a key theme in environmental geography. It is thus fundamentally important to understand and think about what environmental actions and policies can slow human impacts before it is too late.

Beginning in 2000, scientists studying global environmental change began to promote the idea that human impact on Earth was so significant, obvious, and detrimental that it was permanently changing the planet (Crutzen, 2002). These scientists provide numerous examples that show Earth has entered the Anthropocene epoch shaped primarily by humans.

Evidence is clear that the current rate and scale of global environmental change are unprecedented. This includes the fact that atmospheric CO_2 and methane concentrations have increased faster and at a higher level than at the beginning of other epochs. Humans produce materials that are new to the Earth's ecosystem; for example, the annual introduction of hundreds of tons of plastic and half of all concrete ever made was produced in just the last 20 years. At the same time nature has been pushed into a corner; with wildlands reduced to just 25% of ice-free lands on Earth, down from 50% just 300 years ago. This has caused high rates of species extinction; some scientists say this is a new mass extinction. Another example of obvious human impact on natural systems is the presence of isotopes from nuclear weapons testing beginning in the 1940s, which are clearly seen in the geologic record (Barras, 2018; Vaughn, 2016).

When did the Anthropocene begin? There are several theories about which year best signals this shift:

1. The year 1610 denotes the era when colonialism was so widespread that plants, animals, and people were forever changed and, to some extent, "local" environments could no longer remain isolated.
2. The year 1800 marks industrialization as the turning point of human influence as fossil fuel usage began to soar and atmospheric concentrations of CO_2 began to rise.
3. The year 1950 is noted as the time where there is an acceleration of change, with huge growth in human population, the use of nuclear bombs, vast changes to natural systems, and the creation of numerous new materials such as plastics that last forever in the environment.

Regardless of the exact year that marks a new epoch or whether the term is accepted among geologists, the key point is that humans have significant impacts on the planet, more than any other species, but also have the capacity to reduce these negative impacts. This makes humans unique. Sure, cyanobacteria completely changed life on Earth 2 billion years ago by oxygenating the atmosphere, but these microscopic bacteria did not realize it. Humans are the first species that has become influential at the global scale and is aware of this fact (Stromberg, 2013).

Geographers have been talking about these concepts for a long time. In fact, studying human impacts on the environment is at the core of environmental geography. Keep in mind, however, that the Anthropocene takes it up a notch by suggesting that human activities are now as powerful as the global forces of nature. Earth is now changing due to the daily activities of billions of people across the planet. This goes beyond the idea of people affecting the environment; rather, human actions are so potent that they are merged into and can actually alter the planet's inherent biophysical systems. This is not the same thing as historical ideas of human domination over nature, however, because that would imply that humans are in control and understand how their actions affect nature. Now, in the Anthropocene, people do not completely understand how they are affecting the biophysical processes of the planet as a whole.

Using the new term "Anthropocene" is important because it should be another clear indication to the general public that humans now have global environmental impacts. This may counteract what many scientists conclude: "humanity is not taking the urgent steps needed to safeguard our imperiled biosphere" (Ripple et al., 2017). People are changing the Earth, but they also have the knowledge, technology, and capacity to preserve it. The question is whether people have the political willpower to do so.

See also: Section 1: Case Study 2, Case Study 3; *Section 2:* Case Study 6

FURTHER READING

Barras, Collin. 2018. "The Best Place on Earth to Mark the Anthropocene's Dawn." BBC. February 6. http://www.bbc.com/future/story/20180205-the-unexpected-signposts-of-our-new-era.
Crutzen, Paul. 2002. "Geology of Mankind." *Nature* 415: 23.
Lewis, Simon L. and Mark A. Maslin. 2015. "Defining the Anthropocene." *Nature* 519: 171–180.
Ripple, William J., Christopher Wolf, Thomas M. Newsome, Mauro Galetti, Mohammed Alamgir, Eileen Crist, . . . and 15,364 Scientist Signatories from 184 Countries. 2017. "World Scientists' Warning to Humanity: A Second Notice." *BioScience* 16(12): 1026–1028.
Stromberg, Joseph. 2013. "What Is the Anthropocene and Are We in It?" *Smithsonian Magazine.* January. https://www.smithsonianmag.com/science-nature/what-is-the-anthropocene-and-are-we-in-it-164801414/.
The Economist. 2011. "The Anthropocene: A Man-Made World." *The Economist.* May 26. http://www.economist.com/node/18741749.
USGS. 2007. "Age of the Earth." U.S. Geological Survey. pubs.usgs.gov/gip/geotime/age.html.
Vaughn, Adam. 2016. "Human Impact Has Pushed Earth into the Anthropocene, Scientists Say." *The Guardian.* January 7. https://www.theguardian.com/environment/2016/jan/07/human-impact-has-pushed-earth-into-the-anthropocene-scientists-say.

Drought

Most natural hazards hit suddenly and cause immediate disaster. In contrast, droughts are slow to develop, complex to define, and vary significantly from place to place. Although they develop slowly, droughts have widespread social, economic, and environmental impacts. According to the United Nations, droughts displace more people and cause more deaths than any other natural disaster (UNCCD, 2018).

The definition of drought is not based on one universal, numerical value because a drought is a relative term based on geography. Drought conditions in a tropical rainforest are very different from an arid grassland. In addition, the consequences of drought are felt very differently in various geographic regions. Concerns of food security, for example, vary significantly between areas with cutting-edge technology for irrigated agriculture versus a region reliant on rain-fed farming. Because of all this complexity, the planning, policies, and mitigation related to droughts are more difficult than for other natural hazards.

A basic definition of drought is a water shortage caused by deficient precipitation over an extended time. Geographers are also concerned about the impacts of droughts, which are the consequence of both nature (low precipitation) and human society (decisions and cultures that determine water demand). Because drought is not a purely environmental process, it is typically described in both conceptual and operational terms.

Droughts, as evidenced here in the Afar region of Ethiopia, Africa, may increase as anthropogenic climate change disrupts precipitation patterns. Women and girls walk hours each day to carry water from increasingly scarce sources. (Sjors737/Dreamstime.com)

Conceptual definitions are broad and general, which helps people understand the overall concept of drought. One such definition, from the National Drought Mitigation Center, is that "drought is a protracted period of deficient precipitation resulting in extensive damage to crops, resulting in loss of yield" (NDMC, 2018). This type of definition is important because it can help people establish drought policies. In Australia, for example, the term "exceptional drought circumstances" is used to indicate when drought conditions are more extreme than normal risk in agricultural management.

Operational definitions, on the other hand, use data to denote the onset, severity, and end of droughts. This provides a broader understanding of drought characteristics, the probability of drought recurrence, and forecasting of drought severity. Operational definitions vary geographically, based on a multitude of factors related to both humans and the environment. Mathematical models are often used to determine when to initiate emergency conservation measures and drought response actions. Variables indicating that a drought has begun include the level of difference from the average of precipitation and/or other relevant climatic variables over a distinct period of time. Typically, scientists will use data from long-established weather stations to compare the current situation with the historically recorded average. Thresholds are set, based on expert input, to identify the onset; so scientists might say that when a region has received only 50% of its average 30-year precipitation over one growing season, this is defined as a drought (NDMC, 2018).

This type of information is essential for planning and implementing mitigation strategies, developing drought response, and creating preparedness plans. Keep in mind, however, that these operational definitions, although very helpful, require a significant amount of climate data on a monthly, weekly, or even hourly basis and also detailed agricultural impact data regarding crop impacts.

In the United States, for example, it is recognized that drought has significant economic and social impacts. From 1980 to 2000, for example, heat waves and major droughts cost the nation over $100 billion in losses. In one year, 2012, over 60% of the continental United States experienced chronic drought. Due to climate change, severe droughts are projected for the next several decades, so government agencies have worked to gather extensive data to model drought conditions. This includes the real-time "Where is drought this week?" website that provides numerous maps and information (NIDIS, 2018). This program is a collaborative effort between seven federal agencies working with local, state, tribal, and the private sectors to manage drought risks and impacts.

In contrast, many regions of the world lack funding and do not have this level of data collection, or have not had funding to gather data over a long period. Yet many of these regions are most vulnerable to droughts. The impacts of drought are widespread: by 2025, 1.8 billion people will be living under absolute water scarcity, and 66% of people on Earth will be experiencing water-stressed conditions (UNCCD, 2018). Geographic distribution of drought shows highest losses in sub-Saharan Africa, but also high-risk areas in Central and South America, southern Europe, the Middle East, and southern Australia.

The United Nations Sustainable Development Goal 6 is to "ensure availability and sustainable management of water and sanitation for all" through three key pillars of integrated water resources management. This includes drought resilience based on drought monitoring and early warning systems, vulnerability and risk assessment, and drought risk mitigation measure (UNCCD, 2016).

Key mitigation actions include support for local food production based on drought-tolerant, native seed varieties and short-cycle crops; building reservoirs to store water in case of rain deficiency; implementing conservation agriculture techniques to decrease evaporation, use rainwater efficiently, and reduce water runoff; developing water harvesting structures and sustainable irrigation systems that boost reuse and recycling of water; and building alternative jobs and opportunities to provide income in high-risk areas.

Droughts can cause humanitarian disasters, threatening the lives of millions of people. In some places, droughts are part of the normal climate cycle, but in the past, droughts were not always so disastrous. Now, however, rain patterns are becoming more erratic and droughts more frequent and—equally important—social and economic factors have caused greater vulnerabilities. Without mitigation, planning, and action, droughts will continue to have destructive impacts on high-risk populations (FAO, 2018).

See also: Section 1: Water Scarcity; *Section 3:* Case Study 15, Water Conservation

FURTHER READING

Climate.gov. 2018. "Data Snapshot Details: Drought Monitor." National Oceanic and Atmospheric Administration. https://www.climate.gov/maps-data/data-snapshots/data-source-drought-monitor.
FAO. 2018. "Drought." Food and Agriculture Organization of the United Nations. http://www.fao.org/emergencies/emergency-types/drought/en/.
NDMC. 2018. "What Is Drought?" National Drought Mitigation Center. http://drought.unl.edu/DroughtBasics/WhatisDrought.aspx.
NIDIS. 2018. "Where Is the Drought This Week?" National Integrated Drought Information System. https://www.drought.gov/drought/.
NOAA. 2018. "Global Drought Map." National Centers for Environmental Information. https://gis.ncdc.noaa.gov/maps/ncei/drought/global.
UNCCD. 2016. "Drought: Reducing Impacts and Building Resilience." United Nations Convention to Combat Desertification. http://knowledge.unccd.int/topics/drought-reducing-impacts-and-building-resilience.
UNCCD. 2018. "Land and Drought." United Nations Convention to Combat Desertification. http://www2.unccd.int/issues/land-and-drought.

Earthquake

The Earth beneath our feet is composed of the crust and mantle, which are like pieces of a puzzle that fit together on the surface of the planet. These pieces are called tectonic plates and they slowly move around, sometimes bumping into each other. The force of the moving plates sends energy radiating out as seismic waves that shake the ground. Along these edges, called plate boundaries, are faults and this is where most earthquakes happen.

An earthquake happens when two tectonic plates, typically less than 80 km (50 miles) underground, suddenly slip past each other releasing energy. This force is most noticeable directly above it on the Earth's surface, and this is called the epicenter. Before an earthquake, there are sometimes smaller quakes, called foreshocks. Aftershocks can continue for minutes, weeks, or months after the main earthquake.

Each year there are approximately 500,000 earthquakes detected by scientific equipment. Of these, about 100,000 can be felt by people, and 100 of them cause damage (USGS Facts, 2018).

Since 1900, there has been about $5 trillion in earthquake damage. The most economically damaging single event was the Tohoku, Japan earthquake-tsunami-nuclear sequence on March 11, 2011, which caused $335 billion in direct damage, homelessness for 450,000 people, and about 18,500 deaths. It is estimated that between 1900 and 2015, the total number of deaths due to earthquakes is around 2.32 million. Most (about 60%) of people died due to the collapse of buildings, and about 30% due to secondary effects such as a landslide or tsunami (KIT, 2016). But natural disasters hit poor people harder, as they tend to live in crowded, poorly constructed buildings that are vulnerable to more damage, and thus injuries and death tolls are higher. After a disaster, poor regions have more difficulty rebuilding, so health and sanitation conditions remain dangerous for a lengthier time.

The United Nations Educational, Scientific and Cultural Organization (UNESCO) works to help nations prepare for and reduce risk from natural hazards, including a program on geohazard risk reduction. Globally, UNESCO still promotes four priorities for action as stated within the Sendai Framework for Disaster Risk Reduction 2015–2030: Priority 1: Understanding disaster risk; Priority 2: Strengthening disaster risk governance to manage disaster risk; Priority 3: Investing in disaster risk reduction for resilience; and Priority 4: Enhancing disaster preparedness for effective response and to build back better in recovery, rehabilitation, and reconstruction. Specifically, UNESCO helps nations with planning and education to handle natural disasters, policy development to reduce natural hazard vulnerability, networking so leaders can learn from one another, and the implementation of early warning systems and response plans. Most recently, UNESCO has taken the lead on securing building safety and earthquake resilience.

As with other natural hazards, environmental geographers are focused on preparing for an earthquake and ensuring that policies promote an appropriate emergency response if an event occurs. In addition to policies, insurance is an adaptation strategy. Wealthier people can afford to buy earthquake insurance, but it is quite expensive, particularly in regions with higher seismic activity. Buying such insurance is not required, except by banks giving people a loan (mortgage) to buy a house in some locations. Mostly, such insurance is voluntary, and so most people do not have it, which leads to real hardship after a destructive earthquake.

Cities can help citizens prepare for earthquakes by implementing building codes or regulations that specify minimum requirements to protect occupants' safety during earthquakes. These rules are called seismic codes because they specifically address how to build a structure to limit seismic risk, based on best available scientific and engineering research. Building codes vary widely from place to place, as each city, state, and nation has differing rules. It is proven that there are fewer deaths and injuries when building codes are stringent and builders carefully adhere to them (Sanchez, 2014).

National emergency management agencies provide information to educate citizens about the best actions to take during an earthquake. Each family can assess their home's structure and take steps to make it safer during an earthquake and less likely to experience damage. Beyond the building itself, all unsecured objects and furniture are a potential safety concern, as they can move, fall, and break when an earthquake shakes a building (FEMA, 2016). The American Red Cross issues detailed earthquake preparedness information, both for indoors and outside. Staying safe indoors is based on "DROP, COVER and HOLD ON!" Move as little as possible because during earthquakes most injuries happen when people move around, fall, and injure themselves. Overall, stay inside until the shaking stops and only exit if it seems safe. On the other hand, if people are outside, go to a location away from buildings, trees, and power lines and drop to the ground (Red Cross, 2018).

See also: Section 2: Case Study 10, Mitigation, Natural Hazards

FURTHER READING

FEMA. 2016. "Earthquake Safety at Home." U.S. Federal Emergency Management Agency. https://www.fema.gov/earthquake-safety-home.

KIT. 2016. "Natural Disasters Since 1900: Over 8 Million Deaths, 7 Trillion US Dollars." Karlsruhe Institute of Technology. April 18. *ScienceDaily*. www.sciencedaily.com/releases/2016/04/160418092043.htm.

Red Cross. 2018. "Earthquake Safety." American Red Cross. http://www.redcross.org/get-help/how-to-prepare-for-emergencies/types-of-emergencies/earthquake#During.

Sanchez, Ray. 2014. "Experts: Strict Building Codes Saved Lives in Powerful Chile Earthquake." CNN. April 3. http://www.cnn.com/2014/04/02/world/americas/chile-earthquake/index.html.

USGS Facts. 2018 "Earthquake Facts." U.S. Geological Survey. https://earthquake.usgs.gov/learn/facts.php.

UNESCO. 2017. "Geohazard Risk Reduction." United Nations Educational, Scientific and Cultural Organization. http://www.unesco.org/new/en/natural-sciences/special-themes/disaster-risk-reduction/geohazard-risk-reduction/.

Ecosystem Services

In the 1990s, scientists began emphasizing the concept that ecosystem services are the processes and activities that natural ecosystems and other species do to sustain and benefit humans. Indeed, ecosystems support diverse human cultures and "provide aesthetic beauty and intellectual stimulation that lift the human spirit" (Daily, 1997: 4). Ecosystems give both direct and indirect benefits to humans on Earth, which allow people to survive and prosper.

There are several categories of services that ecosystems provide, including provisioning, habitat, regulating, and cultural (BISE, 2018). Provisioning services are the products people get from ecosystems, such as fresh water, wood, food, medicines, etc. Many aspects of human economic systems are based on ecosystems' numerous resources. Habitat services are the complex, integrated, biologically rich landscapes that allow species diversity, migration, and viable gene pools. This includes the complex activities of predators and parasites that work to control populations and halt the potential spread of pests and disease. Regulating services are the life-supporting ecosystem processes that provide relatively stable planetary conditions—things like water purification, waste management, pollination, pest control, and climate. These services are often taken for granted because they are mostly invisible, but when damaged they are extremely difficult to restore. Cultural services are the benefits humans gain from ecosystems through recreation and experiencing natural beauty. Interestingly, these aspects tend to be overlooked, but are often essential in fulfilling people's lives.

The Food and Agriculture Organization (FAO) of the United Nations notes the global importance of ecosystem services to Earth's environment and its ability to provide food for humans. First, ecosystems remove and sequester carbon dioxide from the atmosphere and store it in plants and trees as they grow. Vegetation covers the ground, and this prevents soil erosion and also plays a role in soil fertility through natural biological processes of decomposition. Insects and wind work to pollinate plants and trees so that fruits, seeds, nuts, and vegetables can grow. In fact, pollinators such as bees, birds, and bats influence over 35% of global crop production and actually help increase the production of 75% of the main food crops worldwide. The services of predators and parasites in ecosystems act to control populations of potential pest and disease vectors. Finally, food crops are heavily dependent on water, because its timing, quality, and quantity must be appropriate for crops to grow. Ecosystems work to regulate water flow through various landscapes and watersheds with a complex system known as the hydrologic cycle. As climate change makes an ever-increasing impact on human food production, scientists note that ecosystem services are a significant influence on local weather and on global climate systems, which determine food production capacity (FAO, 2018).

Ecosystems also play a major role in moderating extreme events, as they act to buffer against natural hazards by reducing damage from storms and weather extremes. For example, wetlands, which are defined as areas where water covers the soil part or all of the time, are important because they protect water quality, store floodwaters, guard against droughts, and provide wildlife habitat. Healthy wetlands are vital to protecting people's lives and livelihoods against natural disasters caused by unpredictable weather patterns and extreme climate events. But when wetlands are destroyed, these ecosystems lose their ability to regulate natural hazards. This occurred in the 2004 Asian Tsunami, where healthy wetland mangroves protected coastal areas, but regions where mangroves had been removed suffered severe damage (Nagabhatla and Smakhtin, 2017). Healthy wetlands provide a natural system of conveying storm water and reducing flooding. Indeed, it is estimated that degraded wetlands cause $20 trillion worth of losses in ecosystem services worldwide (Costanza et al., 2014).

In 2000, the United Nations Secretary General Kofi Annan called for a Millennium Ecosystem Assessment to assess ecosystem changes and their consequences for human well-being. Thus it was established to provide an up-to-date scientific appraisal of the world's ecosystems, and over 1,000 experts from across the globe contributed to this report, composed of six synthesis reports and five technical volumes.

The main finding of the global Millennium Ecosystem Assessment was that in the preceding 50 years, humans had changed ecosystems more rapidly and extensively than in any other period. "This has resulted in a substantial and largely irreversible loss in the diversity of life on Earth" (MEA, 2005). The changes that people made to ecosystems have led to substantial gains in economic development and human well-being for some people, but have also caused degradation of many ecosystem services, risk of additional unpredicted changes, and increased poverty for other people. If these problems are not addressed, ecosystem services will be significantly decreased for future human generations. According to this report, the only way to avoid further degradation of ecosystems is to immediately implement policies and update institutions to conserve and boost healthy ecosystem services.

The very concept of ecosystem services, however, has received a great deal of criticism. First, critics say that it downplays the ecological complexity of natural systems and the actual effort needed to effectively and sustainably manage ecosystems. Second, the ecosystem services approach is inappropriate because it mistakenly assumes that there will always be equilibrium between global ecological systems and place-by-place resource use. Third, ecosystem services have taken attention away from the more important question of how humans must make major changes in economic and policy institutions in order to reduce human degradation of ecosystems. Overall, the concern is that relying on ecosystem services is a simple concept where resources flow as services with economic value, and this "is blinding us to the ecological, economic, and political complexities of the challenges we actually face" (Norgaard, 2010: 1219).

In addition, there is an ethical question of whether the complexities of nature can be fully described as a simple monetary value. There is also a counterargument by ecological economists who claim that the best way to protect nature is to figure out how much it is worth. Money talks, as they say. So providing a specific dollar value gives people a better understanding of the many important benefits that nature provides and thus the necessity for humans to protect ecosystems. In 1997, an article by ecological economist Robert Costanza and several colleagues made headlines around the world because they showed the estimated annual net worth of ecosystem services was $33 trillion (Costanza et al., 1997). This value was higher than all of the economies of the world's nations combined. Repeating the study in 2014, the authors pegged this value at $145 trillion (Costanza et al., 2014). Despite the critique of ecosystem services, Dr. Costanza thinks that it actually opens a dialogue between environmentalists and policy makers to try to work in unison toward better environmental management. He notes: "You can say how important you think water supply is as a service until you're blue in the face, but until you say 'It's worth this much in dollars,' it's hard to get people to pay attention" (Harris, 2003). To Costanza and others, putting an actual monetary value on an ecosystem service convinces people of the need for better environmental management because it proves to them just how valuable nature is to humans.

See also: Section 2: Case Study 10; *Section 3:* Case Study 15, Sustainable Development, Water Conservation

FURTHER READING

BISE. 2018. "Ecosystem Services." Biodiversity Information System for Europe. European Union. http://biodiversity.europa.eu/topics/ecosystem-services.
Costanza, Robert, Ralph d'Arge, Rudolf De Groot, Stephen Farber, Monica Grasso, Bruce Hannon, . . . Marjan van den Belt. 1997. "The Value of the World's Ecosystem Services and Natural Capital." *Nature* 387, 253–260.
Costanza, Robert, Rudolf de Groot, Paul Sutton, Sander van der Ploeg, Sharollyn J. Anderson, Ida Kubiszewski, . . . R. Kerry Turner. 2014. "Changes in the Global Value of Ecosystem Services." *Global Environmental Change* 26: 152–158.
Daily, Gretchen, 1997. *Nature's Services: Societal Dependence on Natural Ecosystems.* Washington, D.C.: Island Press.
FAO. 2018. "Ecosystem Services & Biodiversity (ESB)." Food and Agriculture Organization. http://www.fao.org/ecosystem-services-biodiversity/background/regulating-services/en/.
Harris, Lissa. 2003. "Ecological Economist Robert Costanza Puts a Price Tag on Nature." Grist. April 9. http://grist.org/article/what2/.
MEA. 2005. "Overview of the Millennium Ecosystem Assessment." https://www.millenniumassessment.org/en/About.html.
Nagabhatla, Nidhi and Vladimir Smakhtin. 2017. "World Wetlands Help in Dealing with Natural Disasters." United Nations University. May 10. https://unu.edu/publications/articles/world-wetlands-and-natural-disasters.html.
Norgaard, Richard B. 2010. "Ecosystem Services: From Eye-Opening Metaphor to Complexity Blinder." *Ecological Economics* 69: 1219–1227. http://esanalysis.colmex.mx/Sorted%20Papers/2009/2009%20USA%20-3F%20Interd%201.pdf.

Ecotourism

Ecotourism is vacation travel for the specific purpose of experiencing nature. These activities, when undertaken by thousands or millions of people over time, can actually harm the local environment and people. That is why ecotourism advocates are working to promote sustainable ecotourism. This includes concepts of unifying sustainable travel with community engagement and long-term conservation.

Specifically, the people and businesses that participate in sustainable ecotourism must address several goals, including to increase both environmental and cultural awareness; ensure positive experiences for both visitors and host communities; give direct financial support for conservation; provide monetary benefits for both local people and private industry; ensure that tourists are educated about host countries' social and environmental conditions; plan, build, and operate facilities

Boats anchor off Bartolome Island in the Galapagos, Ecuador, where increased tourism threatens the sensitive environment. (National Oceanic and Atmospheric Administration)

with low environmental impact; recognize the rights and beliefs of all people in the host community; and work to empower indigenous people (TIES, 2018).

From a tourist's perspective, the negative aspects of ecotourism might not be very obvious, but these environmental and cultural impacts can devastate local places.

Ecological degradation from ecotourism is particularly surprising, since these tourists are specifically seeking experience in the natural environment. But in fact, ecotourism can cause devastating environmental degradation, particularly as a destination becomes more famous. As more tourists visit an area, the transportation systems must keep up with demand. Air and vehicle travel increase, so airports, roads, and highways expand—often reducing or degrading natural landscapes. Tourists want to see areas of pristine beauty, but natural attractions can actually suffer from overuse. Visitors can disrupt wildlife feeding and mating habits. Human impact can sharply degrade native ecosystems by trampling vegetation, leaving garbage, or causing noise, for example.

The Galapagos Islands, located 1,000 kilometers (about 600 miles) west of Ecuador, are a famous ecotourism destination precisely due to their historical impact in the 1800s, as the location that inspired Charles Darwin's theory of natural selection that underlies the concept of evolutionary change. But fast-forward 200 years, and a massive influx of tourists, new residents (25,000 people total), and suppliers has brought devastating ecological damage, including pollution from sewage and tourist boat oil, and also many new alien species, including rats, cats, mosquitoes, and fire ants which are threatening to change the native ecosystems forever (Sachs, 2016).

At every site, tourism should help the local economy. Unfortunately, this is often not the case, as developers from outside and international corporations buy up land in popular tourist destinations. These businesses have headquarters outside the local area, so profits from the hotels, stores, and restaurants take money away from locally owned businesses. Plus, local people often experience higher prices for food and housing, once the local economy is altered by outside tourist development. In addition, there are often corrupt government officials who take illegal profits from ecotourism. Overall, local communities often see few real benefits from an influx of visitors for ecotourism.

Cultural degradation is seen in many geographical settings. Beach hotels have displaced numerous fishing communities along the coast of Thailand. Sacred indigenous burial grounds have been destroyed by resorts in Hawaii. Native people in the Amazon have been disrupted by tour operators and even suffered illness and death from new diseases they brought. In the North American context, indigenous people work low-wage jobs for outside tourist companies exploiting their cultural traditions.

Tourism can have a negative impact on local cultures that are stretched to meet demands of an invasion of visitors. They often lose their traditional jobs, as cropland is taken over by high-rise hotels. They might get new jobs, but these are often low-paying service work at hotels, and the local cost of living increases at the same time. The historical cultural practices can be degraded as simple tourist shows, which erases traditional cultural meaning and community cohesion.

World leaders believe, however, that tourism has the potential to make a significant contribution to the three pillars of sustainable development, bringing environmental, social, and economic benefits. But tourism must be implemented properly: "we need to call for the right policies, the adequate investment and the proper business practices that can advance us towards fairer, more people-centered, inclusive growth" noted the secretary general of the United Nations World Tourism Organization, Taleb Rifai. And certainly, tourism can have a strong multiplier effect, meaning "that for every job created in tourism, many more jobs are created in other sectors" of the economy according to Supachai Panitchpakdi, secretary general of the United Nations Conference on Trade and Development.

Overall, the United Nations emphasizes sustainable tourism directives in order to create decent jobs, stimulate cultural exchange, and help eliminate poverty. Linkages between local communities and tourism attractions must be improved "in order to make tourism a more effective tool in the fight against poverty and to advance increased awareness among tourists of their obligation to respect and protect the environment considering that it is tourism's prime interest and responsibility to protect natural resources" (UNWTO, 2012). Overall, the tourism industry "has a real interest in protecting the environment and a huge potential for the green economy as its assets are the ones we need to preserve and enhance" (UNWTO, 2012).

In 2012, the United Nations General Assembly unanimously adopted an important resolution titled "Promotion of Ecotourism for Poverty Eradication and

Environment Protection" that calls on all United Nations member countries to adopt policies that promote ecotourism because of its "positive impact on income generation, job creation and education, and thus on the fight against poverty and hunger." But the resolution notes that ecotourism must be done well because "sustainable tourism has a vital role to play in a fairer and sustainable future for all." Indeed, it is recognized that ecotourism, in and of itself, is not necessarily good. Instead "ecotourism creates significant opportunities for the conservation, protection and sustainable use of biodiversity and of natural areas by encouraging local and indigenous communities in host countries and tourists alike to preserve and respect the natural and cultural heritage" (UNWTO, 2013). But opportunities may not always translate into actual sustainability.

The United Nations and other organizations emphasize the need for tourism development plans to account for local needs. Local people and their cultures must be respected within the environmental education that is the basis of authentic ecotourism. Further, each country should establish national laws to promote small and medium-sized tourism enterprises, locally owned cooperatives, and micro-loans specifically for local and indigenous people to build a business in geographic areas with ecotourism potential. These efforts would support a more equitable distribution of profits from ecotourism.

See also: Section 1: Case Study 3; *Section 2:* Case Study 8, Protected Areas and National Parks; *Section 3:* Case Study 13

FURTHER READING

Carr, Anna, Lisa Ruhanen, and Michelle Whitford. 2016. "Indigenous Peoples and Tourism: The Challenges and Opportunities for Sustainable Tourism." *Journal of Sustainable Tourism* 24(8–9): 1067–1079.

Sachs, Andrea. 2016. "Out of the Blue: The Galapagos Islands Are Host to an Evolving Species of Land-Dwellers, They're Called Tourists." *Washington Post.* May 27. www.washingtonpost.com/graphics/lifestyle/galapagos/.

TIES. 2018. "What Is Ecotourism?" The International Ecotourism Society. http://www.ecotourism.org/what-is-ecotourism.

UNWTO. 2012. "Tourism Can Contribute to the Three Pillars of Sustainability." United Nations World Tourism Organization. June 22. http://media.unwto.org/en/press-release/2012-06-22/tourism-can-contribute-three-pillars-sustainability.

UNWTO. 2013. "UN General Assembly: Ecotourism Key to Eradicating Poverty and Protecting Environment." United Nations World Tourism Organization. January 3. http://media.unwto.org/press-release/2013-01-03/un-general-assembly-ecotourism-key-eradicating-poverty-and-protecting-envir.

Wood, Ann. 2017. "Problems with Ecotourism." *USA Today.* http://traveltips.usatoday.com/problems-ecotourism-108359.html.

Flood

Globally, there are more than 150 flood disaster events, which kill an average of 25,000 lives, affect 500 million others, and cost over $60 billion each year. This does not even consider the historical and cultural assets or natural resources that are destroyed (UNESCO, 2018).

Since 2005, the United Nations established the International Flood Initiative (IFI) housed in Tsukuba, Japan. The goal of the IFI is to cut in half the annual number of deaths and greatly reduce the damage caused by water-related disasters. They seek to do this by focusing on water research and then educating, training, and building a group for sharing information and networking among managers. Finally, communities will be given information, technical assistance, and guidance in planning, preparedness, and emergency management.

The reason flood damage has risen to such high levels is due to the rapid growth in human population, much of which is concentrated in high-value, densely concentrated areas. "Climate change and global warming are exacerbating the situation, and are likely to further increase the frequency of water-related disasters" (UNESCO, 2018).

Floods are a naturally occurring phenomena that contribute to sustainable ecosystems and biodiversity; thus, water resource management cannot and should not halt floods. Instead, management goals focus on long-term benefits while reducing negative consequences of flooding through integrated approaches that include public participation, working across economic sectors, and implementing technology that is based on collaboration among multiple disciplines in the social sciences, engineering, and natural sciences.

There are two types of floods, general floods and flash floods, both of which can cause landslides and debris flow in their aftermath. General floods can be predicted in advance, as atmospheric scientists can estimate the amount of flow that a given amount of rainfall will cause within a watershed and model how much this will raise river levels and thus cause flooding. Flash floods, however, are sudden and have caused an extremely large volume of water to inundate an area quickly. This type of flood is difficult to predict; thus people often have no warning and must escape their homes quickly, leaving everything behind. For all floods, emergency response is challenged by logistics, transportation, and distribution of aid because vast geographic areas can be covered in water. Although the initial activities of post-flood management are difficult, with evacuating to higher ground and seeking emergency shelter, often the later stages are even more challenging. Even as flood waters recede and people begin to go home and assess the damage, there are often challenges during prolonged weeks of cleanup (IFRC, 2018).

Floods are the most common and costly natural disaster in the United States. The U.S. National Flood Insurance Program (NFIP) was established in 1968 to allow property owners in participating communities to purchase flood insurance. This insurance thus serves two purposes: it protects the insured citizens against flood losses, and it also requires local and state organizations to develop and enforce management regulations. More than 20,000 communities participate in the NFIP, thus allowing their citizens to buy this federally backed insurance at a more reasonable rate. Private insurance companies sell and service the NFIP flood insurance policies. Since its inception, the NFIP has helped hundreds of thousands of families recover from destructive flooding by paying out billions of dollars in flood insurance claims (FEMA, 2018). Recently, however, increasingly frequent massive severe weather events have caused extensive flooding in many regions so that the federal insurance program runs out of money on an annual basis, causing some to question how NFIP will continue in the future (Leefeldt, 2017). Still, even if NFIP survives amid tightening national budgets, it costs the average homeowner $500 per year, which may not be affordable to lower-income residents.

Social vulnerability refers to the varying social, economic, and cultural variables that affect people's responses to natural hazards. In other words, some communities are harder hit by a flood hazard even though it is the same magnitude of flooding across a wide area. Thus from an emergency manager's perspective, it is important to understand the varying impacts of flood hazards caused by both social vulnerability and differences in risk level. An emergency response must consider the "number of people potentially affected, but also those characteristics of the population that contribute to the social burdens of risk" (Cutter, 2010: 23). Unfortunately, very few emergency management policies actually look at social vulnerability to assess the consequences of flooding and determine the best ways to reduce people's risk. Most management and flood risk policies focus only on economic variables and not the inequitable social and geographic impacts on the communities that are affected.

See also: Section 2: Case Study 9, Case Study 10, Mitigation, Natural Hazards

FURTHER READING

CDC SVI. 2018. "The Social Vulnerability Index." Agency for Toxic Substances & Disease Registry. Centers for Disease Control. https://svi.cdc.gov/.

Cutter, Susan L. 2010. "Social Science Perspectives on Hazards and Vulnerability Science," in *Geophysical Hazards.* T. Beer, ed. Dordrecht, The Netherlands: Springer.

Cutter, Susan L., Christopher T. Emrich, Melanie Gall, and Rachel Reeves, 2017. "Flash Flood Risk and the Paradox of Urban Development." *Natural Hazards Review* 19(1): 1–13.

FEMA. 2018. "National Flood Insurance Program." U.S. Federal Emergency Management Agency. https://www.fema.gov/national-flood-insurance-program.

IFRC. 2018. "Hydrological Hazards: General Floods and Flash Floods." International Federation of Red Cross and Red Crescent Societies. http://www.ifrc.org/en/what-we-do/disaster-management/about-disasters/definition-of-hazard/floods/.

Leefeldt, Ed. 2017. "Federal Flood Insurance May Never Be the Same." CBS News. October 9. https://www.cbsnews.com/news/federal-flood-insurance-may-never-be-the-same/.

UNESCO. 2018. "Floods." United Nations Educational, Scientific and Cultural Organization. https://en.unesco.org/themes/water-security/hydrology/programmes/floods.

USGS. 2018. "Data and Tools: Water." U.S. Geological Survey. https://www.usgs.gov/products/data-and-tools/real-time-data/water.

Global North and Global South

To understand geographic patterns and variations in how the environment affects humans, researchers and policy makers use terms to describe contrasts in countries' economic development. Over time, terms have been used such as More Developed Countries (MDC) and Less Developed Countries (LDC); First World, Second World, and Third World; and more recently Global North and Global South. It is important to understand what the various terms mean, how society and policy makers employ the terms, and how actual conditions vary for people living in these different geographic regions.

The terms MDC and LDC refer to the industrial development of a country. Most often researchers have used the term lesser developed to indicate countries with high rates of population growth, unemployment, and poverty; greater environmental degradation; and tendency for political instability. These places were mostly colonies that supplied natural resources to outside empires, and still are dependent on the economies of the wealthier countries and corporations. These terms, however, promote the concepts that sharply divide the world: rich versus poor, MDC versus LDC and core versus periphery. The notion of "development" sounds like a place is lacking progress and needs to improve. This underscores inequalities, but also downplays the mobility of many countries at the threshold between the two categories. In fact, some countries have moved from less to more developed in just a decade or two. Other countries, however, are caught in the less developed category due to long-term historical relationships, resource exploitation, and economic structures.

The ranked terms First, Second, and Third World are a product of the Cold War, a time when the world was geopolitically divided between the United States and the Union of Soviet Socialist Republics (USSR). All countries were aligned with either the United States or the USSR from post–World War II until the early 1990s, when the USSR broke apart into individual countries that shifted to capitalist economies. Countries were labeled First World if they were capitalist democracies and Second World if they were communist-controlled economies. The Third World was all the remaining, typically poorer, countries. There are many problems with these categories, because they imply a ranking where first is best and third is clearly worst. The terms further became obsolete when the Second World dissolved at the end of the last century.

Other terms came into use beginning in the late 1970s, when the Independent Commission on International Development Issues (ICIDI) was established and led by former German Chancellor Willy Brandt, and thus became known as the Brandt Commission. Their most famous report was "North–South: A Programme for Survival," which stated that their mission was "to study the grave issues arising from the economic and social disparities of the world community and to suggest ways of promoting adequate solutions to the problems involved in developing and attacking absolute poverty" (Brandt, 1980: 296). Indeed, the report developed what became known as the Brandt Line, which is a North-South divide between richer and poorer economies, based on per capita gross domestic product (GDP). Geographically, it is roughly at a latitude of 30° North, passing between North and Central America, north of Africa and India, but dipping south to include Australia and New Zealand above the line. This began the use of the terms North and South to describe variations among countries.

Other researchers added the term Global—so Global South and Global North— to emphasize that the inequalities were linked to global economic structures and

Environmental Crime

Environmental crime is a new area of criminal activity that has grown very rapidly in the last few decades. It is now the world's fourth-largest crime sector, increasing two to three times faster than the global economy. The United Nations Environmental Programme and INTERPOL (the world's largest police organization with 190 member countries) estimate that each year $91 to $258 billion of natural resources are stolen by criminals and traded illegally to earn money for international crime networks and armed rebels (Nellemann et al., 2016). This devastates ecosystems, causes species extinction, and fuels more violence and conflict.

The reason environmental crimes are on the rise is twofold: security forces are underfunded and there are weak environmental laws in many places. Indeed, the amount of money "earned" by criminals through environmental crime is 10,000 times more than the $20 to $30 million spent by international organizations trying to combat it (Nellemann et al., 2016).

Examples of environmental crime include: illegal trading in wildlife, illegal logging and corporate crime in the forestry sector, illegal mining and sale of gold and other minerals, illegal fishing and overfishing, and trafficking and mislabeling of hazardous waste.

One appalling example of illegal wildlife trading is seen in the demand for ivory that is decimating elephant populations. Despite international laws banning ivory trade, over 25% of the world's elephant population has been killed in the past decade. In Tanzania alone, for example, poachers kill 3,000 elephants each year. This earns about $10.5 million in illegal ivory—five times greater than Tanzania's national budget for its entire wildlife division (UNEP, 2016).

The topic of environmental crime receives little world attention, despite its rapid growth and vast scale. Huge sums of money flow through these crime networks and terrorist groups to cause violence and suffering around the world. "The result is not only devastating to the environment and local economies, but to all those who are menaced by these criminal enterprises. The world needs to come together now to take strong national and international action to bring environmental crime to an end," according to UNEP Executive Director Achim Steiner (UNEP, 2016).

Further Reading

Nellemann, C., R. Henriksen, A. Kreilhuber, D. Stewart, M. Kotsovou, P. Raxter, ... S. Barrat, eds. 2016. "The Rise of Environmental Crime—A Growing Threat to Natural Resources Peace, Development and Security." United Nations Environment Programme and RHIPTO Rapid Response—Norwegian Center for Global Analyses. www.rhipto.org.

UNEP. 2016. "Value of Environmental Crime Up 26%." Regional Office: North America. United Nations Environmental Programme. http://www.rona.unep.org/news/2016/unep-interpol-report-value-environmental-crime-26.

---

processes and not just a matter of geographical location. Adding the word "global" indicates the importance of globalization, which is the shift away from powerful nation-states toward more multinational networks and the diffusion of cultural, social, and economic activities. The global economy, though, is not equitable; rather it is linked to problems like poverty and environmental degradation in the Global South and consumerism and high resource use in the Global North.

Although the terms are often used by geographers and other researchers, politicians, and international organizations, most people have never heard of the Global

South or the Global North. So it is not really serving the purpose of educating people about social, economic, and environmental management variations across the planet (Hollington et al., 2015). Most recently, other terms have gained popularity: Majority World denotes that 80% of people on Earth live on about $3,000 a year, compared to the tiny minority of humans who have shelter, food, education, and money for consumer goods (Silver, 2015).

The fact is that words matter, and there are important disparities in the social and environmental conditions among countries around the world. Terms used to describe these variations must be realistic in order to inform people about inequalities and the need to help others, but also careful not to trap a country in one category indicating that people there will always be poor. Geographers tend to use the terms Global North and Global South. Perhaps in the future, better terms will be found. In the meantime, the real issue is understanding the causes of poverty, environmental problems, and people's vulnerability to natural hazards—and working to improve the lives of all people on Earth.

See also: Section 1: Case Study 5; *Section 2:* Case Study 7, Case Study 10; *Section 3:* Case Study 15

FURTHER READING

Brandt, Willie. 1980. *North-South: A Programme for Survival*. London: Pan.
Hollington, A., T. Salverda, T. Schwarz, and O. Tappe. 2015. "Concepts of the Global South." Global South Studies Center, University of Cologne, Germany. http://kups.ub.uni-koeln.de/6399/.
Murphy, James T. 2006. "Representing the Economic Geographies of 'Others': Reconsidering the Global South." *Journal of Geography in Higher Education* 30(3): 439–448.
Rigg, Jonathan. 2007. *An Everyday Geography of the Global South*. New York: Routledge.
Ritzer, G. 2007. *The Globalization of Nothing*." Thousand Oaks, CA: Pine Forge Press.
Silver, Marc. 2015. "If You Shouldn't Call It the Third World, What Should You Call It?" National Public Radio. January 4. https://www.npr.org/sections/goatsandsoda/2015/01/04/372684438/if-you-shouldnt-call-it-the-third-world-what-should-you-call-it.
UNISDR. 2016. "Poverty & Death: Disaster Mortality 1996–2015." United Nations Office for Disaster Risk Reduction and Centre for Research on the Epidemiology of Disasters. https://www.unisdr.org/we/inform/publications/50589.

Habitat and Wildlife

Humans need animals. Many ecological processes on Earth, on which humans depend, are shaped by other species. Insects pollinate food crops and compost waste to create soil. Trees and plants clean the air and water. Animals like dogs and cats are our companions; other animals stimulate our senses, like birds singing. Some animals inspire our imagination of faraway places, like bright blue butterflies in a tropical rainforest.

Tragically, the number of wild animals has dropped by half over the past 40 years. "If half the animals died in London Zoo next week it would be front page news,"

Known for its high biodiversity, Chitwan National Park in Nepal was established in 1973 and covers an area of 932 square kilometers (360 square miles). (Marina Pissarova/Dreamstime)

said Professor Ken Norris, Science Director of the Zoological Society of London, "but that is happening in the great outdoors. This damage is not inevitable but a consequence of the way we choose to live" (Carrington, 2014).

People cause pollution and destroy wildlife habitat. They also kill animals for food, profit, or pleasure in unsustainable numbers. In addition, human actions have greatly affected other species' ecological balance by introducing numerous species in new geographic areas. Two of the most significant human impacts on wildlife, which have geographical, social, and economic linkages, are invasive species and illegal poaching.

An exotic species is one that is not native to an area, so their population grows and becomes an invasive species, which "are plants or animals that do not belong where humans have intentionally or accidentally brought them" (WWF, 2017). They cause immense, irreversible damage to the native ecosystem. Hundreds of species have gone extinct because of invasive species that thrive in a new area where native species are unable to defend themselves against the invaders and where the new species has no predators. Problems have intensified as global trade and interactions contribute to the mixing of wildlife across wide biogeographical boundaries. Today biologists and the International Union for Conservation of Nature (IUCN) actively document and fight to control and eradicate the invaders.

Regulation of invasive species is mostly at the local or state level, so these tend to be reactive—taking action only after people realize there is a problem. The International Plant Protection Convention (IPPC) has been in place since the 1950s, but is not effective. This law primarily works to quarantine pests in international trade. It creates an international program that attempts to prevent spread of plant pests by using certificates between importing and exporting countries' national plant protection offices. But storage containers, packing materials, and transportation facilities often fall outside these regulations, and thus invasive species continue to be transmitted.

Other international laws attempt to manage wildlife, to greater or lesser success. This includes wildlife trafficking, which is the poaching (killing) and taking of protected or managed species and illegally trading this wildlife and their related parts and products. This is increasingly recognized as a new and dangerous area of crime networks and also a significant threat to many animal and plant species. Profits from wildlife trafficking, estimated at $10 billion each year, fund terrorist organizations and other criminal groups that harm people and habitats. The United Nations Office on Drugs and Crime published a comprehensive research overview in 2016: "World Wildlife Crime Report: Trafficking in Protected Species." There is a global database of wildlife seizures, most of which occur at border crossings. Although there are some gaps in data, they already indicate over 164,000 seizures from 120 countries (CITES, 2018). The data are based on information provided by countries in the Convention on International Trade in Endangered Species of Wild Fauna and Flora, which was signed in 1973, and the Marine Mammal Protection Act of 1972.

Wildlife crime is big money, and all regions of the world participate as a source, transit, or destination for illegal wildlife. Specific types of wildlife trafficking are associated with each geographic region: birds are strongly connected with Central and South America, mammals with Africa and Asia, reptiles with Europe and North America, and corals with Oceania. As just one example, the illegal killing of elephants for their ivory is devastating. "Every 15 minutes, an elephant is killed for its ivory. That's nearly 100 elephants a day—almost 40,000 elephants a year—a gruesome practice that could wipe out this species within a generation" (Conservation International, 2017).

How can this type of illegal wildlife trade be addressed? First, every country could implement a national law that prohibits the possession of wildlife that is illegally traded or obtained from anywhere in the world—this would reduce the demand. Second, research and monitoring could be increased and used to inform conservation policy and enforcement strategies. National laws and the international community could support local law enforcement with technical assistance and better funding. This includes tracing and recovering the illegal money created through these activities. Nations can and should do more to target suspicious transactions and dispose of the illegal merchandise to reduce contraband marketing. Plus the international community could support, with monetary and policy initiatives, the establishment of more protected conservation areas to address habitat loss.

Indeed protecting biodiversity as a whole is the most effective way to address declining animal habitat. This is not a new concept, as the Convention on Biological Diversity was adopted at a United Nations meeting in 1992 to promote conservation of biodiversity. Specific strategies and relevant policies include control of invasive species, restoration of threatened species, and protection of endangered species. Another shift in international environmental law is the concept of potentially harmful processes, so instead of managing dwindling wildlife numbers, authorities are encouraged to manage the processes that potentially harm the endangered species.

The global impact of biodiversity loss is detailed in a meta-analysis of nearly 200 research projects led by biologist David Hooper. They found that "the primary drivers of biodiversity loss are, in rough order of impact to date: habitat loss, overharvesting, invasive species, pollution and climate change," Hooper explains (Biello, 2012). Biodiversity loss affects habitats causing species decline. This is evidenced in loss of wildlife species, but in the future may directly affect the human species.

See also: Section 1: Biodiversity Loss, Deforestation, Urbanization; *Section 2:* Case Study 8, Human Modification of Ecosystems; *Section 3:* Case Study 12, Environmental Policy

FURTHER READING

Biello, David. 2012. "How Biodiversity Keeps Earth Alive." *Scientific American.* May 3. https://www.scientificamerican.com/article/how-biodiversity-keeps-earth-alive/.
Carrington, Damian. 2014. "Earth Has Lost Half of Its Wildlife." *The Guardian.* September 30. https://www.theguardian.com/environment/2014/sep/29/earth-lost-50-wildlife-in-40-years-wwf.
CITES. 2018. "Convention on International Trade in Endangered Species of Wild Fauna and Flora." https://www.cites.org/eng/disc/what.php.
Conservation International. 2017. "Wildlife Poaching and Trafficking." https://www.conservation.org/what/Pages/wildlife-trade-and-trafficking.aspx.
Howard, Jules. 2016. "Imagine a World without Animals. You'll Soon See How Much We Need Them." *The Guardian.* October 31. https://www.theguardian.com/commentisfree/2016/oct/31/world-without-animals-pollinating-crops.
UNODC. 2016. "World Wildlife Crime Report: Trafficking in Protected Species 2016." United Nations Office on Drugs and Crime. https://www.unodc.org/documents/data-and-analysis/wildlife/World_Wildlife_Crime_Report_2016_final.pdf.
WWF. 2017. "Impact of Invasive Alien Species." WWF Global. http://wwf.panda.org/about_our_earth/species/problems/invasive_species/.

Human Modification of Ecosystems

The geographic place where an organism lives is its habitat. Many individuals of the same species are called a population, and the combination of all populations of many different species that live together in a habitat is called a community. An ecosystem, then, is a community and its habitat combined.

Ecosystems are composed of all the living species and nonliving components in a specific local geographic environment. Ecosystems include water, air, sunlight, soil, plants, microorganisms, insects, and animals. Ecosystems may be water based (aquatic) or land based (terrestrial), and their size ranges widely—from a vast expanse of forest to a small puddle. Natural ecosystems vary greatly with specific vegetation, animal life, soils, and climate, each distinct from one another. Examples of natural ecosystems include tropical rainforest, temperate forest, taiga, grassland, coral reef, wetland, and desert.

A natural ecosystem is how most people typically think of nature. A forest ecosystem, for example, consists of a forest habitat composed of trees, vines, and plants, plus squirrels, deer, spiders, and other animals that fit within the local climate and soil conditions. In the past, scientists also approached their research by thinking of the natural environment as a native ecosystem.

"Ecologists traditionally have sought to study pristine ecosystems to try to get at the workings of nature without the confounding influences of human activity. But that approach is collapsing in the wake of scientist's realization that there are no places left on Earth that don't fall under humanity's shadow" (Gallagher and Carpenter, 1997: 485).

Humans have significantly modified the natural environment through actions that have both intentional and unintended consequences. These actions are related to specific geographical factors, human population growth, and the use of technology. Human behavior is drastically different from other species because the actions of humans are so powerful and widespread that they can actually change the Earth's environment, which then results in the need for further human adaptation and behaviors that often lead to even more environmental change. Think about the geographic interrelationships between human decisions and activities in one part of the world that affect many other regions, both close by and far away. For example, increasing corn production in the United States reduced the acreage of soybeans, which led to deforestation in the Amazon, where farmers cut down trees to grow soybeans for the world market.

If human modification of the environment is judged as an evolutionary strategy, then "our success at commandeering resources and transforming the landscape to meet our needs has been phenomenal" (Western, 2001: 5460). The human population has grown from fewer than 4 million about 10,000 years ago to over 7 billion today, indicating very successful survival of the species.

This deliberate human modification of ecosystems, however, has caused many negative impacts on Earth's ecosystems, including degradation of soils, heavy use of synthetic chemical pesticides, high natural resource extraction, homogeneity in habitat and landscape, huge use of synthetic chemical nutrient supplements, massive importation of nonsolar (fossil fuel) energy, modified hydrological cycles, and unstable simplified food chains. "All of these characteristics are deliberate strategies to boost production and reproduction. Survival rates have risen, lifespan increased, and other indices of welfare improved in the evolutionary blink of an eye" (Western, 2001: 5460). But there are negative consequences, as humans are distinct from other species because of their global dominance.

The Earth's ecosystems are experiencing the negative side effects of human actions. These unintended consequences include accelerated erosion; broken ecological gradients; ecological impact of toxins and carcinogenic emissions; genetic loss of wild and domestic species; habitat and species loss; loss of soil fauna; overharvesting of renewable natural resources; nutrient leaching and eutrophication; pollution from domestic and commercial wastes; proliferation of invasive exotic species; proliferation of resistant strains of organisms; new and virile infectious diseases; side effects of

Global Environmental Agreements

"Most environmental problems have a transboundary nature and often a global scope, and they can only be addressed effectively through international co-operation" (EU, 2016). Indeed, issues such as climate change, protecting the oceans, or preserving biodiversity cannot be managed by individual countries; instead, they demand international collaboration. Developing such treaties and agreements is very difficult, however, as numerous viewpoints, goals, and interests must be reconciled at each geographic scale. First, each country seeks to promote its own best interest, but even within one national government there may be different goals between agencies (for example, a ministry of agriculture may have different views than a ministry of the environment, which may differ from military objectives). Second, the individual goals of each country may be in conflict. Third, there are outside groups: industrial leaders and environmental organizations that may have very different interests when negotiating an environmental agreement (Susskind and Ali, 2015).

There are thousands of international environmental agreements, considering multiparty (many nations signing) and bilateral treaties (an agreement between just two countries). Key international environmental agreements tend to address these topics: climate change, environmental impact assessments, hazardous chemicals, ocean protection, ozone layer, protection of biodiversity, public participation and information, and waste management (Finland, 2018). Here is a sample of 10 key environmental agreements and their main goals:

- 1972 Convention on the Prevention of Marine Pollution by Dumping of Wastes and Other Matter—known as the London Convention, this makes dumping waste in the ocean illegal.
- 1973 Convention on the International Trade in Endangered Species of Wild Flora and Fauna (CITES)—this makes international trade of endangered species illegal.
- 1979 Geneva Convention on Long-Range Transboundary Air Pollution—collaboration to reduce and prevent air pollution.
- 1987 Montreal Protocol on Substances that Deplete the Ozone Layer—protects the upper-level ozone layer by phasing out substances such as chlorofluorocarbons (CFCs) that cause ozone depletion.
- 1992 United Nations Conference on Environment and Development (UNCED)—called the Earth Summit, it was held in Rio de Janeiro to address numerous global environmental and social issues in unison.
- 1992 United Nations Framework Convention on Climate Change—framework that established goals for the future climate treaties.
- 1997 Kyoto Protocol—international treaty that commits countries to reduce greenhouse gas emissions.
- 1998 UNECE Convention on Access to Information, Public Participation in Decision-Making and Access to Justice in Environmental Matters—known as the Aarhus Convention, guarantees citizens have access to environmental information to keep the regulation process transparent and reliable; signed by nations in Europe and central Asia.
- 2001 Stockholm Convention on Persistent Organic Pollutants—aims to eliminate or restrict the production and use of persistent toxic chemicals.
- 2016 Paris Climate Agreement—agreement negotiated by 196 nations to reduce and mitigate greenhouse gas emissions.

> The United Nations Environmental Programme states that environmental governance is a key factor when seeking to achieve sustainable development. Global economic and social goals can only succeed if strong and effective environmental institutions exist both at the national level and across international borders (UNEP, 2018).
>
> **Further Reading**
>
> EU. 2016. "International Issues." European Union. http://ec.europa.eu/environment/international_issues/agreements_en.htm.
> Finland. 2018. "International Environmental Agreements." Finland Ministry of the Environment http://www.ym.fi/en-US/International_cooperation/International_environmental_agreements.
> Susskind, Lawrence and Saleem H. Ali. 2015. *Environmental Diplomacy: Negotiating More Effective Global Agreements.* New York: Oxford University Press.
> UNEP. 2018. "Environmental Governance." United Nations Environmental Programme. https://www.unenvironment.org/explore-topics/environmental-governance.

synthetic chemical fertilizers and pesticides; simplified predator-prey relationships; unsustainable herbivore-carnivore interactions; unbalanced host-parasite networks; air and water pollution; and overall global changes in land, water, atmosphere, and climate (Western, 2001). Two decades ago, experts already noted:

> Human alteration of Earth is substantial and growing. Between one-third and one-half of the land surface has been transformed by human action; the carbon dioxide concentration in the atmosphere has increased by nearly 30% since the beginning of the Industrial Revolution; more atmospheric nitrogen is fixed by humanity than by all natural terrestrial sources combined; more than half of all accessible surface fresh water is put to use by humanity; and about one-quarter of the bird species on Earth have been driven to extinction. By these and other standards, it is clear that we live on a human-dominated planet. (Vitousek et al., 1997: 494)

The fact that humans can manipulate species and entire ecosystems, sometimes without intending to, is remarkable. In fact, human activities have actually changed the fundamental climate system on planet Earth. If humans have the societal will to acknowledge the importance of this issue, policies may be set up to protect native ecosystems. Implementation of proactive policies today could help slow future ecosystem declines.

Human modification of the Earth's complex ecosystems causes feedback loops and new linkages at multiple scales. These modifications reverberate through entire, vast ecosystems that include humans. The most significant example of this is human-induced climate change. Human actions have affected the Earth's atmosphere to such an extent that now all ecosystem modifications are intensified and future outcomes are uncertain (Pecl et al., 2017). Studying a natural ecosystem, outside of human control, is no longer an option. Indeed, science and policy must encompass existing and future human modifications of Earth's ecosystems.

See also: Section 1: Case Study 3, Biodiversity Loss, Climate Change, Water Pollution; *Section 2:* Case Study 6, Habitat and Wildlife

FURTHER READING

BBC. 2018. "Ecology Definitions." https://www.bbc.co.uk/education/guides/zmyj6sg/revision.
Biello, David. 2012. "How Biodiversity Keeps Earth Alive." *Scientific American*. May 3. https://www.scientificamerican.com/article/how-biodiversity-keeps-earth-alive/.
Gallagher, Richard, and Betsy Carpenter. 1997. "Human-Dominated Ecosystems." *Science* 277(5325):485.
Pecl, Gretta, Miguel B. Araujo, Johann D. Bell, Julie Blanchard, and Timothy C. Bonebrake. 2017. "Biodiversity Redistribution under Climate Change: Impacts on Ecosystems and Human Well-Being." *Science* 355(6332): 9214.
Vitousek, Peter M., Harold A. Mooney, Jane Lubchenco, and Jerry M. Melillo. 1997. "Human Domination of Earth's Ecosystems." *Science* 277(5325): 494–499.
Western, David. 2001. "Human-Modified Ecosystems and Future Evolution." *Proceedings of the National Academy of Science* 98(10): 5458–5465.

Human Population

The number of humans on Earth has increased rapidly in recent decades, which has a profound effect on the Earth's environment. Looking simply at numbers of people does not tell the whole story, however, as one U.S. citizen uses about as many natural resources as 35 people in India (*Scientific American*, 2018). Thus environmental geographers are interested why population growth rates vary from place to place and how population growth affects the availability of resources at local, national, and global scales. Further, geographers investigate examples of how countries can achieve environmental sustainability and what role population control must play.

Given that Earth's complex natural systems typically achieve equilibrium, how can one species have such a significant impact on the planet? Humans are different because they have succeeded in eliminating many of the factors that typically control a species' population. Predators have been killed or controlled by humans' weapons. Food security is maintained by producing crops at an industrial scale. Medicines have been developed to keep people alive and extend lifespans. Humans build structures that protect them from weather extremes. All these actions have led to an explosion in human population, which puts increased pressure on the planet's natural resources.

Over time, several key theories of human population growth and resource use have posed warnings. The first was Thomas Malthus's "An Essay on the Principle of Population" in 1798. This essay stated that human population would continue to grow until it overshot food supplies, and then human population would experience a devastating crash. Several centuries later, after the consequences of the Industrial Revolution were becoming obvious, population theories were again debated. Starting in the late 1960s, environmental and population topics were popularized throughout society with the publication of two key books: *The Population Bomb* (1968) and *The Limits to Growth* (1972). These books presented shocking numbers to support a dire warning about population growth: humans were rapidly reaching a stage of overpopulation, which would lead to a crash. These ideas are called neo-Malthusians, because they continue the argument that human population will outstrip resources. Several organizations also emphasize that society must rapidly implement population policies to control growth, but such ideas are often controversial due to religious beliefs and personal decisions. The Earth Policy Institute, Population Reference Bureau, and Worldwatch Institute, to name a few, have worked to educate people and politicians about the need for policies to control population and slow consumption of natural resources. The book *Collapse: How Societies Choose to Fail or Succeed* (2005) was a recent best-seller that documented historical examples in which overpopulation and resource degradation led to societal collapse.

In biology, a population crash occurs when a species grows beyond the capacity of its environment to sustain it, and once nature loses its basic ability to support the species, their numbers plummet. Humans have caused this to happen to many other species by destroying or degrading their habitat. Today, for example, 37% of all fish species, 29% of all amphibians, 21% of all mammals, and 12% of all birds are threatened with extinction due to human-induced population crashes (IUCN Red List, 2018). The question now is whether a crash might happen to the human species.

Human overpopulation occurs when the population exceeds the planet's carrying capacity or amount of resources needed to support them. Two factors are important: 1) the number of people and 2) the amount of resources each person demands.

First, the world's human population is 7.6 billion and expected to reach 9.7 billion in 2050 and 11.2 billion in 2100, according to the United Nations Department of Economic and Social Affairs. Currently, China and India are the two most populous countries in the world, at more than 1 billion people each and comprising 19% and 18% of the world's population, respectively. The ten largest countries in the world today are distributed with one in Africa (Nigeria), five in Asia (Bangladesh, China, India, Indonesia, and Pakistan), two in Latin America (Brazil and Mexico), one in North America (United States), and one in Europe (Russian Federation). From 2015 to 2050, however, half of the world's population increase is projected to come from just nine nations: India, Nigeria, Pakistan, Democratic Republic of the Congo, Ethiopia, United Republic of Tanzania, United States, Indonesia, and Uganda (UN, 2015). This shows geographic variation from the present locations, as future growth will shift to some African nations, where basic resources like water and food must grow to meet this demand.

Second, there is geographic variation in terms of population and resource use. In fact, countries that consume the greatest amounts of natural resources per person, like the United States, are not the most densely populated. To be clear, the United States comprises 4% of the world's population, but uses 23% of its coal, 27% of its aluminum, and 25% of its oil (*Scientific American*, 2018). This is just one example of resource use in an industrialized country, but overall it is estimated that about 80% of the world's resources are consumed by the highly industrial countries of North America, Europe, Australia, and Japan. This is changing, however, as the consumer class is expanding in many parts of the world. People across the planet now seek the lifestyle, diet, and material items—cars, appliances, electronics, and more—that will consume more and more resources. The Earth only has 1.7 global hectares (4.2 acres) of biologically productive land per person to supply resources and absorb wastes, but the average ecological footprint per person is currently 2.7 hectares (6.7 acres), which is unsustainable and indicates a situation of overpopulation (GFN, 2018). Policy solutions must include control of both population growth and resource consumption.

See also: Section 1: Case Study 1, Case Study 4, Case Study 5, Urbanization; *Section 2:* Human Modification of Ecosystems; *Section 3:* Sustainable Development

FURTHER READING

Conway-Gomez, Kristen, Karen Barton, Min Wang, Dongying Wei, Matt Hamilton, and Marta Kingsland. 2010. "Population and Natural Resources Conceptual Framework: How Does Population Growth Affect the Availability of Resources?" American Association of Geographers. http://cgge.aag.org/PopulationandNatural Resources1e/CF_PopNatRes_Jan10/CF_PopNatRes_Jan10_print.html.

Diamond, Jared. 2005. *Collapse: How Societies Choose to Fail or Succeed.* New York: Viking Press.

Ehrlich, Paul R. 1968. *The Population Bomb.* New York: Sierra Club/Ballantine Books.

GFN. 2018. "Country Overshoot Days." Global Footprint Network. https://www.foot printnetwork.org/.

IUCN Red List. 2018. "Overview of the IUCN Red List." International Union for Conservation of Nature. http://www.iucnredlist.org/about/introduction.

Meadows, Dennis, Donella Meadows, Jørgen Randers, and William W. Behrens III. 1972. *The Limits to Growth.* New York: Universe Books.

PRB. 2017. "World Population Data Sheet." Population Reference Bureau. http://www.prb.org/.

Scientific American. 2018. "Use It and Lose It: The Outsize Effect of U.S. Consumption on the Environment." https://www.scientificamerican.com/article/american-consump tion-habits/.

UN. 2015. "World Population Projected to Reach 9.7 Billion by 2050." United Nations Department of Economic and Social Affairs. July 29. http://www.un.org/en /development/desa/news/population/2015-report.html.

Worldwatch Institute. 2011. "The State of Consumption Today." State of the World 2011: Innovations that Nourish the Planet. http://www.worldwatch.org/node/810.

Hurricane

A hurricane is a large-scale, low-level atmospheric circulation system that has strong winds, low pressure, and heavy rains that rotates counterclockwise in the Northern Hemisphere and clockwise in the Southern Hemisphere. It is defined by maximum sustained surface winds of at least 119 kilometers per hour (74 miles per hour). The storm system is powered by heat released when moist air rises and the water vapor within it condenses. It is sometimes called a "warm core" storm system because the water temperature must be over 26°C (79°F). With ocean temperatures rising due to climate change, scientists fear that hurricanes may become more frequent and/or more powerful (Sneed, 2017).

The term hurricane is used for storms in the western Atlantic and eastern Pacific, whereas the term cyclone is used in the Indian Ocean and South Pacific, and the term typhoon is used in the western Pacific (Zimmermann, 2017). The time of year

Streets in some New Orleans neighborhoods remained flooded for weeks after Hurricane Katrina struck the city in 2005. (U.S. Department of Defense)

with the highest occurrence of storms is called hurricane season, and for the Caribbean, Gulf of Mexico, and Atlantic, this is from June 1 to November 30. Cyclone season is May to October, and typhoon season in the Central Pacific Basin runs from June 1 to November 30 (Ready.gov, 2018).

The Saffir-Simpson Hurricane Wind Scale is a rating from 1 to 5 based on a hurricane's sustained wind speed, which estimates the level of damage it will cause (NHC, 2017). Category 3 or higher are major storm events with significant property damage and potential for human injuries and deaths. Even Category 1 and 2 storms are dangerous and require emergency management. The term "super typhoon" is used in the western North Pacific to indicate a storm with winds over 241 kilometers per hour (km/h) (150 miles per hour or mph) (Zimmermann, 2017). Specifically, the five categories indicate storm characteristics and the levels of damage that emergency managers must expect:

Category 1 has 119 to 153 km/h (74 to 95 mph) sustained winds that produce some damage to roofs and siding; snapping of tree branches; damage to power lines; and likely power outages lasting a few days.

Category 2 has 154 to 177 km/h (96 to 110 mph) sustained winds that cause extensive damage to roofs and siding; shallowly rooted trees to be snapped or uprooted blocking some roads; and near-total power loss, with outages lasting several days.

Category 3 has 178 to 208 km/h (111 to 129 mph) sustained winds that create devastating damage: well-built homes incur major damage or removal of roofs;

many trees can be snapped or uprooted, blocking numerous roads; and electricity and water are unavailable for several days to weeks.

Category 4 has 209 to 251 km/h (130 to 156 mph) sustained winds that lead to catastrophic damage: well-built framed homes likely have severe damage with loss of roofs and some exterior walls; most trees will be snapped and uprooted, blocking roads to isolate some residential areas; power poles are downed so power outages will last weeks to possibly months; and most of the area will be uninhabitable for weeks or months.

Category 5 hurricanes have 252 km/h or higher (>157 mph) sustained winds that lead to catastrophic damage: most homes will be destroyed, with total roof and wall collapse; fallen trees and power poles will isolate most residential areas; power outages will last for weeks to possibly months; and most of the area will be uninhabitable for weeks or months.

Atmospheric scientists gather data to inform emergency managers about hurricane storm systems forming over the ocean and moving toward land. Threats to humans from hurricanes include high winds, heavy rainfall, coastal and inland flooding, rip currents, tornadoes, and storm surges. A storm surge is the abnormal rise in sea level accompanying a major hurricane making landfall. This can be the most dangerous aspect of the storm, causing a high death toll and major coastal flooding, particularly when a storm surge coincides with normal high tide, resulting in storm tides reaching six meters (20 feet) or more. Already, sea levels have risen due to climate change; thus scientists agree that increasingly higher storm surges will be hitting coastlines (Sneed, 2017).

Hurricanes can be forecasted several days in advance, so there may be a warning period announced for a region, and for severe storms, an evacuation. In areas of hurricane risk, evacuation routes are clearly marked with street signs indicating the way (Ready.gov, 2018). This may be a voluntary or mandatory evacuation, meaning that residents are either encouraged to leave or required to leave the area. Some people do not have transportation to leave, and others may think that they need to stay to care for their home.

Accurate predictions of exactly where a hurricane will reach landfall is typically only a few hours' notice for residents. Because people often wait until the very last minute before abandoning their home and possessions, high tides and sudden flooding often cause injuries and deaths. Emergency managers recognize that the initial hit of the hurricane is often extensive and very destructive. Overall, hurricane disasters are usually more destructive than typical floods in terms of recovery (IFRC, 2018). At the onset, sudden high winds cause major damage to buildings and infrastructure, especially weak structures. Then heavy rains and floods come and, in flat coastal areas, high storm surge waves.

With education and preparation, most people realize the level of threat and take recommended actions prior to a hurricane. A Hurricane Watch indicates that atmospheric conditions make it possible for a hurricane to occur within the next 48 hours. Residents should review their evacuation routes, listen to local officials, and be sure they have a disaster supply kit. A Hurricane Warning means that conditions are ideal for a hurricane and one is expected within 36 hours. Residents should take action immediately and follow evacuation orders from local officials (Ready.gov, 2018). Emergency managers may give more specific instructions, depending on when the storm is anticipated to hit and the impact that is projected for a given geographic area.

See also: Section 2: Case Study 9, Case Study 10, Mitigation, Natural Hazards

FURTHER READING

IFRC. 2018. "Meteorological Hazards: Tropical Storms, Hurricanes, Cyclones and Typhoons." International Federation of the Red Cross. http://www.ifrc.org/en/what-we-do/disaster-management/about-disasters/definition-of-hazard/tropical-storms-hurricanes-typhoons-and-cyclones/.
NHC. 2017. National Hurricane Center. http://www.nhc.noaa.gov/aboutgloss.shtml.
Ready.gov. 2018. "Be Prepared: Hurricanes." Department of Homeland Security. https://www.ready.gov/hurricanes.
Sneed, Annie. 2017. "Was the Extreme 2017 Hurricane Season Driven by Climate Change?" *Scientific American.* October 26. https://www.scientificamerican.com/article/was-the-extreme-2017-hurricane-season-driven-by-climate-change/.
Zimmermann, Kim Ann. 2017. "Hurricanes, Typhoons and Cyclones: Storms of Many Names." Live Science. September 5. https://www.livescience.com/22177-hurricanes-typhoons-cyclones.html.

Mitigation

Mitigation is an action taken to prevent or reduce the severity, seriousness, or painfulness of some phenomenon. Dictionaries commonly link this to environmental examples, such as "the emphasis is on the identification and mitigation of pollution," and indicate that usage of the word has significantly increased since the 1990s (Google.com, 2017). In fact, there are two distinct concepts related to the meaning of mitigation in environmental geography: the broader application to natural hazards and the specific imperative related to climate change action.

First, mitigation of natural hazards is an historical concept that geographers have been addressing for over a century. Researchers and environmental managers have learned from expert geographers, such as Gilbert F. White beginning in the 1940s, that purely technological solutions are rarely sufficient; rather, mitigation of natural hazards must also include social concerns and policy actions. For example, to reduce damage from flooding, it is not enough to simply build levees. Instead, policies must prohibit development in floodplains and society must support more logical land-uses such as recreation and wetland preservation in high-risk areas to reduce—or mitigate—losses.

Today, mitigation is a key environmental management approach promoted by the U.S. Federal Emergency Management Agency (FEMA), which suggests a holistic approach that includes implementation of local ordinances, land-use choices, regulations, and building practices that seek to reduce long-term risks from hazards. The overall goal is to reduce the threats to humans and their property (current structures and future construction) both before and in post-disaster settings (FEMA, 2017).

U.S. Global Change Research Program

The U.S. Global Change Research Program (USGCRP) was established by President Reagan and legally recognized by Congress in the Global Change Research Act (GCRA) of 1990 to "assist the Nation and the world to understand, assess, predict, and respond to human-induced and natural processes of global change." This includes providing an assessment report every four years (USGCRP, 2012).

In 2016, 60 experts teamed up to form a Federal Advisory Committee, which wrote the National Climate Assessment. This summarizes the impacts of climate change on the United States, now and in the future. More than 300 scientists also assisted, and the report was reviewed by a panel of the National Academy of Sciences, along with several long periods open for public comments (USGCRP, 2018).

Each National Climate Assessment collects, integrates, and assesses observations and research from around the United States. This allows the report to provide "real-world" information about what is actually happening and how it could affect peoples' lives, livelihoods, and the future. The reports include information on several sectors (human health, water, energy, transportation, agriculture, forests, and ecosystems). It also integrates these sectors and describes how impacts may be felt at the national level. Finally, the report also assesses key impacts on specific regions of the country: Northeast, Southeast and Caribbean, Midwest, Great Plains, Southwest, Northwest, Alaska, and Hawai'i and Pacific Islands, as well as coastal areas (NCA, 2014).

Emphasizing the overwhelmingly broad impacts of future climate change, 13 U.S. government agencies participate in the USGCRP: Department of Agriculture, Department of Commerce, Department of Defense, Department of Energy, Department of Health and Human Services, Department of the Interior, Department of State, Department of Transportation, Environmental Protection Agency, National Aeronautics & Space Administration, National Science Foundation, Smithsonian Institution, and the U.S. Agency for International Development.

Looking toward the future, the USGCRP developed a Strategic Plan with four goals: 1) advance science in order to understand climate and global change; 2) inform decisions on adaptation and mitigation through science; 3) conduct sustained assessments to help the U.S. respond to impacts and vulnerabilities; and 4) communicate and educate to develop the scientific workforce of the future (USGCRP, 2012).

Further Reading

NCA. 2014. "National Climate Assessment 2014." http://nca2014.globalchange.gov/.
USGCRP. 2012. "The National Global Change Research Plan 2012–2021: A Strategic Plan for the U.S. Global Change Research Program." U.S. Global Change Research Program. https://www.globalchange.gov/browse/reports/national-global-change-research-plan-2012%E2%80%932021-strategic-plan-us-global-change.
USGCRP. 2018. "About USGCRP." U.S. Global Change Research Program. https://www.globalchange.gov/about/legal-mandate.

Globally, the United Nations Educational, Scientific and Cultural Organization (UNESCO) notes that with the increasing frequency and severity of natural disasters, it is particularly important to prevent and mitigate their effects. Of all environmental problems, natural hazards are, to some extent, the most manageable because risks are clearly identified, science provides effective mitigation measures, and "the benefits of vulnerability reduction greatly exceed the costs" (UNESCO,

2007). Yet disaster prevention, in the form of mitigation, is usually low on the list of government priorities. Instead, post-disaster relief and recovery receives most funding and focus in disaster risk management, leaving a meager budget for prevention. This is a faulty approach, because cost/benefit analyses consistently show that wise investments in prevention can greatly reduce the negative effects of disasters. In fact, "one dollar invested in disaster preparedness and mitigation will prevent four to eight dollars in disaster losses," and this does not even consider the emotional costs in loss of human life (UNESCO, 2007). Overall, effective natural hazards mitigation must include preventive actions based on science and technology that are implemented through local community involvement, education, and public awareness, united with national policy and global collaboration.

The second common use of the term "mitigation" is related to combating climate change, as society tries to grapple with its current and future effects. As noted by the European Environment Agency: "Climate change is already happening: temperatures are rising, rainfall patterns are shifting, glaciers and snow are melting, and the global mean sea level is rising. Most of the warming is very likely due to the observed increase in atmospheric greenhouse gas concentrations as a result of emissions from human activities. To mitigate climate change, we must reduce or prevent these emissions" (EEA, 2017).

In the United States, the National Aeronautics and Space Administration (NASA) provides "robust scientific data needed to understand climate change" and makes this information available to U.S. citizens and the global community, including planning agencies, policy makers, and other scientists around the world. NASA notes two key approaches in responding to climate change: 1) adaption to the effects that climate change is already bringing to the Earth's environment and 2) mitigation, which entails stabilizing and reducing the levels of heat-trapping greenhouse gases (GHG) in the Earth's atmosphere (NASA, 2018).

Specifically, climate change mitigation must reduce fossil fuel emissions from electricity, heating, industry, and transportation and increase GHG "sinks" (such as healthy forests, soils, and oceans) that gather and store these gases. The goal of mitigation, as noted by the United Nations Intergovernmental Panel on Climate Change, is to reduce or eliminate human intervention with Earth's climate and stabilize "greenhouse gas concentrations in the atmosphere at a level that would prevent dangerous anthropogenic interference with the climate system. Such a level should be achieved within a time frame sufficient to allow ecosystems to adapt naturally to climate change, to ensure that food production is not threatened and to enable economic development to proceed in a sustainable manner" (IPCC, 2014: 4). This statement, endorsed by thousands of climate science experts worldwide, is a warning and call for action.

Europe has taken the lead in climate change mitigation policies, especially related to the 2015 Paris Climate Agreement's main goal of keeping the global temperature rise in this century below 2°C (3.6°F) above average pre-industrial levels. The European Union (EU) adopted a GHG emissions reduction target of 20% below 1990 levels by 2020, which will be met by individual national emissions targets and an emissions trading system that is based on the "cap and trade" principle where a maximum cap of GHG emissions is set and "allowances" are auctioned off to all emitters, such as power plants and factories. If an installation exceeds their allowable amount, they must buy more allowance from another site that has reduced their emissions and can sell their extra allowance. This reduces overall emissions without broad government intervention, because each polluter tries to find the cheapest, least-emitting method of production. The EU has also adopted strong legislation to promote energy-efficient appliances and household items and to encourage more widespread adoption of renewable energy, such as wind, solar, hydro, and biomass, in personal and utility-scale energy production. In fact, the EU has further committed to cut European GHG emissions to 40% below 1990 levels by 2030, which is a policy-driven mitigation effort to address climate change (EEA, 2017).

See also: Section 1: Case Study 3; *Section 2:* Case Study 6, Case Study 7, Case Study 9, Natural Hazards; *Section 3:* Case Study 15

FURTHER READING

EEA. 2017. "Climate Change Mitigation." European Environment Agency. https://www.eea.europa.eu/themes/climate.

FEMA. 2017. "What Is Mitigation?" Federal Emergency Management Agency. https://www.fema.gov/what-mitigation.

Google.com. 2017. "Mitigation." Google Search Dictionary. https://www.google.com/search?site=async/dictw&q=Dictionary#dobs=mitigation.

IPCC. 2014. "Summary for Policymakers," in *Climate Change 2014: Mitigation of Climate Change. Contribution of Working Group III to the Fifth Assessment Report of the Intergovernmental Panel on Climate Change.* Edenhofer, Ottmar, Ramón Pichs-Madruga, Youba Sokona, Ellie Farahani, Susanne Kadner, Kristin Seyboth, . . . Jan C. Minx (eds.). Cambridge, MA: Cambridge University Press.

NASA. 2018. "Responding to Climate Change." National Aeronautics and Space Administration, Propulsion Laboratory and the California Institute of Technology. https://climate.nasa.gov/solutions/adaptation-mitigation/.

NFCCC. 2017. "The Paris Agreement." United Nations Framework Convention on Climate Change. http://unfccc.int/paris_agreement/items/9485.php.

UNESCO. 2007. "Disaster Preparedness and Mitigation UNESCO's Role." United Nations Educational, Scientific and Cultural Organization. http://unesdoc.unesco.org/images/0015/001504/150435e.pdf.

Water Encyclopedia. 2017. "Gilbert White." http://www.waterencyclopedia.com/Tw-Z/White-Gilbert.html.

Wynes, Seth and Kimberly A. Nicholas. 2017. "The Climate Mitigation Gap: Education and Government Recommendations Miss the Most Effective Individual Actions." *Environmental Research Letters* 12(7).

Natural Hazards

Natural hazards are naturally occurring environmental events that can happen either with a slow onset or a sudden occurrence. They can be geophysical (such as earthquakes, landslides, tsunamis, and volcanic activity), hydrological (such as avalanches and floods), climatological (such as drought, wildfires, and heatwaves), meteorological (such as hurricanes and storm surges,) or biological (such as insect infestations and animal plagues) (IFRC, 2018).

Natural hazards exist with or without humans, but they become a natural disaster when people are severely affected. Thus, a natural hazard may culminate in a natural disaster if there is emotional suffering, severe injuries, substantial property damage, or death. These disasters can be so extreme as to destroy entire communities.

Listing types of natural disasters is complicated because humans can be partly responsible. For example, some of the world's deadliest floods and famines were at least partly caused by political actions taken by hostile or negligent regimes (CBC, 2010). In other examples, loss of property and life can be exacerbated by human activities such as developing land in high-risk areas and policies such as insurance programs that encourage such actions.

Environmental geographers are concerned about policies, planning, and management to best assist people before, during, and after a natural disaster. Indeed, environmental geographers realized nearly a century ago that in some locations even minor events can cause extreme consequences and loss due to social variables (White, 1936). From the management perspective, the disaster cycle is composed of four phases: mitigation, preparedness, response, and recovery.

At each stage, disaster management has mostly been focused on improving infrastructure and technology. Certainly, technological solutions, like building a levee to reduce flood risk, are helpful in preparing for natural disasters. But many variables must be part of the management solution because vulnerability to disasters varies depending on an individual's socioeconomic status and a community's ability to fund planning and preparedness (Juntunen, 2005).

Social vulnerability describes the social, cultural, and demographic characteristics of a population that "helps to explain why some communities experience a hazard differently, even though they are affected by the same magnitude or severity" (Cutter et al., 2013: 340). So a flood might harm people more if they have no car and cannot evacuate prior to a storm. This is important because knowing how and where socially vulnerable communities are located allows emergency managers to more efficiently allocate resources.

The concept of social vulnerability within disaster management is a more recent concept developed because managers realized that time and again, "people living in a disaster-stricken area are not affected equally. For example, evidence indicates that the poor are more vulnerable at all stages—before, during, and after—of a catastrophic event" (Flanagan, et al., 2011: 2). Over two recent decades (1996–2015) the United Nations Office for Disaster Risk Reduction notes that 7,056 disasters occurred worldwide, causing over 1 million deaths, and 90% of these people were in low- and middle-income countries (UNISDR, 2016).

In the United States, the Federal Emergency Management Agency (FEMA) worked with the National Institute of Building Sciences to develop HAZUS software for disaster managers "to support risk-informed decision making efforts by estimating potential losses from earthquakes, floods, and hurricanes and visualizing the effects of such hazards" (FEMA, 2017). But this key software only began to include social vulnerability information recently and still focuses on selected social issues like shelter requirements.

Recently, there have been efforts to develop social vulnerability models to help emergency managers with more than just technological solutions. One social vulnerability model was developed by the U.S. government specifically to help local and state agencies. It uses variables from the U.S. Census of Population and Housing in four categories: socioeconomic status, household composition and disability, minority status/language, and housing and transportation (Flanagan, et al., 2011). Another model, the Social Vulnerability Index (SoVI) was developed by geographers and provides maps indicating social vulnerability at the county level in the United States using 42 socioeconomic and built-environment variables that "influence or shape the susceptibility of various groups to harm and that also govern their ability to respond," including political power, industrial development, unemployment, property values, family structure, transportation, medical services, social dependence, and special needs populations (Cutter et al., 2003: 260). Vulnerability to hazards is influenced by many factors, including those listed here, but also by less tangible things like neighborhood cohesion and social networks. It is important to consider all these factors when preparing for and responding to a natural disaster.

Environmental geographers have long argued for "an equal emphasis on technologically-oriented solutions to natural hazards problems, along with emphases on the social, economic, and political factors that lead to the adoption or non-adoption of alternatives for managing disaster risk (Cutter 2017: 1). Effective planning and responding to disasters require social and environmental data enabling local emergency managers to determine the needs of all residents. Local planners can recognize neighborhoods that may require specific response services in the disaster recovery phase. Preparation for natural disasters must be based on understanding both environmental and social systems, with the ultimate goal of resilience from the local to global scale.

There is a paradox that modern society has technology to control nature, yet humans have increased their exposure and vulnerability to natural disasters because of development in risky areas and actions that intensify climate change and related hazards. "The changing context of hazards necessitates more progressive approaches to disaster risk management. This means a reframing of current programs and policies away from response to a more proactive and longer term emphases on enhancing resilience at local to global scales" (Cutter, 2017: 1).

See also: Section 2: Case Study 7, Case Study 9, Case Study 10, Mitigation

FURTHER READING

CBC. 2010. "The World's Worst Natural Disasters: Calamities of the 20th and 21st Centuries." CBC/Radio Canada. August 30. http://www.cbc.ca/news/world/the-world-s-worst-natural-disasters-1.743208.

Cutter, Susan L. 2016. "Resilience to What? Resilience for Whom?" *The Geographical Journal* 182(2): 110–113.

Cutter, Susan L. 2017. "The Changing Context of Hazard Extremes: Events, Impacts, and Consequences." *Journal of Extreme Events* 3(2).

Cutter, Susan L., Bryan J. Boruff, and W. Lynn Shirley. 2003. "Social Vulnerability to Environmental Hazards." *Social Science Quarterly* 84(2): 242–261.

Cutter, Susan L., C. Emrich, D. Morath, and C. Dunning. 2013. "Integrating Social Vulnerability into Federal Flood Risk Management Planning." *Journal of Flood Risk Management* 6(4): 332–344.

FEMA. 2017. "HAZUS." Federal Emergency Management Agency. https://www.fema.gov/hazus/.

Flanagan, Barry E., Edward W. Gregory, Elaine J. Hallisey, Janet L. Heitgerd, and Brian Lewis. 2011. "A Social Vulnerability Index for Disaster Management." *Journal of Homeland Security and Emergency Management* 8(1): article 3.

Grocholski, Brent. 2017. "Natural Hazards." Science. http://www.sciencemag.org/topic/natural-hazards.

IFRC. 2018. "Types of Disasters: Definition of Hazard." International Federation of Red Cross and Red Crescent Societies. http://www.ifrc.org/en/what-we-do/disaster-management/about-disasters/definition-of-hazard/.

Juntunen, L. 2005. "Addressing Social Vulnerability to Hazards." *Disaster Safety Review* 4(2): 3–10.

UNISDR. 2016. "Poverty & Death: Disaster Mortality 1996–2015." United Nations Office for Disaster Risk Reduction and Centre for Research on the Epidemiology of Disasters. https://www.unisdr.org/we/inform/publications/50589.

White, Gilbert F. 1936. "Notes on Flood Protection and Land-Use Planning." *Planners Journal* 3(3): 57–61.

Parks and Urban Green Space

Green space is land in parks, green paths, sports fields, river banks, greenways, trails, conservation areas, and community gardens that has vegetation such as grass or trees. The broader term "open space" is also used by urban planners to indicate any parcel of land that is open to the public and is "undeveloped," meaning there are no buildings or structures on it; so this can include cemeteries, schoolyards, public plazas, and even vacant lots.

Historians note that urban parks have evolved through five stages, each with distinct goals and activities, from the mid-1800s to present: pleasure ground, reform park, recreation facility, open space system, and sustainable park. Pleasure ground parks were prominent from 1850 to 1900, encompassing very large land areas that promoted strolling, pushing a baby carriage, and picnicking. The reform park was popular from 1900 to 1930, and these were small sites tucked into neighborhoods with supervised activities like plays, dancing, and classes. Parks as a recreation facility were common during 1930 to 1965, with mostly medium-sized land parcels in the suburbs established as a site for specific sports like basketball, tennis, baseball, and swimming. From 1965 to 1990, parks within an open space system were many small sites linked together as a network to provide visitors with stress relief, music, and free-form play. Most recently, the sustainable park stage encompasses the goals of parks from 1990 to the present day, which seek to promote both human and ecological health, with activities like hiking, biking, bird watching, and environmental education (Cranz and Boland, 2004).

Overall, green space and parks can offer benefits to individuals through improved physical and mental health, and to communities through economic and

Established in 1857, Central Park in New York City is now one of the most popular urban parks in the U.S., welcoming over 40 million visitors annually to its 341 hectares (843 acres). (Alyaksandr Stzhalkouski/dreamstime.com)

environmental improvements. Studies show that conveniently located parks play a significant role in increasing the activity levels of community members. Such physical activity can decrease heart disease, diabetes, high blood pressure, some cancers, and feelings of anxiety among park participants. Further, parks and nature provide numerous psychological benefits to people of all ages and backgrounds by improving their moods, reducing stress, and increasing a reported sense of wellness (Bedimo-Rung et al., 2005). Communities benefit from parks because they create a sense of pride and cohesion and stimulate social interactions. Businesses may spring up nearby as a park brings people to an area and encourages them to linger. Urban parks also contribute environmental benefits by providing wildlife and plant habitats, help control runoff, and improve air quality.

In the past, urban parks were simply grass and a few trees set up to provide relief from overcrowded city streets and tiny apartments. Today, parks play a broader social and ecological role, as they are integrated into urban areas, bringing nature within the city and providing benefits to the community as a gathering place. Land values increase where a well-maintained park is integrated into a neighborhood. In addition, modern urban planning has a new emphasis on integrating parks and green infrastructure (APA, 2017). For example, building a walking path along a canal for storm water runoff provides both recreational opportunities and necessary engineering solutions to reduce flood risk.

It is globally recognized that urban areas must develop and maintain parks, as noted in the United Nations Sustainable Development Goal number 11: "Make cities inclusive, safe, resilient and sustainable." There is interest among city planners and managers around the world in increasing the number and size of urban parks.

Happiness and Sustainability

Sustainable happiness is a relatively new concept that seeks to unify goals of sustainability and the psychological understanding of happiness. It is defined as "happiness that contributes to individual, community and/or global well-being without exploiting other people, the environment or future generations" (O'Brien, 2010: 1).

Unfortunately, sustainable happiness has only recently been discussed, because society promotes a short-term view of happiness, often based on how much "stuff" a person is able to gather (De Graaf et al., 2014). Society pushes people to have high expectations for material things and in fact values this "stuff" above friends, family life, and community well-being. Think about sports stars and bloggers on social media who emphasize all the things people should buy.

Consumerism is everywhere, from social media, to commercials, to corporate sponsorship of schools and sports. People are "continually bombarded by commercial messages" that suggest a "happy life results from the acquisition of wealth and the possessions that convey the right image and high status." But there are personal, social, and ecological costs to this consumerism (Kasser, 2006: 200).

The term "affluenza" was conceived in the 1980s to describe how affluence—people's constant desire to obtain more and more stuff—is like an illness with symptoms, causes, and a cure. Shopping has become a hobby that has no benefit in terms of health or well-being. To treat this illness, people must build their immunity so that they do not judge themselves based solely on how much stuff they have. The best way to do this is to rebuild families and communities and respect the Earth (De Graaf et al., 2014). In other words, sustainable happiness can be achieved by building sustainable social and environmental relationships (O'Brien, 2008).

To get started, people can monitor their own behavior and experience for one day and assess the impact of their activities. Make a Sustainable Happiness Baseline Chart that includes these categories: time, activity, emotional experience, impact on self, impact on others, and impact on natural environment. In addition, people can reflect on "areas where I could improve my own well-being, or the well-being of others and/or the natural environment" (O'Brien, 2010: 18). This helps identify what activities truly provide happiness and sustainability for each individual.

According to researchers, spending time in nature; talking to friends, parents, grandparents, or neighbors; or participating in a community environmental volunteer project are most likely to lead to sustainable happiness (De Graaf et al., 2014; O'Brien, 2010).

Further Reading

De Graaf, John, David Wann, and Thomas H. Naylor. 2014. *Affluenza: How Overconsumption Is Killing Us and How to Fight Back*. San Francisco: Barrett-Koehler.

Kasser, T. 2006. "Materialism and Its Alternatives," in *A Life Worth Living: Contributions to Positive Psychology*. Csikszentmihalyi, Isabella Selega and Mihaly Csikszentmihalyi, eds. Toronto: Oxford University Press.

O'Brien, Catherine. 2008. "Sustainable Happiness: How Happiness Studies Can Contribute to a More Sustainable Future." *Canadian Psychology* 49(4): 289–295.

O'Brien, Catherine. 2010. "Sustainability, Happiness and Education." *Journal of Sustainability Education*. April 28. http://www.susted.com/wordpress/content/sustainability-happiness-and-education_2010_04/.

For example, World Urban Parks (WUP) is a large international collaboration of organizations and individuals interested in promoting and advocating for urban parks, conservation, recreation, and green city initiatives to build livable, healthy, and sustainable communities. They note that by 2050, more than 70% of the world's population will live in urban areas, and with rapid urbanization, parks are increasingly important for both nature and humans. The WUP envisions "a world where people value and have easy access to quality urban parks, open space and recreation" (WUP, 2017).

Despite the numerous potential benefits of urban parks, there can also be both environmental and social drawbacks. First, if parks are poorly managed they can cause negative environmental impacts like erosion from trails that are overused or not maintained; further, parks can actually destroy wildlife habitat by bringing more people into an area (EPA, 2017). Second, parks can be expensive to establish and maintain, especially if brought into a region that has been neglected for decades; it would be cheaper to simply make a natural area that was protected from human use. Parks are not designed for purely environmental benefit; rather, managers must accommodate human visitors and navigate political constraints within an urban governmental structure. Sometime parks can actually increase negative urban blight and crime if they are ignored, unsafe, and left in disrepair. Finally, there can be significant social equity problems when parks are built by outsiders in inner-city regions.

Although there seem to be obvious benefits to reclaiming former industrial sites for green space, the unforeseen consequence can be that long-time, low-income residents are pushed out because land values rise and they cannot afford to remain in their own neighborhood. This is the negative aspect of environmental gentrification, where a new park is developed and pushes out poorer residents, because the wealthier now want to live in this greener location. On the other hand, if parks are established with local community input and participation, some problems can be avoided. There are good examples of parks that included community programs like youth training, gardening, and skills training that make parks meaningful, assessable, and thus beneficial to the original community members (Walker, 2004).

See also: Section 1: Urbanization; *Section 2:* Protected Areas and National Parks; *Section 3:* Case Study 12, Sustainable Cities

FURTHER READING

APA. 2017. "City Parks Forum." American Planning Association. www.planning.org/cityparks/.

Bedimo-Rung, A., A. Mowen, and D. Cohen. 2005. "The Significance of Parks to Physical Activity and Public Health: A Conceptual Model." *American Journal of Preventive Medicine* 28(2): 159–168.

Cranz, Galen and Michael Boland. 2004. "Defining the Sustainable Park: A Fifth Model for Urban Parks." *Landscape Journal* 23(2): 102–120.

EPA. 2017. "What Is Open Space/Green Space?" U.S. Environmental Protection Agency. April 10. https://www3.epa.gov/region1/eco/uep/openspace.html.

Walker, Christopher. 2004. "The Public Value of Urban Parks." The Urban Institute. June 21. https://www.urban.org/research/publication/public-value-urban-parks/view/full_report.

Wolch, Jennifer R., Jason Byrne, and Joshua P. Newell. 2014. "Urban Green Space, Public Health, and Environmental Justice: The Challenge of Making Cities 'Just Green Enough.'" *Landscape and Urban Planning* 125: 234–244.

WUP. 2017. "Welcome, Bienvenue, Willkommen, Bienvenidos, 歓迎" World Urban Parks Organization for Open Space and Recreation. http://www.worldurbanparks.org/en/.

Protected Areas and National Parks

A globally accepted definition of a protected area is: "A clearly defined geographical space, recognized, dedicated and managed, through legal or other effective means, to achieve the long-term conservation of nature with associated ecosystem services and cultural values" (Dudley, 2008: 8).

The International Union for Conservation of Nature (IUCN), established in 1948, has become the global authority on ecological data and objective, scientifically based, environmental management recommendations. With a global network of over 10,000 experts and 900 staff in more than 50 countries, IUCN is the only environmental group to have earned official United Nations Observer Status (IUCN, 2018). To meet one of their key mission areas, IUCN formed the World Commission on Protected Areas (IUCN-WCPA).

The IUCN-WCPA has developed a list of categories of protected areas based on their management objectives. These categories are recognized by international organizations and by many national governments; thus, they serve two purposes. First, the categories are useful in providing a deeper understanding of the types of

Known for its many geothermal features including Old Faithful geyser, seen here, Yellowstone (Wyoming) was established in 1872 as the first U.S. National Park. (Aneese/iStockphoto.com)

parks and preserves gained through specific "best practice" management. Second, the IUCN-WCPA categories are based on data and scientific expertise, so these concepts have been used as the basis for legislation and policies to preserve land in many countries and communities (IUCN-WCPA, 2018).

Specifically, IUCN-WCPA Protected Area Categories, each with distinct management goals to both protect nature and provide access to visitors, include the following:

Category Ia: Strict Nature Reserves have rigorous rules to protect biodiversity, so human visits and use are limited and these areas provide valuable data for monitoring and scientific research.

Category Ib: Wilderness Areas are undisturbed by significant human activity, so there is some public access, but only at a level that will maintain the wilderness qualities for present and future generations.

Category II: National Parks are large natural or near-natural areas that have been reserved to protect key ecological processes, species, and ecosystems, but also provide visitor opportunities for recreation, education, science, and spirituality that are culturally compatible with the surrounding region.

Category III: Natural Monuments or Features are small areas set aside to protect a specific natural element such as a landform, cavern, or geological feature that has high visitor value.

Category IV: Habitat/Species Management Areas are typically small, fragmented areas set aside specifically to protect particular species or habitats, so their management is targeted and often requires active interventions to protect against further human-induced degradation.

Category V: Protected Landscapes/Seascapes are established for nature conservation in areas where interactions between human culture and nature through time have created a distinct local character with ecological, scenic, and social value, which may be threatened by outside influences.

Category VI: Protected Areas with Sustainable Use of Natural Resources are typically large areas protected in order to conserve ecosystems in unison with traditional indigenous natural resource management systems.

Across the planet, the number of protected areas has increased substantially in the last 100 years, as people and governments realize their important contribution to ecological conservation, ecosystem services, and sustainable development. Keep in mind that the concept of a protected area varies by category, but also differs considerably from place to place. "Protected" can mean strict rules prohibiting human activities or rather lax general statements about preference for less human impact. Still the United Nations Environment Programme—World Conservation Monitoring Centre publishes detailed analyses of global protected areas, as defined by IUCN criteria. Their latest global data indicate over 114,000 protected areas totaling over 19 million square kilometers (7.3 million square miles) exist, covering 12.9% of the Earth's land surface (Chape et al., 2008: 13).

As noted earlier, national parks are a specific type of protected area. There are over 1,200 national parks in 100 different countries around the world (NPS, 2018).

Each country establishes its own management goals, administrative framework, and priorities for ecological and social sustainability in their national parks.

In the United States, the first national park was Yellowstone, established in 1872. Several more national parks and monuments were set aside, mostly in the federal lands of the U.S. West, and they were administered by the U.S. Department of the Interior, the U.S. War Department, and the U.S. Forest Service of the Department of Agriculture. It was a bit haphazard, because there was no single U.S. agency or clear management goals for the federal parklands. Then in 1916, President Woodrow Wilson signed the act creating the new National Park Service, in the Department of the Interior. The purpose of the national parks "is to conserve the scenery and the natural and historic objects and the wild life therein and to provide for the enjoyment of the same in such manner and by such means as will leave them unimpaired for the enjoyment of future generations." Later, in 1970, the U.S. Congress signed the General Authorities Act of 1970, which clarified that the National Park System had "grown to include superlative natural, historic, and recreation areas in every region." Indeed, the U.S. National Park System now includes more than 400 areas covering more than 84 million acres in all 50 states and many U.S. territories (NPS, 2018).

In all parks and protected lands, there are competing demands from human visitors, neighboring economic desires, habitat requirements, and ecological integrity, to name a few. Sometimes, a park can be "over-loved," and too many visitors actually causes environmental degradation, harming the very ecosystems that visitors wish to see and managers seek to protect. In these sites, careful management is crucial. For example, when narrow mountain roads within a national park become overcrowded, park managers will prohibit private cars and implement mandatory bus transportation in the park, at least during high season. Others have discussed even more strict rules, such as capping or limiting the number of visitors allowed in a park each day (Zuckerman, 2017).

As noted by Parks Canada, the agency that manages 38 national parks, 171 national historic sites, and a handful of national marine conservation areas, they work to "protect and present nationally significant examples of Canada's natural and cultural heritage and foster public understanding, appreciation and enjoyment in ways that ensure their ecological and commemorative integrity for present and future generations" (Parks Canada, 2017). This statement indicates the complexity faced by managers within the mission of national parks. The difficulty is balancing ecological protection with visitor needs and enjoyment. In protected areas, managers must fill many complex and sometimes competing roles: guardian of diverse cultural and recreational resources, environmental educator and advocate, partner in community revitalization, leader in preservation, and protector of natural ecosystems.

See also: Section 2: Case Study 8, Parks and Urban Green Space; *Section 3:* Case Study 13, Environmental Policy

FURTHER READING

Chape, Stuart, Mark Spalding, and Martin Jenkins. 2008. *The World's Protected Areas: Status, Value and Prospects in the 21st Century.* Los Angeles: University of California Press.

Dudley, Nigel, ed. 2008. "Guidelines for Applying Protected Area Management Categories." International Union for Conservation of Nature. https://portals.iucn.org/library/sites/library/files/documents/PAG-021.pdf.
IUCN. 2018. "About." International Union for Conservation of Nature. https://www.iucn.org/about.
IUCN-WCPA. 2018. "Protected Area Categories." International Union for Conservation of Nature-World Commission on Protected Areas. https://www.iucn.org/theme/protected-areas/about/protected-area-categories.
NPS. 2018. "History." National Park Service. https://www.nps.gov/aboutus/history.htm.
Parks Canada. 2017. Parks Canada Agency. https://www.pc.gc.ca/en/agence-agency.
Zuckerman, Laura. 2017. "U.S. National Parks, Led by Utah's Zion, Weigh Limits on Visitors." Reuters. August 3. https://www.reuters.com/article/us-usa-nationalparks/u-s-national-parks-led-by-utahs-zion-weigh-limits-on-visitors-idUSKBN1AJ31N.

Tornado

Tornadoes are the most violent type of atmospheric storm (Ready.gov, 2018). Specifically, a tornado is a rapidly rotating column of air extending from a thunderstorm to the ground. Wind is invisible, so it can be difficult to see a tornado unless a funnel of dust, debris, and water condensation has formed. In a big thunderstorm, which can produce a tornado, there is typically lightning, hail, high winds, and heavy rains.

Globally, the mid-latitudes that are between 30° and 50° North or South tend to have the appropriate conditions for tornadoes to form. These are regions where cold, polar air meets warm, subtropical air. And precipitation can form as air rises along the boundaries of colliding air masses. Plus, in these mid-latitudes, there tend to be air masses moving in different directions at varying speeds at diverse elevation levels, and this causes rotation to develop inside a storm cell. Overall, where there is a higher occurrence of convective storms, combined with temperature variations, this provides favorable conditions for producing tornadoes (NCDC, 2018).

Tornadoes have been documented on every continent except Antarctica. Although many countries experience tornadoes, the United States has the most, averaging over 1,200 each year. Second is Canada with about 100, followed by concentrations in Bangladesh and Argentina. Also northern European countries, western Asia, Australia, New Zealand, China, Japan, and South Africa have tornadoes. Some places, like the UK, with a smaller land area, actually have a relatively high number of tornadoes per square kilometer, but they tend to be fairly weak (NCDC, 2018).

For emergency managers, location and timing are key. In the United States, there is a region of the central plains popularly known as a "tornado alley," which has a higher occurrence of tornadoes, but in fact every state is at risk. It is true that most tornadoes occur east of the Rocky Mountains with concentrations in the central and southern plains, the Gulf Coast, and Florida. Given that cold and warm air masses must meet, tornadoes occur more during some times of the year, and this is called tornado season. It varies by geographic location, with peak tornado season in the northern Great Plains in June or July, but along the Gulf Coast, it is earlier in the spring. Tornadoes can happen any time of day or night, but most happen between 4 and 9 PM, as the air is the warmest, resulting in a greater difference in temperature when the warm air meets cold air masses (NSSL, 2018).

Likewise, massive thunderstorms can create numerous tornadoes over a large area, so damage can be severe in one location or in several, and moderate damage may be experienced across a vast geographic area. Because thunderstorms accompanied by tornadoes can cause high winds and heavy rains, this type of natural disaster can lead to extensive damage. Buildings, roads, electricity, sewage, water, communications, and other services can be destroyed. There are often floods after the event (Red Cross, 2017). Overall, emergency managers must prepare for significant damage following a tornado.

The intensity of a tornado is ranked by the Enhanced Fujita (EF) scale that is based on engineering concepts and damage assessment. Weak tornados are classified as EF1 and EF2. They have winds of 105 to 177 kilometers per hour (65 to 110 miles per hour) causing light to moderate damage. These tornadoes make up 69% of all tornadoes and cause 5% of the deaths. Strong tornadoes (29% of tornadoes, causing 30% of all deaths) are EF2 and EF3s. They have 179 to 266 kph winds (111 to 165 mph) and cause considerable to severe damage. Violent tornadoes are EF4 and EF5 with winds of 267 to 323 kph (166 to 201+ mph) that comprise only 2% of tornadoes but 70% of all tornado deaths (NWS, 2009).

Weather forecasting has improved in recent years, so that now analysts and storm spotters recognize certain features within a thunderstorm that make the formation of a tornado likely. For example, some patterns in Doppler radar images can be identified through computer modeling, and this makes it easier for forecasters to identify dangerous tornado activity. Plus, storm spotters are trained to recognize tornado conditions on the ground and report back to the National Weather Service, which can then issue an alert.

The path of a tornado can be over 1.6 kilometers (1 mile) wide and stay on the ground for 80 kilometers (50 miles). Each tornado varies by size, intensity, and the amount of time it travels on the ground, but all can cause destruction to buildings and injuries from flying debris. Tornadoes often strike quickly, with little warning, so emergency management is based on preparedness, particularly making people aware of the tornado hazard and necessary actions to reduce risk. A Tornado Watch means that weather conditions are right for possible tornados to form, so people should stay alert. A Tornado Warning means that a tornado has been sighted or indicated by radar, so people in its path must take shelter immediately. The average lead time for tornadoes in the United States is 11 minutes, which means that when a warning is issued, it is imperative that people immediately take shelter (NWS, 2009).

To be prepared for tornado season, families should identify safe rooms or storm shelters in their home, office, or school and have a plan for getting to this location quickly when a warning siren is sounded. Overall, people should be alert to weather conditions and look for approaching storms. Signs of danger include a dark, often greenish sky; large hail; low clouds that might be rotating; and a loud roar that sounds like a train. During a tornado, people should take shelter in a small, windowless interior room in the basement or lowest level, and particularly protect their head from flying debris (Ready.gov, 2018).

See also: Section 2: Mitigation, Natural Hazards

FURTHER READING

NCDC. 2018. "U.S. Tornado Climatology." National Climatic Data Center. https://www.ncdc.noaa.gov/climate-information/extreme-events/us-tornado-climatology.

NSSL. 2018. "Severe Weather 101: Tornadoes." National Severe Storms Laboratory. http://www.nssl.noaa.gov/education/svrwx101/tornadoes/.

NWS. 2009. "Tornadoes: Quick Facts." National Weather Service. http://www.crh.noaa.gov/Image/dvn/downloads/quickfacts_Tornadoes.pdf.

Ready.gov. 2018. "Be Informed: Tornadoes." U.S. Department of Homeland Security https://www.ready.gov/tornadoes.

Red Cross. 2017. "Tornado Safety." American Red Cross. http://www.redcross.org/get-help/how-to-prepare-for-emergencies/types-of-emergencies/tornado#About.

Volcano

From myth, to ancient history, to modern times, volcanos have evoked awe and fear. One of the most famous volcanoes is Mount Vesuvius, in southern Italy, which erupted in 79 BCE, killing 2,000 people and burying the ancient Roman city of Pompeii in volcanic ash (History, 2018). The city was abandoned and forgotten until the mid-1700s when archeologists discovered that beneath a thick blanket of volcanic ash and debris, Pompeii was mostly still intact, including skeletons of some people running from the volcano, frozen in time. Today scientists, called

In Iceland, Mount Eyjafjallajokull erupted in 2010, creating a 11 kilometer (7 miles) column of steam and ash that winds forced over northern Europe; many countries closed airports for several days to avoid the dangers of flying through volcanic ash clouds. (Johann Helgason/Dreamstime.com)

volcanologists, provide emergency managers more warning time prior to a major volcanic eruption.

A volcano is an opening on the Earth's surface where hot molten rock, called magma, comes from the planet's interior to the surface. Above ground, this material is called lava, and when it escapes it causes an eruption, which can be like an explosion or like a calm, hot flow. The erupted material includes lava, ash, cinders, and/or gas. Volcanoes also occur on other planets and even moons, with different materials erupting from their interiors. The Earth was built by volcanoes, with about 80% of its above and below sea-level surface formed by volcanoes (CNN, 2017). In volcanic areas, mountains usually form from the numerous layers of accumulated ash and rock. Scientists classify volcanoes as active, dormant, or extinct. An active volcano is one that has erupted since the last ice age, so in the past 10,000 years. A dormant volcano has not erupted in 10,000 years, but is expected to erupt again. An extinct volcano is one that scientists do not expect to erupt ever again (USGS, 2017).

Several terms are used to describe the hazards caused by volcanoes: lava flows, tephra, pyroclastic flows, and lahars. Lava flows are the molten rock that pours or oozes across the Earth's surface, burning everything in its path. Tephra is the term for all the pieces of rock that have been blasted into the air by the eruption. Most falls back close to the volcano, enlarging its slopes. But ash is the lighter, smaller pieces that are less than 2 millimeters (1/10 inch) in size, and these can travel thousands of miles, causing problems far from the volcano. In fact, airborne ash endangers jet aircraft and telecommunications, and once it settles, it threatens water supplies, damages machinery and electronics, and causes breathing problems for people and animals. Pyroclastic flows are the hot avalanches of volcanic gas and lava fragments caused by the eruption and fallout. Volcanic ash landslides are called lahars, which are fast-moving slurries of rock, mud, and water that look like flowing concrete (USGS, 2017).

Volcanic eruptions are measured with the Volcanic Explosivity Index (VEI), which is a classification system based on a scale from 1 to 8, and each succeeding VEI is 10 times greater than the last (CNN, 2017). The VEI is based on the volume of pyroclastic material ejected by the volcano, the eruption duration and cloud height, and observations of the eruption and its aftermath.

A supervolcano is a VEI 8, which has measured deposits that are greater than 1,000 cubic kilometers (240 cubic miles). As its name implies, this type of eruption causes major changes on the landscape. The most recent supervolcano was over 27,000 years ago at Taupo, on the north island of New Zealand. Another well-known supervolcano, occurring 640,000 years ago, is what helped create what is now Yellowstone National Park, specifically the park's huge crater, measuring 48 by 72 kilometers (30 by 45 miles across). According to most volcanologists, the chance of a supervolcano eruption happening today is slim, with odds of 1 in 700,000 every year (Live Science, 2016). Other research shows that super eruptions range from 5,200 to 48,000 years, "with a best guess value of 17,000 years" (Hignett, 2017).

Although the massive volcanoes are now relatively rare on Earth, there are still more than 500 volcanoes that have erupted at least once, and 50 of those are located in the United States (CNN, 2017). Of the active volcanoes today, nearly 75% encircle the Pacific Ocean, and this is known as the Ring of Fire or the Circum-Pacific

Belt. The Ring of Fire marks borders between several major tectonic plates on the Earth's crust, so it also contains other geological features: mountain trenches, ocean trenches, and sites of earthquake activity. In fact, approximately 90% of all earthquakes occur along the Ring of Fire (USGS, 2017). Geographically, the ring is more of a half-circle that starts in New Zealand, goes north along the eastern coasts of Asia and Russia, across to the Aleutian Islands of Alaska, and south down the western coasts of North, Central, and South America.

One example of a volcanic eruption along the Ring of Fire is the well-known Mount Pinatubo in the Philippines. In 1991, it had a massive, VEI 6 eruption that ejected more than 5 cubic kilometers (1 cubic mile) of debris into the air and created a column of ash and gas that went up 35 km (22 miles) into the atmosphere. The eruption blocked enough solar radiation to drop mean global temperatures by 0.5°C (1°F) during the following year (Live Science, 2016). In May 1980, Mount St. Helens in the U.S. state of Washington erupted as a VEI 5, and despite widespread evacuations, 57 people were killed. Due to its geographic location, air, highway, and railway transportation was regionally disrupted for an extended time (CNN, 2017).

In the United States, scientists from the Geological Survey's Volcano Observatories assess and monitor volcanic activity and issue warnings of predicted eruptions through a formal notification system. For communities in at-risk locations, there are usually indications that a volcano may erupt, but time before the actual eruption may be days, weeks, or months. Government officials are responsible for developing emergency response plans.

From an emergency management perspective, volcanoes cause large numbers of people to move quickly as they flee the lava flow. The response must be adjusted to meet the needs of each specific volcano, eruption, and geographic setting. In general, response must include temporary shelters, safe water, basic sanitation, food supplies, and the short-term provision of basic health services and supplies (IFRC, 2018). The major threats to human health from a volcanic eruption are first during the initial disaster and later in the aftermath. With a sudden, dangerous eruption, people can die from the blast impact or from suffocation. Later exposure to ash and gases can be harmful. Most volcanic gases dissipate quickly, but heavier gases can linger and cause breathing problems in both healthy people and people with respiratory problems. At high exposure levels, gases can cause headache, dizziness, and suffocation (CDC, 2012). When residents follow warnings and evacuate, the chances of adverse health effects from a volcanic eruption are very low.

See also: Section 2: Case Study 10, Natural Hazards

FURTHER READING

CDC. 2012. "Key Facts about Volcanic Eruptions." Natural Disasters and Severe Weather, Centers for Disease Control and Prevention. https://www.cdc.gov/disasters/volcanoes/facts.html.

CNN. 2017. "Volcanoes Fast Facts." September 20. http://www.cnn.com/2013/07/25/world/volcanoes-fast-facts/index.html.

Hignett, Katherine. 2017. "Catastrophic Supervolcano Eruptions Happen Way More Often than Scientists Thought." *Newsweek*. November 29. http://www.newsweek.com/catastrophic-civilization-ending-supervolcano-eruption-725618.
History. 2018. "Pompeii." History.com. http://www.history.com/topics/ancient-history/pompeii.
IFRC. 2018. "Geophysical Hazards: Volcanic Eruptions." International Federation of Red Cross and Red Crescent Societies. http://www.ifrc.org/en/what-we-do/disaster-management/about-disasters/definition-of-hazard/volcanic-eruptions/.
Live Science. 2016. "The 11 Biggest Volcanic Eruptions in History." Live Science. https://www.livescience.com/30507-volcanoes-biggest-history.html.
USGS. 2017. "Volcano Hazard Program." U.S. Geological Survey. https://volcanoes.usgs.gov/index.html.

Wildfire

A wildfire is an unplanned, unwanted fire that typically starts in a natural area, like a grassland or forest. In recent decades, people have increasingly developed into these areas, building homes in places that are vulnerable to wildfires. In general, the United States has weak zoning laws; thus, housing often encroaches on wildlands in many local areas. These built-up areas are called the wildland urban interface (WUI), and they are the subject of extensive concern for emergency managers and firefighters.

Wildfires can start anywhere; sometimes they originate in remote wilderness areas, in parks, or even in somebody's backyard. Historically, wildfires mostly started from natural causes, such as lightning. But due to human expansion into wildlands, most are now caused by humans, either intentionally or unintentionally, from actions like leaving campfires burning unattended, tossing a cigarette out the window, or burning leaves outdoors. Wildfires occur throughout the year, but there is higher occurrence during seasons when grass, trees, and low vegetation are especially dry and thus burn easily. High winds also make fires more dangerous, unpredictable, and rapidly spreading.

Wildfires can have a huge impact over a large geographic area, with sparks and cinders moving more than a mile from the flames, and smoke causing health problems for people much farther away. Wildfires can damage or destroy buildings; disrupt infrastructure like electricity, communications, and transportation; and most tragically, cause injuries and even death. Long-term ecological problems occur after a big fire, because the scarred landscape is likely to have increased mudslides, erosion, and even flooding.

Environmental geographers are particularly focused on the location, preparedness, and recovery from wildfires. Emergency management provides extensive information for residents in the wildland urban interface. Homes in these areas are most often ignited by small flames or embers, which are pieces of wood or dry vegetation that are blowing in the air. Homeowners can reduce their risk by preparing the site up to 60 meters (200 feet) around their home, an area known as the home ignition zone (HIZ). Fire research has shown that the HIZ is divided into the immediate zone (0 to 1.5 meters or 0 to 5 feet), the intermediate zone (1.6 to 9 meters or 6 to 30 feet), and the extended zone (10 to 60 meters or 31 to 200 feet).

The homeowner's goal in the immediate zone should be to maintain a noncombustible area, so proper construction materials and removal of all flammable vegetation are essential. In the intermediate zone, all landscaping must be carefully planned to create breaks with paved areas that decrease fire activity, and grass should be mowed to lower than seven centimeters (four-inches), trees should be spaced farther from the house, and vegetation near the ground should be trimmed so fire cannot move up a tree and spread to other trees within the canopy (Meyer, 2017). In the extended zone, the goal is to keep flames small and on the ground, so all dry trees and vegetation must be dug out, and all small trees growing between larger trees should be removed (NFPA, 2018).

Many regions of the globe experience severe wildfires (NASA, 2018). It was a particularly severe fire season in 2017. Satellite imagery shows huge fires burned in many regions across the globe: Siberia in Russia, Chile in South America, regions of sub-Saharan Africa, Southeast Asia, and New Zealand. In western and northwestern Canada fires were extreme, with the province of British Columbia experiencing the largest fires ever recorded there. In the western United States, too, the wildfire season severely affected many states, including Oregon, Washington, Montana, Nevada, Alaska, and California (Pierre-Louis, 2017). In California, "at the peak of the wildfires there were 21 major wildfires that, in total, burned over 245,000 acres, 11,000 firefighters battle the destructive fires that at one time forced 100,000 to evacuate, destroyed an estimated 8,900 structures (as damage assessment continues, this is the latest count), and sadly, took the lives of 43 people" (CAL FIRE, 2017).

Past forest management practices are partly to blame because the focus was on fire suppression, meaning that all fires, even naturally occurring fires, were extinguished immediately. This intensifies vulnerability and risk, because many landscapes gather too many flammable materials like dried brush and dead trees. But this alone does not explain the notable increase in wildfires, especially for fires located in places where there is no direct forest management.

"Wildfires are a complex and challenging disaster type which we can expect to see more of as heatwaves and drought become more frequent and intense as a result of climate change and other factors" according to the United Nations Secretary General's Special Representative for Disaster Risk Reduction (United Nations, 2016). Indeed, people are moving into fire-prone areas. Plus, climate change is predicted to increase the number and extent of wildfires; thus, recent years are occurring as scientists have predicted.

Several factors related to climate change act to extend wildfires both in locational extent and seasonal time frame. Overall, warmer temperatures cause more evaporation and drier soils, which create circumstances that promote more fires. Some regions, then, have earlier snow melt that leads to drier conditions for longer periods, and this means a longer fire season (Meyer, 2017). Climate change has shifted and increased the insect pests that harm trees, thus causing more dead vegetation for fires to engulf. Plus, human introduction of invasive species has led to more vegetation and dry fuel. These changes mean that when a fire does ignite from either natural causes, like lightning, or human causes, like bonfire embers, it now spreads faster, farther, and at a higher temperature than in the past (Pierre-Louis, 2017).

See also: Section 1: Climate Change, Urbanization; *Section 2:* Human Modification of Ecosystems, Natural Hazards

FURTHER READING

CAL FIRE. 2017. "California Statewide Fire Summary." October 30. http://calfire.ca.gov/communications/communications_StatewideFireSummary.

Meyer, Robinson. 2017. "Has Climate Change Intensified 2017's Western Wildfires?" September 7. https://www.theatlantic.com/science/archive/2017/09/why-is-2017-so-bad-for-wildfires-climate-change/539130/.

NASA. 2018. "Global Maps: Fire." Earth Observatory, National Aeronautics and Space Administration. https://earthobservatory.nasa.gov/GlobalMaps/view.php?d1=MOD14A1_M_FIRE.

NFPA. 2018. "Public Education: Wildfire." National Fire Protection Association. http://www.nfpa.org/Public-Education/By-topic/Wildfire.

Pierre-Louis, Kendra. 2017. "This Is How Much of the World Is Currently on Fire." Popular Science. August 4. https://www.popsci.com/global-wildfire-maps.

UN. 2016 "Wildfires Rage in a Changing Climate." United Nations Office for Disaster Risk Reduction. https://www.unisdr.org/archive/49866.

Section 3
Introduction: Sustainable Management of Natural Resources

How many natural resources does each American consume in one day? And how much garbage does each one produce in a day? In a week? In a year? Is this comparable to a citizen in a different country? Geographers are very interested in these questions, as they seek to understand the spatial patterns of how people use, manage, and protect natural resources and, ultimately, planet Earth.

In modern society, consumerism and technology affect how natural resources are used and who is using them. There is an uneven distribution of resource use across the planet. One American consumes as many goods and services as 35 people from India or 53 Chinese people. And "a child born in the United States will create thirteen times as much ecological damage over the course of his or her lifetime than a child born in Brazil" (Earth Talk, 2018).

The amount of energy a person consumes and in what form—a warm bath powered by a coal-powered electricity plant or a small solar array on their rooftop—makes a big difference around the globe. How do we use energy? What foods—meat or plant based—does each person consume? Overall, how many things—the latest cell phone, fashionable clothing, and recent car model—does a person buy? These are the key questions when figuring the carrying capacity of the Earth. Rather than just considering simple population numbers, it matters more how each person uses environmental resources. Indeed the Ecological Footprint model shows that if all people used resources the way Americans do, we would need about eight planet Earths to sustain this consumer lifestyle (WWF, 2017).

But nature cannot be stretched so thin for long without showing some degradation that will affect human populations well into the future. Climate change is now an example of this abuse and continued degradation. Human society will take action to mitigate hazards and shortages, but natural cycles may be permanently altered, which may drastically affect carrying capacity in the future. Although climate change science has been prominent in recent decades, expert warnings came well before this.

The book *Limits to Growth* was published by the Club of Rome, written by Donella Meadows and her three co-authors, in 1972. The Club of Rome is a group of scientists, economists, and public servants with "common concern for the future of humanity." They are active in more than 30 countries, and they seek to understand the interconnectedness of global challenges through scientific analysis, communication, and advocacy. Although *Limits to Growth* was written nearly 50 years ago, the facts remain unchanged: Earth cannot support the rate of resource use, even with advanced technology. These five areas must be limited if humans are to survive on Earth: population increase, agricultural production, nonrenewable resource depletion, industrial output, and pollution generation. They used computer models to show that humans must rethink and take action to lower their environmental impact in order to survive. Earth and its resources are finite and ultimately determine human life (Meadows et al., 1972).

SUSTAINABLE DEVELOPMENT

The concept of sustainability is fairly new. In the 1980s, the United Nations had important meetings and established the World Commission on Environment and Development, led by G. H. Brundtland. She is a highly respected politician from Norway, and her leadership was so important that the group became commonly called the Brundtland Commission. They worked for many years and eventually wrote a report called *Our Common Future* in 1987. This was the first time the concept of sustainable development was defined: "Development that meets the needs of the present without compromising the ability of future generations to meet their own needs." From here on, sustainability would be described as the balance of three pillars: economic development, social development, and environmental protection

The goal of the Brundtland Commission was to draw attention to these issues and unite countries to work for a common goal of global sustainable development. At the international level, the UN worked to make environmental issues into key political topics and to encourage world leaders to address environmental and social issues in unison. A few years later, in 1992, the Rio Earth Summit was held in Brazil. Here many international leaders met and agreed to Agenda 21, a plan that noted specific actions to promote sustainable development at national, regional, and international levels.

In the United States, an important next step occurred in 1993, when President Bill Clinton established the President's Council on Sustainable Development (PCSD) to provide information on sustainable development and create "bold, new approaches to achieve our economic, environmental, and equity goals" (PCSD, 1993). Again, notice the theme of the three pillars that work in unison. This council worked to build a "national sustainable development action strategy" that would "foster economic vitality" (*Federal Register*, 1999). These policy initiatives bring up a few topics for elaboration: How is development defined and how is it linked to economics and industrialization?

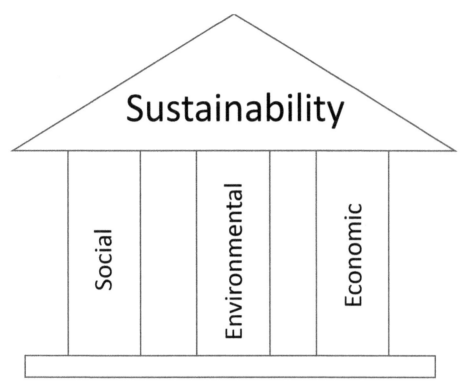

The Three Pillars of Sustainability. (Duram, 2017)

INDUSTRIALIZATION

Geographers are interested in understanding patterns of activities on the Earth. Environmental geographers are particularly interested in factors that influence people's use of natural resources. A key factor is economic variation, specifically the level of industrialization. Many researchers have written about and formed theories that describe differences among countries and regions. This section presents a few key terms that help categorize countries in order to explain how different economies use natural resources and thus have varying impacts on the Earth.

People who live off the land, simply using natural resources as needed for themselves, tend to have basic economic systems. They might trade with their neighbors if they have an extra crop of grain and the neighbor has an extra crop of vegetables. Through much of human history, these types of societies existed and depended on natural resources like wood for heating and cooking, and soil for farming. But things changed significantly when people developed machinery and factories, so by the early 1800s the economies of Europe and North America were shifting. Instead of people working in agriculture, many people started to work for a factory owner or mining company to earn wages. They spent the money they earned on necessities like food and housing. If there was any extra, the worker could also buy a consumer good or service like a book or a train ticket. The owners of profitable factories had even more extra money (called disposable income) to spend on consumer items like jewelry or larger homes.

There was uneven growth in these types of economies, both among individuals within a country and among different countries. So nations today are at very different levels of economic development. Colonialism was an important influence on these variations, as wealthy powerful countries built their economies by exploiting the resources and people in their colonies. These former colonies still have less developed economies, because their natural resources were taken away without compensation. These resources allowed the colonial powers to amass great wealth and build industries.

The more industrialized countries use a lot of natural resources and consume a lot of goods. The less industrialized countries remained more similar to their historic past, with less developed consumer economies. Some people use the terms More Developed Country (MDC) and Less Developed Country (LDC) to describe these differences. Due to their geographic location, some people use the terms Global North for more industrialized and Global South for less industrialized countries.

International organizations are helpful in providing definitions of these economic variations so that people can understand what caused and maintains the current disparities. The United Nations "classifies all countries of the world into one of three broad categories: developed economies, economies in transition and developing economies," which is intended to "reflect basic economic country conditions" (UN, 2014). The International Monetary Fund ranks some countries as "Advanced Economies," and these include countries in Europe, plus the United States, Canada, Australia, New Zealand, Israel, South Korea, Taiwan, and Japan (IMF, 2016).

LEAST DEVELOPED COUNTRIES

Because they are able to apply for specific types of aid, the "least developed countries" are clearly defined by the UN. Several criteria are used, including average income per person, education levels and literacy, malnutrition, and childhood mortality rates. Accordingly, people in the least developed countries tend to have low incomes, get few opportunities to attend high school, are unable to read as an adult, are more likely to experience hunger, and have a high percentage of children who die before they reach age five (UN DESA, 2018).

As this list implies, there are often tragic situations in these places, where people must work very hard just to survive. This contrasts sharply to countries, probably where the readers of this book live, where people are mostly concerned about getting the newest iPhone. People in these countries have all the basic needs of food, shelter, clean water, and education—and maybe even take them for granted. If readers live in a country where education is available, most people have access to food, and babies born today will live past the age of five, then they do not live in a least developed country. The geographic distribution of the least developed countries in the world is 34 in sub-Saharan Africa, 8 in East Asia, 4 in South Asia, 1 in West Asia, and 1 in Latin America (UN DESA, 2018). Often warfare or corruption is the root cause of human suffering, and these demand global attention and unified solutions to address economic, social, and ecological stability for a population.

There is a significant gap between the richer and poorer countries, as seen in the income categories used by the UN and World Bank. People in low-income countries earn, on average, less than $1,035 each year; in lower-middle-income countries people earn between $1,036 and $4,085; in upper-middle-income countries, people make between $4,086 and $12,615; and people in high-income countries have incomes of more than $12,615 per year. So, globally, the highest-income people earn 12 times more than the lowest. Keep in mind, too, that although the high-income countries are defined with individual earnings over $12,615 per year, the average income in the United States is actually about $52,000, which means that the average American earns 50 times what someone in a low-income country earns (UN, 2014; U.S. Census Bureau, 2018). Certainly, there are wide variations within a country, as seen in the United States and elsewhere. In the United States, for example, the richest 1% of families control 39% of the country's wealth (CNN, 2017).

Looking globally, the richest 1% of Americans earn over $400,000 per year, or 400 times what a person in a low-income country earns. According to the business magazine *Forbes*, the richest family in America is the Waltons, who own Walmart, and they control about $130,000,000,000—yes, those zeros add up to billions with a "B" (Dolan, 2016). This is a whopping 125 million times more than a person in a low-income country earns in a year (UN 2014).

Although there are huge variations around the globe, it might sound negative to say that one country is "less" developed than another, especially because it is not necessarily in their control. Many global economic structures are set in stone; for example, the large multinational banks and automobile corporations are well established. It would be very difficult for a new bank or car company to pop up

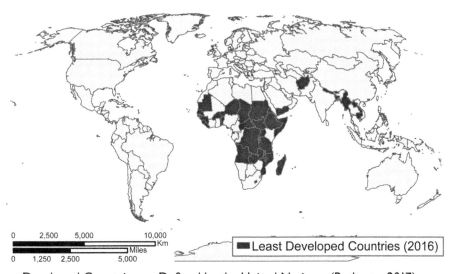

Least Developed Countries, as Defined by the United Nations. (Bathgate, 2017)

and provide profits to citizens in a small, less developed country. There is indeed some controversy with using terms like more and least developed, and so the UN explains that "the designations 'developed' and 'developing' are intended for statistical convenience and do not necessarily express a judgment about the stage reached by a particular country or area in the development process" (UN Statistics, 2018).

Acknowledging these types of variations among places might allow people to better identify regions that need help. In addition, just because a country is in a lower economic category, that does not mean that all people are unhappy and lacking a strong community.

ALTERNATIVE WAYS TO MEASURE DEVELOPMENT

Money does not buy happiness. Yet gross domestic product (GDP) is a typical measurement used when making national comparisons. The GDP is the monetary value of all goods and services produced in a country, so it shows the size of an economy. As of 2015, these are the top 10 rankings of GDP: United States, China, Japan, Germany, United Kingdom, France, India, Italy, Brazil, and Canada (World Bank, 2017).

Using the GDP, however, overemphasizes consumerism and consumption of goods, without considering the well-being of the people. Keep in mind that just because there is a high GDP, that does not necessarily mean that this wealth is evenly distributed among all people. Even if the GDP per person is figured, it does not really describe the well-being of the population.

For example, the GDP includes money spent for "happy" things like building a new house, but does not distinguish between this and the "sadness" of money spent to rebuild a house after a fire destroys it. Also think about how including military

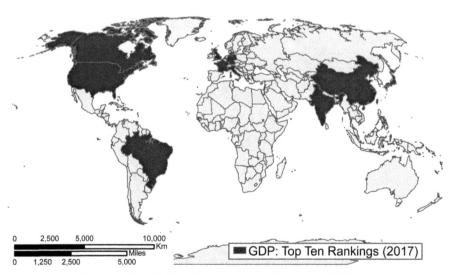

Gross Domestic Product (GDP): Top 10 Countries. (Bathgate, 2017)

Section 3: Introduction

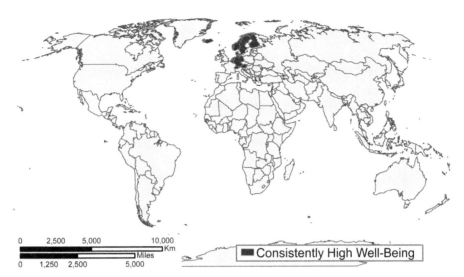

Top 10 Countries that Consistently Maintain Economic Growth and Social Well-Being. (Bathgate, 2017)

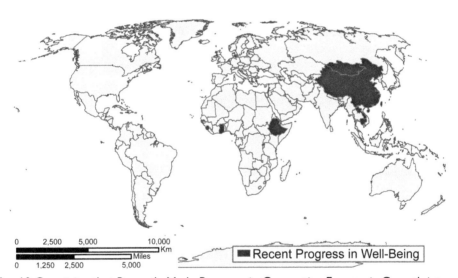

Top 10 Countries that Recently Made Progress in Converting Economic Growth into Social Well-Being (2006–2014). (Bathgate, 2017)

expenses show economic measures, but not happiness factors—we may have factories building bombs and tanks, but these weapons cause so much pain and suffering. Should these things actually be excluded from wealth? And just because a nation has a high GDP, does every citizen have equal access to wealth and opportunity? Overall, is there a better way to measure a nation's wealth?

These questions and ideas are behind several approaches that use alternative methods of measurement. The Sustainable Economic Development Assessment

(SEDA), for example, shows that the typical view of economic growth, based on GDP, is faulty because it does not measure people's well-being. Instead, the key factors should be how national wealth is distributed among all people in a country and how economic growth has been used to improve citizens' lives. The SEDA looked at 160 countries in terms of three elements: economics, sustainability, and investment, with 10 detailed dimensions that include things like income equality, health, education, and infrastructure (Thomson, 2016).

There are two lists related to this ranking. First are the top 10 countries that are best at consistently converting economic growth into well-being: Norway, Netherlands, Finland, Germany, Austria, Denmark, Switzerland, Iceland, Belgium, and Sweden (Thomson, 2016). This is a somewhat predictable list, composed of European countries with well-developed economies and strong workers' rights.

Second are countries that have made the most progress recently. These are the top 10 in converting economic growth into well-being in the years 2006–2014: Ethiopia, China, Rwanda, Mongolia, Qatar, Sierra Leone, Timor-Leste, Cambodia, Laos, and Ghana (Thomson, 2016). These shifts are often due to cessation of war, which increases a nation's ability to meet basic human needs and improve living conditions for its population. Both these lists look very different from the GDP categories shown earlier, and certainly these countries should be commended for striving to provide well-being for their citizens.

There are several other alternative measures that look more broadly than just at the GDP and economic size. The Genuine Progress Indicator (GPI) assesses 26 variables related to economic, social, and environmental progress. In the GPI, the "pain and suffering" that are included in the GDP are omitted. For example, costs related to crime and military are a negative in the equation, as is pollution, nonrenewable energy use, climate change, and environmental resource depletion (Daly and McElwee, 2014). Several U.S. states, including Vermont and Maryland, have opted to adopt policies to use the GPI over the GDP when assessing the well-being of their population (Ceroni, 2014).

Another global measurement is the Better Life Index, which was developed by the Organisation for Economic Cooperation and Development, which measures 11 topics: income, jobs, housing, health, access to services, environment, education, safety, civic engagement and governance, community, and life satisfaction (OECD, 2018). The OECD interactive website at www.oecdbetterlifeindex.org provides detailed information that is worth exploring to see how individual countries fit in. The United States is the 13th happiest country, just below Israel and Austria, and just above Costa Rica and Puerto Rico. Interestingly, although GDP has risen vastly over time, most alternative measures show significant declines. For example, global GPI has been decreasing since 1978, and researchers find that once a modest income per person is achieved, "life satisfaction" levels off, meaning that higher wealth does not bring more happiness (Kubiszewski et al., 2016).

SUSTAINABILITY AND HAPPINESS

The World Happiness Report provides a survey of the state of global happiness (SDSN, 2017). The report has come out annually since 2012 and coincides with the

UN World Happiness Day, on March 20. The report draws from experts in many subjects—from economics to health, psychology, and more—to assess the progress of nations. There is a newly emerging science of happiness, which seeks to understand variations among individuals, communities, and countries. This is increasingly linked to people demanding that happiness, not just economics, be considered in government policy.

Stepping further away from economic views of variations among countries, the Happy Planet Index measures citizens' health and happiness in combination with sustainability. It describes how well countries are achieving long, happy, sustainable lives (NEF, 2018). This index uses the measure of people's well-being and longevity and how equally these are distributed across the population. Then this result is compared to each country's ecological footprint, or "the amount of the environment necessary to produce the goods and services necessary to support a particular lifestyle" (WWF, 2017). The repeat winner of the Happy Planet Index is Costa Rica, due to its high life expectancy (78.5 years) and low environmental impact per person (Bruce-Lockhart, 2016). Happiness and sustainability clearly go together in these measures.

VISUALIZING SUSTAINABILITY

Established in the *Our Common Future* report by the Brundtland Commission, the definition of sustainability is defined as the balance among three pillars: society, environment, and economy (Meadows et al., 1972). All three pillars must be strong or the other two will fall.

Building on the three pillars, sustainability can be used in other contexts. Triple bottom line is a concept that was created among sustainability-seeking business consultants in the 1990s (Elkington, 2004; *The Economist*, 2009). The term "bottom line" is commonly used to mean profit for a company, which is usually the single factor deemed important to a company. But the triple bottom line introduced the idea that a good business looks at not just one "P" of profit, but three "Ps" of profit, people, and planet.

Likewise this shows the complexity of sustainability, because the three topics are interrelated in the bubble diagram (Venn diagram) with overlapping themes. Again, the three key goals are social, environmental, and economic sustainability, and there are times when the three themes come together to create true sustainability. In other cases, only two of the themes overlap, so society and the economy overlap on sustainable issues commonly labeled as equitable. Where the economic and environmental issues intersect, people seek a viable situation. Where society and the environmental overlap, the goal is commonly stated as bearable. These intersecting terms seem to underestimate possible sustainability goals.

Certainly measuring sustainability is a tricky endeavor. Humans seem to have an understanding of the economic aspect of sustainability; as noted earlier, this is often in terms of consumer goods and productivity. The environment can be assessed by scientists who study the planet's ecological systems, but it is difficult

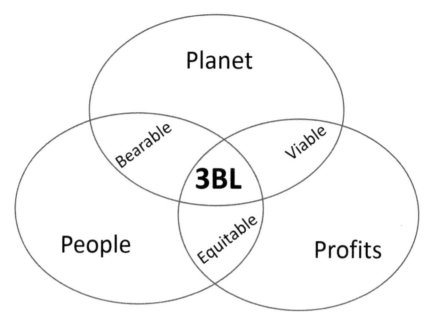

Sustainability: The Triple Bottom Line or 3BL. (Duram, 2017)

to decide what levels of environmental change are acceptable. Of the three pillars, the society category has been the least understood over time because human relationships with the environment and with one another are very complicated. The social component of sustainability can mean everything from just a bland concept like social progress to idealistic goals of social equity and poverty alleviation (Kates et al., 2005). These topics greatly affect people's use of the natural environment and their environmental decision making.

Building on the ideas noted earlier, an updated and more complex graphic is needed. It shows that in the real world, decisions must be made to bring about a healthy relationship between people and the planet, a unity between the planet and prosperity, and equality between people and prosperity. Overall, this view de-emphasizes the monetary value of economic growth and instead encourages a more long-term view of prosperity that is linked to human and environmental well-being.

ASSESSING ENVIRONMENTAL SUSTAINABILITY

It is helpful to visualize sustainability: what it includes and how the various components fit together. This way, people can assess how human actions affect the environment. This is important because what gets measured gets done! In other words, if people want to take action to improve an ecological problem, they need to understand the whole situation: causes and possible solutions. In order to fully assess the environmental sustainability of an activity, people need to qualify the variables, understand the system, consider new policies, and push for appropriate changes. This process of assessment and action can lead to environmental solutions. There are useful tools and concepts for tracking environmental sustainability, such as Life Cycle Assessment (LCA) and the circular economy.

Section 3: Introduction

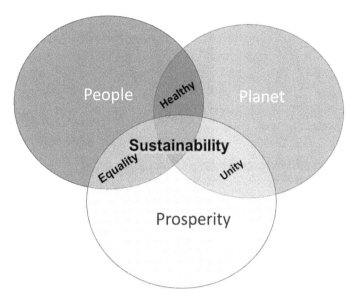

Sustainability in the Real World: A Hopeful Future. (Duram, 2017)

Life Cycle Assessment

LCA is a valuable method for the systematic evaluation of the environmental aspects of a product or service system through all stages of its life cycle (UNEP, 2018). The stages of a product would include materials, manufacturing, distribution, usage, and disposal (EPA, 2017). This includes all the environmental impacts along the way, and even what happens to the product when it is no longer in use.

The International Organization for Standardization (ISO) is the federation of national groups that determine the international accepted standards of measurements. In the case of the Life Cycle Analysis, the ISO developed ISO 14040 (ISO, 2006). According to these standards, LCA shows the possible environmental impacts of a product throughout its life cycle "from raw material acquisition through production, use, end-of-life treatment, recycling and final disposal (i.e., cradle-to-grave)" (ISO, 2006).

LCA studies are very complicated, because there is potentially so much information to gather over a long period. The concept of the entire "life" of a product means that there are linked stages from resource extraction, to producing a consumer item, to selling the item, to throwing it "away." Because such a study is so complex, ISO breaks this down and states that there are four stages for an LCA study: setting goals and defining scope, inventory analysis, impact assessment, and interpretation.

LCA has been used in many different places for comparing many different products and services. One example is the comparison between the LCA of reading a paper newspaper versus a digital newspaper version (Toffel and Horvath, 2004). Data for this LCA would include the impact of cutting trees, milling, water used making paper and ink, printing, distributing, recycling, or creating garbage from the newspaper. This would be compared to the impacts of gathering the raw materials (e.g., metals, petroleum) needed to make the digital device (plastics, electronics), the manufacturing of the device (for example, machinery, water use, chemicals, electricity), the distribution and energy use of the finished device, and its eventual disposal (often in a landfill). One big question in this type of LCA comparison is that the long-term toxicity of the electronic components in the waste stream is not fully known (Moberg et al., 2010). In addition, a new newspaper would be obtained daily and disposed of, whereas the digital device should last multiple years. LCA comparisons emphasize the need for consumers to keep the items they purchase for as long as possible, rather than throwing them "away" and buying new devices before necessary. This is hard to do in a throw-away society, where people are motivated through advertising and culture to buy the latest consumer goods.

Circular Economy

This brings up the idea of a circular economy. In a linear economy, which modern society follows, products are made from raw materials that are mined or newly created (like oil to make new plastics). A quick way to remember the linear economy is the slogan: "take, make, dispose," where manufacturing new consumer goods depletes natural resources and creates waste that mostly goes to landfills.

On the other hand, in a circular economy, all materials for manufacturing would be "used" materials from recycling, recovery, or reuse (Pearce and Turner, 1989). In a circular economy, manufacturers would begin with the concept of designing all manufactured items to be reusable, recyclable, or easy to upgrade, thus creating zero waste. Global transportation of consumer goods would be reconsidered. All bulk product deliveries would be from set pickup points at the same time; in other words, people would rethink the current manufacturing transportation system to be much more efficient and sustainable. The circular economy would require a new sustainability education component, and both government and business should be involved in this process. Consumers must make smarter choices in order to buy and discard products that do not send valuable resources to a landfill. Finally, waste take-back programs would need to be integrated into the manufacturing process so that used materials are returned to their manufacturers or taken to recycling facilities where these resources can be sorted and reused.

The circular economy is related to ideas that took shape in the late 1970s when Walter Stahel worked to develop a closed-loop approach to manufacturing (Stahel, 2013). This contrasts to the current, typical approach of "cradle to grave" (where a product is born, used, and its grave is the garbage can or, ultimately, the landfill). Later William McDonough and Michael Braungart wrote the book *Cradle to Cradle: Remaking the Way We Make Things* (2002), and the documentary film *Waste = Food* (2007) further explains the concept. The central idea is that waste becomes the food or input for manufacturing, so eventually there is no more waste. All items are disassembled and all components are simply used again in a closed loop or circular system.

ENGOs

Imagine a day when there is no garbage can. In this future world, everything people buy is consumed, reused, or recycled. So what is now called "waste" will then be a valuable material in future manufacturing. Sadly, it feels like that day is far in the future. However, sometimes when people learn about a radical new way to use natural resources or to protect the environment, they can sometimes think of a way to make it happen. People must take action and make innovative changes for a sustainable future.

One way is to join together with like-minded people and form an environmental group, often called a nongovernmental organization (NGO). "A non-governmental organization (NGO) is any non-profit, voluntary citizens' group which is organized on a local, national or international level" to work for a common goal separate from government or business (UN NGO, 2017). There are many kinds of NGOs, from school sports booster clubs, to Girl Scouts, to business groups' Chamber of Commerce, just to name a few. But environmental geographers are interested in groups that focus on environmental issues, and these are called ENGOs.

There are hundreds of larger ENGOs, with the total number in the thousands, once every community park and nature group is included. It is really exciting to think that so many people want to participate in environmental issues. Plus it is easy to find a group, so everybody can get involved.

The Sierra Club, founded in 1892 by naturalist John Muir, was one of the first national ENGOs in the United States. The Sierra Club is still one of the best known ENGOs in North America, with over 2 million members today, 64 local chapters, and 20,000+ group outings and hikes each year (www.sierraclub.org). As noted in their mission statement, they seek to "to explore, enjoy, and protect the wild places of the earth." They have also been active in reducing the number of coal power plants in the United States, promoting their "Beyond-Oil" campaign, and pushing for the protection of more wilderness.

Greenpeace began much later (in the 1970s), but has grown to over 3 million members active in 55 countries, with global campaigns to take action on topics of climate, forests, oceans, food, nuclear weapons, and toxins. Greenpeace is a non-violent, direct action organization, which means they protest to get attention and disrupt actions that harm the environment. For example, they block logging trucks' access to a forest road or they hang a banner on a coal plant's smokestack. Greenpeace owns three ships, which is "a unique asset in the battle to save planet Earth and protect the global commons." They say their ships are "at the forefront of Greenpeace campaigning, often sailing to remote areas to bear witness and take action against environmental destruction" (greenpeace.org, 2018).

Today, there are many examples of ENGOs, and they have a great deal of influence on environmental policy, protection, and education. There are hundreds of ENGOs around the world, but here is an example of just a handful: World Wildlife Fund works in over 100 countries to protect biodiversity (wwf.org). Natural Resources Defense Council employs nearly 400 lawyers and scientists to give nature a voice in the legal system (NRDC.org). Nature Conservancy has worked to protect over 1 million acres of land and thousands of miles of rivers around the world (nature.org). Conservation International brings together scientists and communities to promote protection of biodiversity and local cultural sustainability (conservation.org). Connected to concepts of sustainable development, the International Institute for Sustainable Development is an ENGO based in Canada with international programs that link research, citizens, businesses, and politicians to address issues of sustainable economies, clean energy, fresh water, and resilient communities (iisd.org).

Beyond these global ENGOs, it is important for each person to find local groups that address issues of concern in their own community. What is of interest? Hiking, rock climbing, recycling, or teaching kids about nature? People have probably already started a local group that matches this same area of interest. Local libraries, park districts, city governments, and the Internet provide helpful lists of various ENGO groups. Once a person finds their place, they can decide how to best participate and educate others about building a sustainable future from the local, grassroots level up to the national, and even international, settings.

WORLDVIEW

It is tricky to talk about culture, because there are so many historical factors: family influences, individual variations, and melding of ideas through globalization. Rather than the term "culture," which is difficult to describe in a complex place like the United States, think about the concept of worldview, or dominant social paradigm (DSP). This is the culmination of society's beliefs, values, and ideals. The DSP greatly influences a society's collective environmental thoughts. In other words, think about this: What makes America, "America"? What makes the United States unique? Usually people say things like individualism, technology, freedom, religion, progress, consumerism, education, capitalism, democracy, infinite resources, science, etc. These concepts describe society's attitudes toward the environment, and thus Americans' use of natural resources and environmental policies.

When discussing types of environmentalism, there may be "shades of green" (Steffen, 2004). The Light Greens try to do ecologically friendly things because it is trendy and their friends do it. The Dark Greens believe that the capitalist system is inherently built on environmental degradation, so they seek radical political change. The Bright Greens believe that new technologies and social innovations will lead to constructive solutions that may involve significant economic and political changes to bring about a sustainable society (Newman, 2011; Steffen, 2004, 2011). Each person can think about which shade of green best fits with their personal beliefs and actions.

Another important term is "greenwashing," which is the spin or green marketing efforts that deceptively promote the perception that a company's products, aims, or policies are environmentally friendly (Greenwashing Index, 2018). But sadly, sometimes companies spend more effort on advertising that they are "green" than actually taking action to be more environmentally sound. You may notice this in terms of labeling (using a flower or natural-looking images on a bottle of household cleaner that still contains toxic chemicals) or millions of dollars spent on a public relations campaign to advertise how a highly polluting energy company is ecofriendly.

Consumers must be cautious and educated. People can learn about the products they buy, how consumer goods are produced, and the ethics of the companies that produce them. This is perhaps the most important thing individuals can do. If people demand more environmentally friendly products from companies that use environmentally sustainable production methods, this will push the production of products that are better for the environment. People can have a big impact with their wallet.

PARTICIPATE IN POLICY

Whether natural resources are used sustainably is inexorably linked to human culture, society, and worldviews, which determine environmental policies. Many people are interested in participating in local environmental groups, and this can have increasing influence as more people learn about and participate in the environmental policy process. In a democracy, all people should have a voice. If that voice is pro-environment, then the politicians who are elected must follow the wishes of the people. Policies, regulations, and laws must reflect society's environmental goals.

OUR FUTURE

The Earth is in our hands. People can take action to reduce human impact on the Earth through their daily actions. Water consumption, electricity use, diet, and waste production all add up over a lifetime, especially when combined with family members, neighborhoods, and entire communities. This adds up to the global environmental system. People—as individuals, as communities, and as nations—can choose a sustainable future.

CASE STUDIES 11–15

Case Studies in this section provide examples of sustainability in action. Keep in mind that environmental, social, and economic sustainability are very complicated and may require a balance of innovation and tradition.

Indeed, shifting toward a more sustainable future is precisely what Case Study 11 describes. Up until now, China has had a dismal environmental record, having focused on rapid, unregulated industrialization for decades. A few examples of the degradation caused by China's reliance on dirty fossil fuel energy include air pollution, water pollution, and land degradation. In early 2017, however, the Chinese government announced massive new investments in renewable energy. Trillions of yuan (billions of U.S. dollars) will be invested to create millions of jobs with a goal of shifting China into the renewable energy leader. Sustainability demands this type of shift and initiative.

Everybody can get involved with environmental science, as shown in Case Study 12. In this example, citizen science projects related to bird observations are the perfect opportunity for individuals to take action and to benefit the environment. Case Study 13 focuses on the nation of Costa Rica, which dismantled its military in 1949, and instead puts this money toward education, health care, and environmental

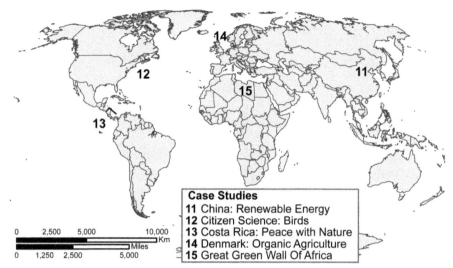

Section 3 Case Studies: Sustainable Management of Natural Resources. (Bathgate, 2017)

protection. Even the Costa Rican Constitution states that each citizen has the right to an ecologically balanced environment.

Case Study 14 describes the national policies of Denmark that promote organic agriculture. Denmark has specific goals and government initiatives: both farmers and consumers benefit from training, information, and education about growing and eating organic foods.

Case Study 15 presents information on the Great Green Wall of Africa. This vast initiative encompasses both a continent-wide concept and many locally driven projects. The integration of geographic scales exemplified here could be a lesson for other initiatives worldwide. People may be best able to address global concerns by understanding and taking action in their local communities.

Overall, these five Case Studies illuminate how people are taking action to create a more sustainable world. The Key Concepts continue along this theme, providing details about environmental policies, sustainability, and the greening of consumers. In addition, several innovations and alternatives are described: Earth Day, electric cars, fair trade, green buildings, Green Party politics, and sustainable cities, to name a few.

FURTHER READING

Bruce-Lockhart, Anna. 2016. "Which Is the Greenest, Happiest Country in the World?" World Economic Forum. July 29. https://www.weforum.org/agenda/2016/07/greenest-happiest-country-in-the-world.

Ceroni, Marta. 2014. "Beyond GDP: US States Have Adopted Genuine Progress Indicators." *The Guardian*. September 23. https://www.theguardian.com/sustainable-business/2014/sep/23/genuine-progress-indicator-gdp-gpi-vermont-maryland.

CNN. 2017. "Record Inequality: The Top 1% Controls 38.6% of America's Wealth." September 27. http://money.cnn.com/2017/09/27/news/economy/inequality-record-top-1-percent-wealth/index.html.

Daly, Lew and Sean McElwee. 2014. "Forget the GDP. Some States Have Found a Better Way to Measure Our Progress." *New Republic*. February 3. https://newrepublic.com/article/116461/gpi-better-gdp-measuring-united-states-progress.\

Dolan, Kerry A. 2016. "Billion-Dollar Clans: America's 25 Richest Families." *Forbes*. June 29. http://www.forbes.com/sites/kerryadolan/2016/06/29/billion-dollar-clans-americas-25-richest-families-2016.

Earth Talk. 2018. "Use It and Lose It: The Outsize Effect of U.S. Consumption on the Environment." *Scientific American*. https://www.scientificamerican.com/article/american-consumption-habits/.

The Economist. 2009. "Triple Bottom Line: It Consists of Three Ps: Profit, People and Planet." November 17. http://www.economist.com/node/14301663.

Elkington, John. 2004. "Enter the Triple Bottom Line," in *The Triple Bottom Line: Does It All Add Up?* Adrian Henriques and Julie Richardson, eds. London: Earthscan Publications.

EPA. 2017. "Design for the Environment Life-Cycle Assessments." U.S. Environmental Protection Agency. April 19. https://www.epa.gov/saferchoice/design-environment-life-cycle-assessments.

Federal Register. 1999. "President's Council on Sustainable Development." https://www.federalregister.gov/agencies/president-s-council-on-sustainable-development.

Greenpeace.org. 2018. "Greenpeace." www.greenpeace.org/usa.

Greenwashing Index. 2018. "About Greenwashing." EnviroMedia Social Marketing and the University of Oregon School of Journalism and Communication. http://greenwashingindex.com.

IISD. 2018. "Evidence, Passion, Sustainability." International Institute for Sustainable Development. http://www.iisd.org/.

IMF. 2016. "World Economic Outlook: Too Slow for Too Long." International Monetary Fund. http://www.imf.org/external/pubs/ft/weo/2016/01/pdf/text.pdf.

ISO. 2006. "Environmental Management-Life Cycle Assessment-Principles and Framework." International Standards Organization. https://www.iso.org/obp/ui/#iso:std:iso:14040:ed-2:v1:en.

Kates, Robert W., Thomas M. Parris, and Anthony A. Leiserowitz. 2005. "What Is Sustainable Development?" *Environment: Science and Policy for Sustainable Development* 47(3): 8–21.

Kubiszewski, I., R. Costanza, C. Franco, P. Lawn, J. Talberth, T. Jackson, and C. Aylmer. 2016. "Beyond GDP: Measuring and Achieving Global Genuine Progress." *Ecological Economics* 93: 57–68.

McDonough, William and Michael Braungart. 2002. *Cradle to Cradle: Remaking the Way We Make Things*. New York: Farrar, Straus and Giroux.

Meadows, Donella H., Dennis L. Meadows, Jorgen Randers, and William W. Behrens III. 1972. *The Limits to Growth*. New York: Universe Books.

Moberg, Åsa, Martin Johansson, Göran Finnvedena, Alex Jonsson. 2010. "Printed and Tablet e-Paper Newspaper from an Environmental Perspective: A Screening Life Cycle Assessment." *Environmental Impact Assessment Review* 30(3): 177–191.

NEF. 2018. "Happy Planet Index." New Economics Foundation. www.happyplanetindex.org.

Newman, Julie. 2011. *Green Ethics and Philosophy: An A-to-Z Guide*. Thousand Oaks, CA: SAGE Publications.

OECD. 2018. "Better Life Index." Organisation for Economic Co-operation and Development. http://www.oecdbetterlifeindex.org.

PCSD. 1993. "Overview." President's Council on Sustainable Development. https://clinton2.nara.gov/PCSD/Overview/.

Pearce, David William and R. Kerry Turner. 1989. *Economics of Natural Resources and the Environment*. Baltimore, MD: Johns Hopkins University Press.

SDSN. 2017. "World Happiness Report." Sustainable Development Solutions Network. http://worldhappiness.report/.
Stahel, Walter R. 2013. "Policy for Material Efficiency—Sustainable Taxation as a Departure from the Throwaway Society." *Philosophical Transactions of the Royal Society A* 371: 20110567.
Steffen, Alex. 2004. "Tools, Models and Ideas for Building a Bright Green Future: Reports from the Team." August 6. Worldchanging.com.
Steffen, Alex. 2011. *Worldchanging, Revised Edition: A User's Guide for the 21st Century*. New York: Harry N. Abrams.
Thomson, Stephanie. 2016. "Which Countries Are Best at Converting Economic Growth into Well-Being?" *World Economic Forum*. July 28. https://www.weforum.org/agenda/2016/07/which-countries-are-best-at-converting-economic-growth-into-well-being.
Toffel, Michael W. and Arpad Horvath. 2004. "Environmental Implications of Wireless Technologies: News Delivery and Business Meetings." *Environmental Science and Technology* 38(11): 2961–2970.
UN. 2010. "Sustainable Development." United Nations. http://www.un.org/en/ga/president/65/issues/sustdev.shtml.
UN. 2014. "Country Classification." United Nations. http://www.un.org/en/development/desa/policy/wesp/wesp_current/2014wesp_country_classification.pdf.
UN DESA. 2018. "LDC Criteria." United Nations, Department of Economic and Social Affairs. https://www.un.org/development/desa/dpad/least-developed-country-category/ldc-criteria.html.
UN NGO. 2017. "What Is an NGO?" United Nations. https://outreach.un.org/ngorelations/content/about-us-0.
UN Statistics. 2018. "Methodology." United Nations Statistics Division. http://unstats.un.org/unsd/methods/m49/m49.htm.
UNEP. 2018. "Life Cycle Thinking and Knowledge." United Nations Environmental Programme. http://web.unep.org/resourceefficiency/what-we-do/assessment/life-cycle-thinking-and-knowledge.
U.S. Census Bureau. 2018. "Income." https://www.census.gov/topics/income-poverty/income.html.
Waste = Food. 2007. 51 minutes. Film by Rob van Hattum. Icarus Films.
World Bank. 2017. "GDP Ranking." http://data.worldbank.org/data-catalog/GDP-ranking-table.
WWF. 2017. "Ecological Footprint." World Wildlife Fund. http://wwf.panda.org/about_our_earth/teacher_resources/webfieldtrips/ecological_balance/eco_footprint/.

Case Studies

Case Study 11: China's Bold Steps toward Renewable Energy—Better Late than Never

Sometimes there are great examples of sustainability that have immediate and obvious beneficial impacts. But sometimes, in the real world, things are a bit tricky. China, for example, is the biggest polluter in terms of greenhouse gas (GHG) emissions that cause climate change (UCS, 2018). They overtook the United States in this not-so-desirable leadership about a decade ago, as their manufacturing sector has grown relentlessly, often with few environmental protections (Forsythe, 2017). More recently, however, China has shifted its outlook and is investing heavily in renewable energy. This case study represents one of those complex situations that environmental geographers love to discuss: it is not an "easy" or absolute example of sustainability, but it is a "do-able" example that shows how national policy can shift to embrace alternatives.

Over 1 million people die from polluted air in China each year, and millions more are forced to breathe poisonous air every day (Vaughan, 2016). This air pollution is caused by coal-burning factories, coal electric power plants, and oil-driven vehicle traffic. In addition to the high death tolls, there are other significant and wide-ranging health impacts, including cancer, birth defects, heart disease, and respiratory diseases.

China is not alone. According to a report by the World Health Organization, 92% of the world's population lives in places with air pollution above recommended levels. Over 3 million deaths each year are caused by outdoor air pollution, and the vast majority of these are in lower- or middle-income countries in Southeast Asia and western Pacific regions (WHO, 2016). These are examples of countries that are forced to promote the immediate benefits of industrial development over long-term sustainability.

To improve environmental quality and human health, these nations must change their economic model and move away from coal consumption. China is a leading example of this shift.

CHINA

China has the largest population of any country in the world, with more than 1.37 billion people, which makes an obvious contribution to the country's massive energy use. China's landmass makes it the fourth largest country, just slightly smaller than the United States, and it shares borders with 14 countries, including Russia, Mongolia, India, North Korea, and Vietnam (CIA, 2018).

The landscape of China varies from hills, plains, and river deltas in the east to mountains, high plateaus, and deserts in the western part of the country. The international border between Nepal and China (Tibet, now a region controlled by China) runs across the summit of Mount Everest, which at 8,850 meters (29,035 feet) in elevation, is the highest mountain in the world. The climate of China is likewise diverse, with subarctic cold in the north and at high elevations and tropical conditions in the south (CIA, 2018).

These physical environmental variations have influenced population distribution: 95% of Chinese live in the eastern one-third of the country. Along the coast and in a few central commercial centers, there is a high level of economic development, which pulls in hundreds of millions of immigrants from the poorer rural interior regions each year. Ethnic Chinese make up about 92% of the population (*National Geographic*, 2004).

China has the world's longest continuous civilization, estimated at over 40 centuries old. Chinese developed art, philosophy, technology, and a vast military power that built enduring empires across Asia (BBC, 2018; *National Geographic*, 2004).

Flash forward to modern times. The People's Republic of China is a communist country with a centrally controlled economy. In 2001, China joined the World Trade Organization, taking on a full role in the global economy. Indeed China now has the largest economy in the world, with a gross domestic product (GDP, figured as purchasing power parity) of $21.27 trillion in 2016; by comparison the U.S. GDP was $18.56 trillion (CIA, 2018). Chinese industrial output is focused on iron, steel, coal, machine building, armaments, and consumer goods; thus it now exports vast amounts of machinery, equipment, textiles, clothing, footwear, toys, sporting goods, and mineral fuels.

CHINA'S SHIFT IN PLANNING

China's rapid industrial growth led to alarming pollution: four of the world's most polluted cities in terms of air pollution are in China. Coal powered the massive growth in China's industrial output, and this caused significant environmental degradation (*National Geographic*, 2004).

In early 2017, China's National Energy Administration announced a plan to invest 2.5 trillion yuan ($363 billion) in renewable energy, with the goal of dominating the world in renewable energy generation and equipment manufacturing (Reuters, 2017). Interestingly, at the same time, the United States under President Donald Trump is backing away from renewable energy and supporting more exploration, tax breaks, and consumption of fossil fuels.

Case Study: China's Bold Steps

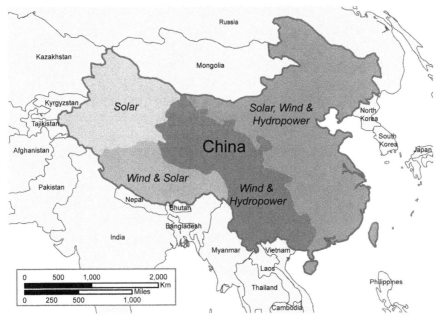

China: Regional Variations in Types of Renewable Energy Potential. (Bathgate, 2017)

The Chinese plan states that about half of new electricity generation will be from renewable sources by 2020. Further, 13 million new jobs will be created in the renewable energy sector by 2020 (Forsythe, 2017). China plans to focus on renewable power supplies like wind, solar, and hydro, which each hold potential in distinct regions (GENI, 2016a, 2016b; Hu and Cheng, 2013).

Specifically, the five-year plan written by China's economic planning agency states that approximately 700 billion yuan ($100 billion) will go toward wind farms, 500 billion yuan ($73 billion) to hydropower, and new investments in tidal power and geothermal will reach about 275 billion yuan ($40 billion). Solar power will get 1 trillion yuan ($145 million), with the goal of increasing solar capacity by five times its current level. To give us perspective, that is like adding about 1,000 major new solar power plants over the next few years. China is already the world's top producer of solar power (Forsythe, 2017; Reuters, 2017). Because demand continues to skyrocket, now stimulated by the national government's new commitments, most experts say the costs of renewables will continue to go down.

A RENEWABLE ENERGY LEADER

It is interesting that China's new policy is more than simply an act to promote environmental sustainability; indeed, the Chinese are basically staking a claim on the renewable energy industry. The economic power of China, already a big player in renewable energy, will now expand rapidly, as they seek to manufacture all components of wind and solar energy production. With the huge investment from the Chinese government, jobs that could have been created in the United States or other countries may well go to Chinese workers. Because of the huge market within China, their companies will be able to produce these products at a huge scale, ramping up production and driving down costs (China Dialogue, 2016).

On top of all the recent investments, the shift has already been happening on the ground. For example, China installed more than one wind turbine every hour, every day, for the whole year of 2015. They were also active with solar installations; China covered the area of one soccer field every hour for all of 2015, too (Greenpeace, 2017).

But this shift to renewable energy faces a big challenge because the Chinese coal industry has a lot of political power. Coal has traditionally been the main supplier of energy to China's electric grid, and indeed the country has continued to build coal-burning power plants. This important announcement by the Chinese National Energy Administration did not mention how or if coal power would be reduced. Still, environmentalists are hopeful that the country will continue to shift away from dirty, nonrenewable sources of energy.

In a country as huge as China, with its powerful industrial economy, it is a huge challenge to make this shift. Even with this important policy commitment, there will still be significant fossil fuels being consumed. By 2020, renewable energy will still only comprise 15% of total Chinese energy consumption (Reuters, 2017).

THE GROUP OF TWENTY

Will other countries follow this example? Yes, there are changes in other industrialized countries' dependence on fossil fuels.

The Group of Twenty (G20) is an organization of key industrialized countries that meet to discuss global financial stability. These nations represent about 80% of global economic output (GDP) and 75% of world trade. China has the world's largest economy, followed by the United States, India, Japan, and Germany. The remaining G20 members are Argentina, Australia, Brazil, Canada, France, Indonesia, Italy, South Korea, Mexico, Russia, Saudi Arabia, South Africa, Turkey, the United Kingdom, and the European Union (G20, 2018).

These countries have a great deal of influence over world energy usage and development. At the G20 Summit in 2009, the leaders discussed how to make economic growth have a positive impact on people all over the Earth. Thus they signed an agreement to promote the "convergence of living standards" between rich and poor countries. The way to do this, as stated by the G20, is to "bring clean affordable energy to the poorest, such as the Scaling Up Renewable Energy Program." These economic powers stated that "inefficient fossil fuel subsidies encourage wasteful consumption, reduce our energy security, impede investment in clean energy sources and undermine efforts to deal with the threat of climate change" (G20 Leaders Statement, 2009).

From this definitive statement, it is clear that the world's leaders are now ready to embrace renewable energy to benefit the global economy, society, and the environment. But how can this really happen? From an environmental geographer's perspective, reducing the subsidies will make fossil fuels more expensive and would encourage the development of renewables.

LOOKING AHEAD

Fossil fuels are such a prominent force in the world economy because they are heavily subsidized. According to the International Energy Agency, globally the level of direct subsidies to fossil fuels was $493 billion in 2014, which is more than triple the subsidies going to renewable energy ($150 billion) (IEA, 2016). These subsidies take the form of tax breaks, grants given to coal or oil producers, investments by state-owned enterprises, government funding of fossil fuel research, and financing from government-owned banks.

Subsidy estimates skyrocket to $5 trillion per year when indirect costs are included: military spending used to support a country's fuel interests, or the health, climate, or environmental costs related to fossil fuels (Bast et al., 2015; China Dialogue, 2016).

For China specifically, direct fossil fuel subsidies are estimated at $17 billion annually (China Dialogue, 2016). Annually, the United States provides more than $20 billion in direct national fossil fuel production subsidies each year; despite calls from President Barack Obama to eliminate industry tax breaks, Congress refused to act (Bast et al., 2015; Oil Change International, 2018). For example, oil companies can deduct from their taxes their costs of cleaning up an oil spill. So the huge oil corporation BP was able to claim $9.9 billion in tax deductions from its $32 billion cleanup after its disastrous Deepwater Horizon accident in the Gulf of Mexico in 2010 (Pandey, 2015).

In 2017, the Bloomberg New Energy Outlook report noted that the cost of wind and solar continues to drop. These two technologies will become the cheapest ways of producing electricity in many countries during the 2020s and in most of the world in the 2030s. The cost of onshore wind costs will fall by 47% and solar photovoltaic costs will drop by 66% by 2040 (Bloomberg, 2017). Geographically, the Asia-Pacific region leads in renewable energy investment, representing 50% of all new investment worldwide, with China as the leading nation.

Specifically, China increased its foreign investment in renewables by 60%, to $32 billion, in 2016 (IEEFA, 2017). China has recently invested in over 10 international projects, each worth more than $1 billion. China has been growing its financial investments regionally, too, by establishing the Asia Infrastructure Investment Bank. Also, China is a key player in the BRICS' (Brazil, Russia, India, China, and South Africa) New Development Bank, which is making loans for renewable energy projects (Jaeger et al., 2017). China outspent the United States by two-and-a-half times in their investments in domestic renewable energy production (Bloomberg, 2017). China is now the global leader in terms of technology, investment, manufacturing, and employment related to renewable energy (IEEFA, 2017).

China has a history of environmental degradation due to its industrialization-at-all-cost goals. It has a very poor environmental record over the last 100 years. Yet China has recently made massive and significant commitments to support renewable energy. Maybe China, and other countries, will make this much-needed shift away from fossil fuels in the near future.

See also: Section 1: Climate Change, Energy, Mining; *Section 3:* Environmental Policy, Green Technology, Renewable Energy

FURTHER READING

Bast, Elizabeth, Alex Doukas, Sam Pickard, Laurie van der Burg, and Shelagh Whitley. 2015. "Empty Promises: G20 Subsidies to Oil, Gas and Coal Production." Overseas Development Institute. http://priceofoil.org/content/uploads/2015/11/Empty-promises_main-report.2015.pdf.

BBC. 2018. "China profile—Timeline" November 8. http://www.bbc.com/news/world-asia-pacific-13017882.

Bloomberg. 2017. "New Energy Outlook." Bloomberg New Energy Finance. https://www.bloomberg.com/company/new-energy-outlook/.

China Dialogue. 2016. "Fossil Fuel Subsidies Reform in China: Current Status and Future Directions." December 15. https://chinadialogue-production.s3.amazonaws.com/uploads/content/file_en/9499/EFC_Fossil_Fuel_Subsidies_Reform_in_China.pdf.

CIA. 2018. World Fact Book. U.S. Central Intelligence Agency. https://www.cia.gov/library/publications/the-world-factbook/geos/ch.html.

Forsythe, Michael. 2017. "China Aims to Spend at Least $360 Billion on Renewable Energy by 2020." *New York Times.* January 5. https://www.nytimes.com/2017/01/05/world/asia/china-renewable-energy-investment.html.

G20. 2018. "G20 Members and Participants." Organisation for Economic Co-operation and Development. http://www.oecd.org/g20/g20-members.htm.

G20 Leaders Statement. 2009. "The Pittsburgh Summit." September 24–25. http://www.g20.utoronto.ca/2009/2009communique0925.html.

GENI. 2016a. "Solar Energy in China." Global Energy Network Institute. http://www.geni.org/globalenergy/library/renewable-energy-resources/world/asia/solar-asia/solar-china.shtml.

GENI. 2016b. "Wind Energy Potential in China." Global Energy Network Institute. http://www.geni.org/globalenergy/library/renewable-energy-resources/world/asia/wind-asia/wind-china.shtml.

Greenpeace. 2017. "China Kept on Smashing Renewables Records in 2016." January 6. http://energydesk.greenpeace.org/2017/01/06/china-five-year-plan-energy-solar-record-2016/.

Hu, Yuanan and Hefa Cheng. 2013. "The Urgency of Assessing the Greenhouse Gas Budgets of Hydroelectric Reservoirs in China." *Nature Climate Change* 3: 708–712.

IEA. 2016. "International Energy Agency Fact Sheet: World Energy Outlook 2016." International Energy Agency. https://www.iea.org/media/publications/weo/WEO2016Factsheet.pdf.

IEEFA. 2017. "IEEFA Report: China Set to Dominate U.S. in Global Renewables Boom; $32 Billion in Overseas Investments in 2016 Alone." Institute for Energy Economics and Financial Analysis. http://ieefa.org/ieefa-report-china-set-dominate-%E2%80%A8global-renewable-energy-boom-expands-lead-u-s/.

Jaeger, Joel, P. Joffe, and R. Song. 2017. "China Is Leaving the U.S. Behind on Clean Energy Investment." Renewable Energy World. http://www.renewableenergyworld.com/articles/2017/01/china-is-leaving-the-us-behind-on-clean-energy-investment.html.
National Geographic. 2004. "China Facts." *Atlas of the World*, 8th ed. Washington, D.C.: National Geographic Society.
Oil Change International. 2018. "Fossil Fuel Subsidies: Overview." http://priceofoil.org/fossil-fuel-subsidies/.
Pandey, Avaneesh. 2015. "US Fossil Fuel Subsidies Increase 'Dramatically' Despite Climate Change Pledge." *International Business Times*. November 12. http://www.ibtimes.com/us-fossil-fuel-subsidies-increase-dramatically-despite-climate-change-pledge-2180918.
Reuters. 2017. "China to Plow $361 Billion into Renewable Fuel by 2020." January 4. http://www.reuters.com/article/us-china-energy-renewables-idUSKBN14P06P.
UCS. 2018. "Each Country's Share of CO2 Emissions." Union of Concerned Scientists. http://www.ucsusa.org/global_warming/science_and_impacts/science/each-countrys-share-of-co2.html#.WKOXwG8rIdU.
Vaughan. Adam. 2016. "China Tops WHO List for Deadly Outdoor Air Pollution." *The Guardian*. September 27. https://www.theguardian.com/environment/2016/sep/27/more-than-million-died-due-air-pollution-china-one-year.
WHO. 2016. "WHO Releases Country Estimates on Air Pollution Exposure and Health Impact." World Health Organization. September 27. http://www.who.int/mediacentre/news/releases/2016/air-pollution-estimates/en/.

Case Study 12: Citizen Science—Helping Scientists Understand Migratory Birds

Citizen science is public participation in the scientific process. Understanding environmental change is complex and requires huge amounts of data from vast areas across the globe. Scientists can gain a lot of information by linking their research with ordinary citizens. Citizen science projects involve the general public in the scientific process in order to gather information, increase monitoring, represent geographic variation, and assist scientists in their important work.

One aspect of citizen science began in the late 1990s and focuses on computers. Specifically, this version of citizen science uses people's computer downtime to help process big datasets. The University of California at Berkeley started the SETI@home project to use volunteers' computers to comb through lots of radio telescope data in the search for alien signals from outer space (Hand, 2010).

Building on this idea, these university computer engineers built computer software that could be used for these sort of generalized projects where people volunteer their help (Berkeley Open Infrastructure for Network Computing or BOINC). By the early 2000s, there were dozens of active BOINC projects and hundreds of thousands of users worldwide (Hand, 2010). But then the researchers realized that the human mind and peoples' skills could be as valuable as their computer processing. So they developed a program where people use their computers and also see shapes and help find solutions for scientific questions (in this case, variations in how amino acids can fold into protein). This idea, that everyday people could help scientists, sparked many exciting opportunities for citizen science (*Scientific American*, 2018).

LOCAL TO GLOBAL RESEARCH

Many citizen science projects have websites that explain the background on the research project and provide specific information on the research protocol, or the specific methods that citizens must take to participate in the program. These research projects vary by topic and geographic scale. Some people might help at the local level by gathering temperature information or collecting water samples from a nearby creek. Others might report in on national or international scientific research.

Across the globe, there are examples of citizen science. Some involve a few people or school groups, whereas others include millions of people collaborating on a project (Scistarter, 2018). The key point is that everyday people volunteer to be involved in the discovery of new scientific information (National Wildlife Federation, 2018). In fact, often public participation in data collection and reporting happens at such a large geographic scale that the research wouldn't otherwise be possible. Citizen science refers to volunteers helping with scientific data collection and then providing community members with relevant scientific information (CitizenScience.Org, 2018). Some types of science are particularly relevant to citizen science, such as wildlife studies that seek to track location, geographic range, and species diversity (Citizen Science Alliance, 2018; Zooniverse, 2018).

CORNELL LAB OF ORNITHOLOGY

The focus of this case study is the Cornell Lab of Ornithology, which manages many citizen science programs related to birds, including the popular Great Backyard Bird Count, eBird, Feederwatch, and NestWatch, each of which is described here (Cornell Lab of Ornithology, 2018).

The Cornell Lab of Ornithology defines citizen science as "[p]rojects in which volunteers partner with scientists to answer real-world questions" (Cornell Lab of Ornithology CC, 2018). They state that public participation in scientific research (PPSR) includes citizen science, volunteer monitoring, and other types of organized research activities. Specifically, members of the public engage in the process of scientific investigations: asking questions, collecting data, and/or interpreting results (Cornell Lab of Ornithology PPSR, 2018).

Ornithologists study birds. They are scientists in a subfield of zoology, or animal biology, which is the study of animals. Avian is another related term, which simply means "bird" or related to birds. So avian communities are groups of interacting birds (Ornithology.com, 2016).

As geographers, we are interested in the location, density, and distribution of various species. In addition, a geographer's special interest in understanding people and the environment is especially relevant to birds, as humans have greatly affected bird habitat and health. Think about everything that humans do, from cutting down trees and building roads, to lighting vast cities and spraying for insects, to having outdoor cats as pets. All of these actions can harm birds. Even more widespread problems are caused by climate change (Brahic, 2008).

A seven-year study published by the National Audubon Society warns that the migratory routes and habitats of more than half of the birds in North America are now or soon will be threatened by climate change (Audubon, 2015). Another report looked at 40 years of climate data and records from bird censuses conducted by the Audubon and the U.S. Geological Survey. Researchers compared changes in climate to shifts in bird migration patterns of 588 bird species in North America. They, too, discovered that migration paths and habitats of over half of these birds are threatened by climate change (NPR, 2014).

To understand the great diversity, geographical locations, and human impact on birds, we need to gather vast amounts of data. To study birds, citizen science provides wonderful opportunities for widespread data collection and reporting.

HOW TO GET INVOLVED

The **Great Backyard Bird Count** was initiated by the Cornell Lab of Ornithology and National Audubon Society in 1998. This was the first citizen science bird project to collect data online and display information in real time. Now, over 160,000 people across the globe participate in this four-day bird count in mid-February each year. Through this work, scientists learn a great deal about the geographic distribution and abundance of various bird species. Anybody can participate by counting the numbers and kinds of birds you see for at least 15 minutes on one or more days of the bird count. Volunteers can count from any location, anywhere in the world.

Participants set up a free online account and begin entering a checklist of bird sighting information. This has grown into such a popular program that in 2016, for example, volunteers in over 130 countries counted 5,689 species of birds on more than 162,000 checklists! Now The Great Backyard Bird Count has merged with eBird, another popular program from the same organizations so volunteers can continue to count birds all year long (GBBC, 2018).

Participants in **eBird** submit their observations through a simple website, available in English, French, and Spanish. They indicate when and where they went birding and then check off all the birds they saw and heard on the trip. Regional bird experts have developed special automatic data quality filters that review all the data entered on the eBird website before they go into the database. Local experts look into records that are flagged by the filters because they seem unusual. These local partner conservation organizations are very important to the broader citizen science effort. Rather than just having one faraway group managing the whole project, eBird's local expertise helps maintain the integrity of the whole project by focusing on specific goals from a local area. But the eBird data are fully integrated within the database, so data can be viewed and analyzed across geographic regions and political borders. In addition to this geographic collaboration from local to national to international scales, the eBird data are accessible to anybody through the eBird website and other related biodiversity websites worldwide (eBird, 2018) The eBird observation data are shared globally, and land managers can make better decisions because they can now understand bird numbers at specific geographic locations. This is particularly important because global assessments by the International Union for the Conservation of Nature indicate that one in eight species of birds is threatened with extinction (Brahic, 2008).

Citizen scientists contribute valuable data to research across the world. (Duram, 2017)

The **Feederwatch.org** website explains that counting birds is as easy as 1-2-3: 1) Install a bird feeder. 2) Count the birds that visit. 3) Enter the data for the scientists. Detailed information notes that volunteers must register online, decide on set days and times to observe their feeders, and learn the bird species from the convenient Feederwatch guide. The observation season is mid-November to early April—specifically, the annual project always starts on the second Saturday of November and runs for 21 weeks. New participants must pay a small fee for their new participant kit, which includes the FeederWatch handbook and instructions, a color poster of common feeder birds, and a Bird Watching Days Calendar to help keep track of count days. The Cornell Lab of Ornithology and Bird Studies Canada are nonprofit organizations that rely on donations to maintain the website and database for Project FeederWatch (Feederwatch, 2018).

NestWatch.org is a program that monitors birds' nests through the help of citizen scientists in order to discover the trends in reproductive biology of birds. Specifically the program tracks information on when nesting occurs, how many eggs are laid and hatched, and how many hatchlings survive. The NestWatch database can be used to study the current bird breeding conditions and to see how they may be changing over time as a result of climate change, habitat degradation and loss, expansion of urban areas, and the introduction of non-native plants and animals. To participate in NestWatch, volunteers must do the following: 1) Get certified as a NestWatch monitor by studying their online resources and taking a test. 2) Search and find active nests to monitor. 3) Visit these nests every three or four days and record what they see. 4) Report their data using the NestWatch.org online data entry tools (Nestwatch, 2018).

LOOKING AHEAD (OR OVERHEAD)

Anybody can be a citizen scientist and volunteer their time to help scientists with their research. Because thousands or even millions of people can get involved, this helps scientists gather a huge amount of data and then spread valuable ideas faster than if they had to work alone.

The Cornell Lab of Ornithology and National Audubon Society began eBird in 2002 with the goal of providing rich data on bird numbers and distribution across geography and time scales. Their slogan is "global tool for birders, critical data for science," which nicely summarizes the program: people record the birds they see, keep track of this information on bird lists, share their sightings with the eBird community, and contribute to science and conservation (eBird, 2018). To help with identifying different bird species, check out the smart phone app called Merlin Bird ID. Just answer five simple questions about a bird, and Merlin will come up with a list of possible matches (Merlin, 2018).

Citizen science projects like the ones described here are truly a local-to-global effort. Because we are all linked through computers and the Internet, we can each gather information and upload it to a central research site, which may be located across the country or even across the globe. The eBird data are used to produce "Bird Cast" real-time bird migration forecasts. These maps and data allow anyone to look up what birds are migrating in their area at a given time. Individual citizen scientists participate by gathering information and then can learn more by tracking trends in their region, continent, or even across the whole planet. These types of tools are particularly relevant to environmental geographers who seek to understand how human influences and ecological factors vary from place to place. Given the urgency of declining bird populations and increasing disruptions from climate change, this is a particularly important time to participate in helping to better understand bird species worldwide (Audubon, 2015).

Anybody can get involved in citizen science. With a bit of training and careful gathering of data, people across the planet can work together to advance the scientific knowledge base. In the examples given here, people can help provide valuable information on birds. Whether a beginner or experienced environmentalist, the citizen science projects want everyone to get involved. Get online and learn how you can participate!

See also: Section 1: Biodiversity Loss, Deforestation, Endangered Species; *Section 2:* Habitat and Wildlife, Parks and Urban Green Space, Protected Areas and National Parks; *Section 3:* Earth Day

FURTHER READING

Audubon. 2015. "Audubon's Birds and Climate Change Report." National Audubon Society. http://climate.audubon.org/.
Brahic, Catherine. 2008. "One in Eight Bird Species Threatened by Climate Change." *New Scientist*. May 19. https://www.newscientist.com/article/dn13935-one-in-eight-bird-species-threatened-by-climate-change/.
Citizen Science Alliance. 2018. "What Is the Citizen Science Alliance?" http://www.citizensciencealliance.org/.

CitizenScience.org. 2018. "The Power of Citizen Science." http://citizenscience.org/.
Cornell Lab of Ornithology. 2018. "Exploring and Conserving Nature." http://www.birds.cornell.edu.
Cornell Lab of Ornithology CC. 2018. "Defining Citizen Science." Citizen Science Central. http://www.birds.cornell.edu/citscitoolkit/about/definition.
Cornell Lab of Ornithology PPSR. 2018. "What Is Citizen Science and PPSR?" Citizen Science Central. http://www.birds.cornell.edu/citscitoolkit/about/defining-citizen-science/.
eBird. 2018. "Welcome to eBird." Audubon and Cornell Lab of Ornithology. http://ebird.org/content/ebird/.
Feederwatch. 2018. "Project Feederwatch." Cornell Lab of Ornithology. http://feederwatch.org/.
GBBC. 2018. "The Great Backyard Bird Count." The Cornell Lab, Audubon, Bird Studies Canada. http://gbbc.birdcount.org/.
Hand, Eric. 2010. "Citizen Science: People Power." *Nature* 466: 685–687.
Merlin. 2018. "Merlin Bird ID App." Cornell Lab of Ornithology. http://merlin.allaboutbirds.org/.
National Wildlife Federation. 2018. "Citizen Science." https://www.nwf.org/Educational-Resources/Wildlife-Guide/Understanding-Conservation/Citizen-Science.
Nestwatch. 2018. "Nest Watch: Where Birds Come to Life." Cornell Lab of Ornithology. http://nestwatch.org/.
NPR. 2014. "More than Half of U.S. Bird Species Threatened by Climate Change." September 9. *Morning Edition*, National Public Radio. http://www.npr.org/2014/09/09/345833757/more-than-half-of-u-s-bird-species-threatened-by-climate-change.
Ornithology.com. 2016. "Ornithology: The Science of Birds." http://ornithology.com/.
Scientific American. 2018. "Citizen Science." www.scientificamerican.com/citizen-science/.
Scistarter. 2018. "What Is Citizen Science?" https://scistarter.com/citizenscience.html.
Zooniverse. 2018. "What Is the Zooniverse?" https://www.zooniverse.org/about.

Case Study 13: Costa Rica's Peace with Nature— Conservation, Biodiversity, and Sustainability

The nation of Costa Rica has made a concerted effort to protect its natural environment. Article 50 of their Constitution states that "[t]oda persona tiene derecho a un ambiente sano y ecológicamente equilibrado." This translates to "[e]very person has the right to a healthy and ecologically-balanced environment" (CostaRica.com, 2015; República de Costa Rica, 2018).

Indeed, this small Central American nation, with rich ecological diversity, is guided by this goal. Costa Rica contains 5% of the world's species biodiversity, even though its land area is only about 0.03% of the planet (Embassy of Costa Rica, 2018). Indeed, it is known globally as an example of a nation with excellent policies that succeed in both protecting the environment and building a strong economy linked to nature.

Although there are many valuable lessons to be learned from Costa Rica, this case study will focus on how long-term commitment to environmental policies can succeed in protecting biodiversity and developing renewable energy. Environmental geographers are especially interested in seeing the outcome of long-term policies that prioritize both people and the environment.

Biodiversity is plentiful in Costa Rica. (Bathgate, 2017)

BACKGROUND

Costa Rica is in Central America, with coasts on both the Pacific Ocean and Caribbean Sea. Nicaragua is to the north, and Panama is on its southern border. Costa Rica is 51,100 square kilometers (19,730 square miles), or about the size of Kentucky, with a population of 4.5 million. National parks and reserves cover about 26% of the nation's land area (13.42 square kilometers or 518 square miles) (Embassy of Costa Rica, 2018).

The climate is tropical and subtropical with a dry season (December to April) and a rainy season (May to November). It is cooler in the interior highlands. The coastal flats are separated by rugged mountains that include over 100 volcanic cones, several of which are active volcanoes.

Costa Rica declared its independence from Spain in 1821 and had a fairly clear path toward the development of a democracy. In 1949, Costa Rica abolished its military and instead decided to use these funds for education, health care, and environmental initiatives. It now has a strong ecotourism industry and growth in technology jobs.

Costa Ricans enjoy a high standard of living, political stability, and an excellent social services system. The government has consistently spent 20% of gross domestic product (GDP) to reach its goals of providing health care, education, safe water, sanitation, and electricity to all people. Costa Rica spends 6% of its GDP on education (Embassy of Costa Rica, 2018), which is quite high compared to many developed countries. Indeed, Costa Rica has high life expectancy, low infant mortality, and a low birth rate. In fact, these demographic numbers are all very similar

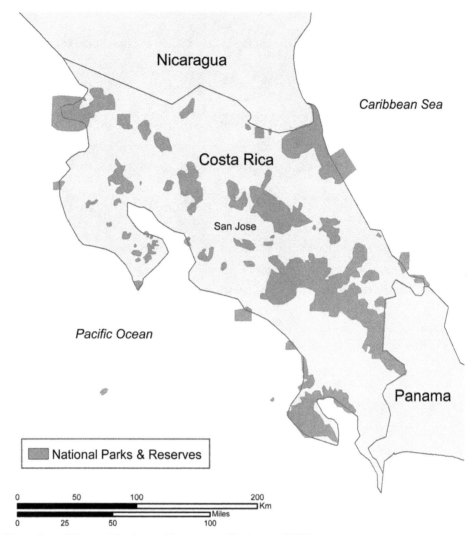

Costa Rica: National Parks and Reserves. (Bathgate, 2017)

to the United States. Costa Rica's poverty rate is lower than other Central American nations and is in line with many developed countries.

A SUCCESS STORY

The Constitution of Costa Rica, implemented in 1949, lays out the framework for national environmental policy. Today, we can see success in several realms.

First, biodiversity is promoted through reforestation incentives such as government-subsidized loans, tax credits, and direct payments that benefit both small and large landowners. In terms of administration, forest resource use and protected areas are merged, so planning involves both timber harvest (cutting trees) and recreation (enjoying rainforests and habitats) within one agency. There are protected areas,

Case Study: Costa Rica's Peace with Nature

Forest preserves are home to diverse wildlife in Costa Rica. (Bathgate, 2017)

with government management offices, in each geographic region of the country, which promotes forest protection. Science, ecology, and forestry are emphasized in schools, and Costa Rican landowners can get a lot of technical support. Because of this education, many conservation efforts are actually the responsibility of individual citizens. Further, Costa Rica has an office that deals with trading carbon emissions in the international market, so other countries or multinational corporations pay Costa Rica to protect its forests (De Camino et al., 2000).

Costa Rica won the 2010 Future Policy Award at the Global Summit on Biodiversity. The prize was given for its 1998 Biodiversity Law Number 7788 (Ley de Biodiversidad, Ley 7788) of 1998, and its 2008 amendments are recognized as among the best national biodiversity laws in the world. These are considered a model for other nations to follow (Watts, 2010).

Costa Rica funds nature reserves from a fuel tax, a car tax, and energy fees. The nation considers this essential to protecting its ecosystem services like biodiversity, clean air, and clean water. Costa Rica set up a national biodiversity collaborative with scientists, government workers, indigenous people, and others working together to develop long-term solutions. With these major actions, Costa Rica was able to increase forest cover from 24% in 1985 to 46% in 2010 (Walker, 2017). It is also worth noting that Costa Rica ranks first in the Happy Planet Index and high in numerous environmental rankings (Watts, 2010).

Second, renewable energy production is soaring. In recent years, Costa Rica has come very close to relying solely on renewable sources for its national electricity grid. In 2015, for example, the country ran on a combination of hydropower, geothermal, wind, and solar energy for 299 days (Costa Rica News, 2016). No other country of similar size has come close to this. Some countries work toward being carbon neutral, which means they achieve net-zero carbon emissions by burning some fossil fuels but then planting trees to offset this usage. On the other hand, Costa Rica's 100% renewable days means that their electricity is created entirely from these infinite resources.

How has Costa Rica done this? The Costa Rican natural environment provides many opportunities. Early on, the government invested in renewable energy for power generation. Because of abundant rainfall, hydroelectric plants produce a lot of energy. Geothermal renewable energy is also present here, due to underground steam from volcanoes. Solar and wind power are an increasingly important part of the national energy portfolio.

To its advantage, the country has a relatively small population and an economy that is not focused on manufacturing. In any case, the Costa Rican government makes renewable energy a priority. There is a new program that promotes solar panels and other renewables by allowing consumers to store their surplus power in the national grid (Arias, 2016). In 2016, the nation set up four new wind farms and built a 305-MW hydroelectric plant that will power 525,000 homes (Science Alert, 2016). Most recently, Costa Rica began construction of three 50-MW geothermal power plants, costing nearly a billion dollars (Costa Rica News, 2016). This is a huge commitment for a small country, but it shows they are investing in the future—and that future is green.

NATIONAL ENVIRONMENTAL POLICY

Costa Rica has been successful in protecting biodiversity and developing a renewable energy electricity grid due to their long-term environmental policies. The nation has made a long-term commitment to environmental sustainability, as seen in both policy implementation and in community participation. Demilitarizing the nation 70 years ago meant that all of these defense funds can be concentrated into education, health care, and environmental protection. As stated in the Costa Rican Constitution, people have a right to a clean environment (CostaRica.com, 2015).

Costa Rica's environmental policies are based on the Organic Environmental Act of 1995 (Ley Orgánica del Ambiente General No. 7554), which spells out specific details about conservation, sustainable use of natural resources, and inclusive economic development. This law also expands public participation in the management of biodiversity and the fair sharing of benefits derived from it. Further, the law states that whoever pollutes or damages the environment is subject to penalty. Law 7554 established the National Environmental Council as an advisory group that reports directly to the president of Costa Rica. This emphasizes the key role that environmental issues play in the national mind-set (Cordero and Ryan, 2016).

According to this environmental act, the lead agency for environmental policy is the Ministry of Environment, Energy and Telecommunications (MINAE), in Spanish, the Ministerio de Ambiente, Energía y Telecomunicaciones. This ministry has the responsibility of managing natural resources and protecting the environment, with 10 agencies under its control, overseeing everything from meteorological data to marine parks, but including forestry and biodiversity protections.

The Costa Rican Ministry of the Environmental (MINAE) divided the country into 11 geographic areas to be managed by the National System of Conservation Areas (SINAC). The powerful Environmental Court system, which is strict against violators, is particularly involved in slowing development and urbanization that can harm habitats high in biodiversity (Cordero and Ryan, 2016). Public participation is encouraged, and the agency works to set policies to achieve sustainable natural resources management. Education plays a key role in conservation efforts in Costa Rica. Because national policies emphasize education, science and conservation are taught early on (De Camino et al., 2000). These are examples of how environmental protection is integrated throughout Costa Rican policies and engrained in society.

INTEGRATED POLICY LEADS TO SUCCESS

In terms of renewable energy, biodiversity, and more, Costa Rica provides examples of environmental policies worth noting. By integrating global initiatives, national goals, and local actions, Costa Rica illustrates the balance environmental geographers seek.

For example, the award-winning Costa Rica Biodiversity Law is closely aligned with the UN Convention on Biological Diversity. The stated objective of this UN law is to protect the "conservation of biodiversity and the sustainable use of the resources as well as to distribute in an equitable manner the benefits and derived costs" (WIPO, 2017).

This is notable in several aspects. Costa Rican law works for its own interests and for global benefit. It promotes incentives and rewards for ecological actions gained from conservation and sustainable biodiversity. It sets up preventive and precautionary principles and states that biodiversity should be used in the public interest, including the interest of future generations. Thus, the Costa Rican government has integrated environmental goals across other policy areas, such as the economy and education. There are even ethical issues to be learned from these environmental policies (Blasiak, 2011).

One Costa Rican official explained: "We are declaring peace with nature" (Watts, 2010). This comment refers to the fact that Costa Rica disbanded its military, and has thus focused on environmental protection and social well-being.

See also: Section 2: Ecotourism, Habitat and Wildlife, Protected Area and National Parks; *Section 3:* Environmental Nongovernmental Organizations, Environmental Policy, Renewable Energy

FURTHER READING

Arias, L. 2016. "Costa Rica's Electricity Mostly from Renewable Sources in 2016." *Tico Times*. April 16. http://www.ticotimes.net/2016/04/16/costa-rica-renewable-energy.

Blasiak, Robert. 2011. "Ethics and Environmentalism: Costa Rica's Lesson." United Nations University. https://ourworld.unu.edu/en/ethics-and-environmentalism-costa-ricas-lesson.

Cordero, Roberto and Lindsay Ryan. 2016. "Environment & Climate Change Law: Costa Rica." International Comparative Legal Guides. http://www.iclg.co.uk/practice-areas/environment-and-climate-change-law/environment-and-climate-change-law-2016/costa-rica.

Costa Rica News. 2016. "Costa Rica Used 100% Renewable Energy for 299 Days." October 7. https://thecostaricanews.com/costa-rica-used-100-renewable-energy-299-days/.

CostaRica.com. 2015. "Costa Rica Constitution." Culture. https://www.costarica.com/culture/costa-rica-constitution/.

De Camino, Ronnie, Olman Segura, Luis Guillermo Arias, and Isaac Pérez. 2000. "Costa Rica: Forest Strategy and the Evolution of Land Use. Evaluation Country Case Study Series." The World Bank. https://ieg.worldbankgroup.org/Data/reports/cstrcacs.pdf.

Embassy of Costa Rica. 2018. "Environment." http://www.costarica-embassy.org/?q=node/12.

República de Costa Rica. 2018. "Constitución Política de 1949 con reformas hasta 2003." Political Database of the Americas, Center for Latin American Studies, Georgetown School of Foreign Service. http://pdba.georgetown.edu/Constitutions/Costa/costa2.html.

Science Alert. 2016. "Costa Rica Has Been Running on 100% Renewable Energy for 2 Months Straight." September 7. http://www.sciencealert.com/costa-rica-has-been-running-on-100-renewable-energy-for-2-months-straight.

Walker, Peter. 2017. "Costa Rica's Electricity Was Produced Almost Entirely from Renewable Sources in 2016." *The Independent*. January 2. http://www.independent.co.uk/environment/costa-rica-renewable-energy-electricity-production-2016-climate-change-fossil-fuels-global-warming-a7505341.html.

Watts, Jonathan. 2010. "Costa Rica Recognised for Biodiversity Protection." *The Guardian*. October 25. https://www.theguardian.com/environment/2010/oct/25/costa-rica-biodiversity.

WIPO. 2017. "Costa Rica Biodiversity Law No. 7788." World Intellectual Property Organization. http://www.wipo.int/wipolex/en/details.jsp?id=896.

Case Study 14: Denmark's Achievements in Organic Agriculture

Denmark is leading the way in organic farming, due to governmental policies that emphasize small-scale, environmentally friendly food production. For decades, Denmark has looked to the future, set goals, and followed through on achieving a significant shift toward organic agriculture.

Denmark has had its own national organic brand for over 25 years, making it well recognized both within and outside the country. Plus, the Danish government has worked to shift agricultural production to organic and sustainable methods for decades. Most recently, in 2015, the Danish Action Plan outlined 67 points, with the goal of doubling organic farming by 2020. Other countries can learn from this forward-thinking example.

Organic farming is based on healthy soils, crop diversity, and integrated methods that do not allow the use of synthetic chemicals for fertilizer and pesticides. It is rooted in ecology, and due to its smaller scale of production, organic farming also promotes the social benefits of linking farmers and consumers. So this topic presents the juncture of ecological and social goals, which is so fascinating for environmental geographers.

DENMARK'S ORGANIC FARMING LAWS

Denmark is a modern, prosperous country in northern Europe, with a population of 5.7 million people. Its land area is 43,094 sq km (16,638 sq mi), making it about twice the size of Massachusetts (CIA, 2018). Although the country's total boundary length is 7,314 km (4,545 mi), only 68 km (42 mi) is the land boundary with Germany—the remainder is water boundaries in the Baltic Sea and North Sea. The climate is temperate, mostly cool, humid, and overcast, with mild, windy winters and cool summers; the landscape is low and flat with gently rolling plains.

This country has the highest percentage of arable land compared to any other nation, with approximately 59% (CIA, 2018). Danish people have a high life expectancy, because of generous government funding of education and health care. According to the World Happiness Report by the United Nations, Denmark ranks first because of its high minimum wage, good economy, extensive health care, broad social programs, and high level of individual freedom (Kramer, 2016).

Denmark was one of the first countries to establish organic farming laws, starting in 1987 (DVFA Website, 2015). European organic regulations were established in 1991, and since that time Denmark has adhered to those laws (EEC, 1991; Organic

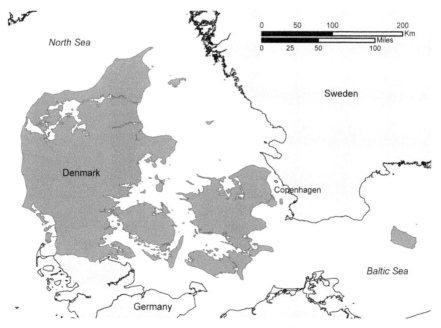

Denmark Is a Country in Northern Europe Composed of Several Islands. (Bathgate, 2017)

Europe, 2015). Danish officials emphasize the importance of "efficient state inspection from stable to table," which ensures that all food that is labeled organic is truly produced with organic methods. The Danish Organic label and logo are well known throughout the country, with nearly 100% of Danes recognizing the red certified Ø-label (Frederiksen, 2016; Organic Europe, 2015). Now, organic products are sold at farm stores, farmers' markets, online, and overwhelmingly at "ordinary places of purchase, such as supermarkets" (DVFA website, 2015).

In 2015, the Danish Ministry of Environment and Food was created by merging the Ministry of the Environment and the Ministry of Food, Agriculture and Fisheries of Denmark. It is notable that for the Danish government, food is very closely linked to environmental quality. The ministry has local offices throughout the country, which oversee environmental protection and farming programs (DMEFD, 2018).

The Danish Agriculture and Food Council wrote an informative and readable report that presents a very positive view of organic food and farming in terms of both environmental concerns and health topics. The report notes that organic products prohibit the use of synthetic colors, flavors, sweeteners, genetically modified organisms (GMOs), and irradiation (DAFC, 2015: 8). In addition, "organic farming is an agricultural production system designed to take the best possible care of the environment, while ensuring that the agricultural system operates as naturally as possible" (DAFC, 2015: 4). This report continues by explaining that organic farmers protect water quality, soil fertility, and biodiversity through planting a wide variety of crops rather than just one type of crop over and over. These approaches help break the cycle of pests and weeds and allow soil to recover its nutrients that are drawn out by any one crop. They do not use chemical pesticides or synthetic fertilizers, which can run off farm fields and poison nearby creeks and lakes. Fields are kept smaller. Natural trees and shrubs line the field to provide natural habitat (DAFC, 2015).

ORGANIC ACTION PLAN FOR DENMARK

Denmark's commitment to organic is shown in its consumption and agricultural statistics. Of the total food consumed in Denmark, 8% is organic, which is the highest proportion of any country. In the United States, for example, only about 4% of food is organic (USDA-ERS, 2017). Specifically, 20% of eggs, 11% of fruit, and 10% of all vegetables consumed in Denmark are organic (DAFC, 2018).

To keep up with this demand, Denmark's organic farms have grown at an unparalleled rate, averaging 22% annual increases since 2003 (Hirsh, 2014). Approximately 7% of Danish farmland is certified organic, and the government plans to more than double this, to 15% by 2020 (DAFC, 2018; DVFA Plan, 2015).

The Danish Ministry of Environment and Food announced a bold new strategy in 2015 with the goal of greatly increasing organic farming production and organic food consumption in their nation. The 67-point plan called Økologiplan Danmark (Organic Action Plan for Denmark) seeks to provide strong government support in order to achieve these goals. "With Økologiplan Danmark, we will strengthen cooperation between municipalities, regions and ministries with a long line of new initiatives. We will commit ourselves to, among other things, have more organic items on the menu in canteens, hospitals and daycare institutions," stated Minister of Food and Agriculture Dan Jørgensen (The Local, 2015).

The Organic Action Plan for Denmark has delineated significant goals for increasing organic farming within its borders that include local, regional, and international actions. Internationally, the plan will promote exports of Danish products abroad. China, for example, is seen as a big potential market. Regionally, the plan encourages Denmark's government institutions, which serve 800,000 meals each day, to go organic (Sørensen et al., 2015). It will also promote public consumption of organic food through sales promotions and promoting more organic food brands.

Locally, the government will provide assistance to organic farmers by investing in organic education, advisory services, and funding an organic production research center. The government will also help farmers make the transition to organic methods by subsidizing farmland with a per-hectare (acre) payment for certified organic land. The Action Plan notes that organic agriculture has more risk, as farmers must learn to work with nature, rather than try to control it with chemicals. This risk can lead to economic uncertainty. So the government will help the organic sector through agro-ecological actions and scientific assistance (DVFA Plan, 2015).

THE EU AND ORGANIC FARMING

Denmark is part of the European Union (EU), so many of its organic, agricultural, and other policies are related to the EU's broader rules and goals.

The EU is a political and economic collaboration between 28 European countries. It was established after World War II, with economic interdependence designed to rebuild a stable continent. So, by 1958, the European Economic Community (EEC) brought economic cooperation between just six countries: Belgium, Germany, France, Italy, Luxembourg, and the Netherlands. Then, in 1973, Denmark, Ireland, and the United Kingdom joined, making it nine member states. As more countries joined, the goals expanded from just economics to broader policy collaborations, including topics like health, security, and the environment. In 1993, the name was changed to the European Union (EU) to reflect its new outlook. This is a unique collaboration, as everything the EU does is founded on treaties that are voluntary and based on representative democracy, with citizens from each country represented at the European Parliament, and each nation is represented at the European Council (European Union Information, 2018).

As part of its health and environmental policies, organic agriculture has been strongly supported by the EU. "Organic has become a way of living" (European Commission, 2016: 4).

In 1991, the Europeans adopted EEC Regulation 2092/91 on organic farming and labeling of organic foods. This was expanded and now covers many things, including both plant and animal products. A few key provisions are the prohibition of genetically modified organisms (GMOs) and a two- to three-year transition to organic farming from farming with synthetic chemicals (EEC, 1991). These rules laid out minimum requirements, but each country could introduce stricter standards if they wished. Because there have been constant updates, the EU organic regulations are very long and complicated (European Commission, 2018).

Organic agriculture has expanded rapidly in recent years. Now 11.1 million hectares (27.4 million acres) are cultivated as organic, which has more than doubled in the past decade, with annual increases of about 500,000 hectares (1.2 million acres). In 2015, there were 185,000 organic farms in the EU, and a total of over 300,000 registered operators, which includes producers, processors, and importers.

But certified organic farming is not the only way to shift agricultural production. Other parts of the world also recognize the importance of smaller-scale, ecologically based agriculture.

ORGANIC FARMING AND FOOD SECURITY

Organic farming uses methods similar to agroecology. According to a special report to the United Nations: "As a set of agricultural practices, agroecology seeks ways to enhance agricultural systems by mimicking natural processes, thus creating beneficial biological interactions and synergies among the components of the agroecosystem" (United Nations, 2010).

Whereas some industrial nations consistently subsidize large-scale, chemically intensive agricultural production, claiming that it is the only way to feed the world's growing populations, there is increasing evidence to the contrary. Indeed, the United Nations has published several reports that underscore the need for sustainable, organic agriculture to "feed the world" (United Nations, 2011).

A 2008 report stated that food security in Africa must be based on organic, small-scale agriculture, where crop yields and production would increase significantly if it shifted toward these methods (United Nations, 2008: 16). The 2010 UN report specifically noted that small-scale farming is the best way to produce enough food for the world's growing population, instead of genetically modified, large-scale industrial operations (UN, 2010). Then the UN published "Trade and Environment Review 2013: Wake Up Before It Is Too Late," a report with contributions from more than 60 experts around the world. This report stated that major changes are necessary: agricultural production, food, and trade must all shift toward smaller, diversified, localized approaches. Specifically, these experts recommend "a rapid and significant shift from conventional, monoculture-based and high-external-input-dependent industrial production toward mosaics of sustainable, regenerative production systems that also considerably improve the productivity of small-scale farmers" (United Nations, 2013: 2).

Thus from the global scale to the local, experts and policy makers see the advantages of agriculture that works with nature. Denmark is wisely focused on organic farming to promote both environmental and social sustainability.

See also: Section 1: Agriculture; *Section 2:* Ecosystem Services; *Section 3:* Alternative Agriculture, Assessments and "Footprints," Composting, Environmental Policy, Green Consumerism, Sustainable Cities, Sustainable Diet

FURTHER READING

CIA. 2018. "Denmark." World Fact Book. https://www.cia.gov/library/publications/the-world-factbook/geos/da.html.

DAFC. 2015. "All about Organics: Danish Food that Matters." Danish Agriculture and Food Council. http://www.agricultureandfood.dk/danish-agriculture-and-food/organic-farming/all-about-organics#.

DAFC. 2018. "Organic Farming." Danish Agriculture and Food Council. http://www.agricultureandfood.dk/danish-agriculture-and-food/organic-farming.

DMEFD. 2018. "The Ministry." Danish Ministry of Environment and Food of Denmark. http://en.mfvm.dk/the-ministry/.

DVFA Plan. 2015. "Organic Action Plan for Denmark." Danish Veterinary and Food Administration. https://www.foedevarestyrelsen.dk/english/SiteCollectionDocuments/Kemi%20og%20foedevarekvalitet/Oekologiplan%20Danmark_English_Print.pdf.

DVFA Website. 2015. "Organic Food." Danish Veterinary and Food Administration. https://www.foedevarestyrelsen.dk/english/Food/Organic_food/Pages/default.aspx.

EEC. 1991. "Council Regulation (EEC) No 2092/91." European Economic Community. June 24. http://eur-lex.europa.eu/LexUriServ/LexUriServ.do?uri=CONSLEG:1991R2092:20080514:EN:PDF.

European Commission. 2016. "Facts and Figures on Organic Agriculture in the European Union." Agricultural and Farm Economics Briefs, Number 14. https://ec.europa.eu/agriculture/sites/agriculture/files/rural-area-economics/briefs/pdf/014_en.pdf.

European Commission. 2018. "Organic Farming." Agriculture and Rural Development. February 20. https://ec.europa.eu/agriculture/organic/index_en.

European Union Information. 2018. "The EU in Brief." https://europa.eu/european-union/about-eu/eu-in-brief_en.

Frederiksen, Anders. 2016. "The Danish Organic Label." Organic Denmark. September 1. http://organicdenmark.com/organics-in-denmark/the-danish-organic-label.

Hirsh, Jesse. 2014. "Ask an Ag Minister: Denmark's Dan Jørgensen." Modern Farmer. March 21. http://modernfarmer.com/2014/03/ask-ag-minister-denmarks-dan-jorgensen/

Kramer, Sarah. 2016. "Here's Why People in Denmark Are Happier Than Anyone Else in the World." *Business Insider*. March 17. http://www.businessinsider.com/denmark-worlds-happiest-country-2016-3.

The Local. 2015. "Denmark Launches 'Most Ambitious' Organic Plan." January 30. https://www.thelocal.dk/20150130/denmark-announces-most-ambitious-organic-plan.

Organic Europe. 2015. "Country Report—Denmark." Knowledge Centre for Agriculture. Research Institute of Organic Agriculture. www.organic-europe.net/country-info/denmark/country-report.html.

Sørensen, N., A. Lassen, H. Løje, and I. Tetens. 2015. "The Danish Organic Action Plan 2020: Assessment Method and Baseline Status of Organic Procurement in Public Kitchens." *Public Health Nutrition* 18(13): 2350–2357.

United Nations. 2008. "Organic Agriculture and Food Security in Africa." UNEP-UNCTAD Capacity Building Task Force on Trade, Environment and Development. http://unctad.org/en/Docs/ditcted200715_en.pdf.

United Nations. 2010. "Report Submitted by the Special Rapporteur on the Right to Food, Olivier De Schutter." December 17. http://www2.ohchr.org/english/issues/food/docs/A-HRC-16-49.pdf.

United Nations. 2011. "Eco-Farming Can Double Food Production in 10 Years." March 8. http://www.srfood.org/images/stories/pdf/press_releases/20110308_agroecology-report-pr_en.pdf.

United Nations. 2013. "Trade and Environment Review 2013: Wake Up Before It Is Too Late." United Nations Conference on Trade and Development. http://unctad.org/en/PublicationsLibrary/ditcted2012d3_en.pdf.

USDA-ERS. 2017. "Organic Market Overview." United States Department of Agriculture-Economic Research Service. https://www.ers.usda.gov/topics/natural-resources-environment/organic-agriculture/organic-market-overview.

Case Study 15: Great Green Wall of Africa

This environmental geography case study exemplifies collaboration across a continent, with specific initiatives tailored to address the crucial combination of environmental degradation and social sustainability in each local place.

The Great Green Wall of the Sahara and the Sahel Initiative (GGWSSI) is a project that began as a plan to plant a wall of trees across Africa at the southern edge of the Sahara Desert to slow or prevent desertification. The goals have expanded over the years, and now the GGWSSI seeks to reduce the detrimental social, economic, and environmental impacts of land degradation felt among people in the Sahel and the Sahara. This is a major African-led initiative that aims to build resilient landscapes to address the challenges of climate change in the region (LPAA, 2018).

According to the African Union (2018: 3): "The overall goal of the Great Green Wall initiative is to strengthen the resilience of the region's people and natural systems with sound ecosystems management, sustainable development of land resources, the protection of rural heritage and the improvement of the living conditions of the local population."

DESERTIFICATION

According to the UN Convention to Combat Desertification, desertification is "land degradation in arid, semiarid and dry subhumid areas resulting from various factors, including climatic variations and human activities" (UNCCD, 2018). This does not mean the actual advancement of sandy deserts, but rather persistent degradation of dryland ecosystems caused by human activities like overgrazing, unsustainable farming, clear-cutting trees, and anthropogenic climate change. Indeed, the Report of the Millennium Ecosystem Assessment (2005) states that land degradation is the reduction of biological or economic productivity of drylands (WRI, 2005).

Drylands take up 41% of Earth's land area, and more than 2 billion people (nearly one-third of the human population) live in these areas. Sadly, some of the world's most poverty-stricken areas are located in these arid regions. An estimated 74% of the world's poor are affected by land degradation. Given such a broad impact, it is not surprising that desertification can be studied at every geographic scale. Indeed, the negative effects of desertification are experienced locally, nationally, regionally, and globally.

Soil is a living thing, composed of animals, plants, fungi, and microorganisms. Once this living soil is disrupted in dryland areas, wind and water erosion worsen problems by carrying away topsoil and leaving just infertile dust and sand. The combination of human actions and environmental processes transforms healthy landscapes into degraded land and eventually to desert that is unable to grow vegetation.

Droughts have increased significantly since the 1970s, with double the land area now experiencing severe droughts. It is estimated that 12 million hectares (29.7 million acres) are lost each year to drought and desertification, and on this land 20 million tons of grain could have been grown. This staggering number indicates that this is clearly both a social concern and an environmental disaster (UN, 2018).

At the global level, the United Nations Convention to Combat Desertification (UNCCD) was established in 1994. It is a legally binding international agreement that focuses on sustainable land management to promote both environmental protection and economic development. The UNCCD countries work together in arid regions to improve soil health, restore lands, and mitigate the effects of drought (UNCCD, 2018). For example, they have experts who can advise and demonstrate innovations that can be tailored to the individual country level.

THE GREAT GREEN WALL

The Great Green Wall is an effort to combat desertification in Africa, specifically along the southern edge of the Sahara Desert, which reaches 7,700 kilometers (4,785 miles) from the Atlantic Ocean on the west to the Red Sea on the east.

Originally, the main effort was concentrated on planting a 15-kilometer-wide (9 miles) band of trees from Djibouti City, Djibouti (in the east) to Dakar, Senegal (in the west) (Powers, 2014). In recent years, the idea of this simple belt of trees has expanded to include related projects in other North and West African countries. Now 20 countries are participating—each with individual local action plans.

This idea took hold in 2005, when the African Union initiated the idea of planting a wide band of trees along the southern part of the Sahara desert to block wind and improve the soil quality by adding nutrients and binding sediment together (Laing, 2016). Soon many African leaders became interested, as they saw the initiative as a possible action to combat some of the social and environmental issues faced in the Sahel-Sahara region where over 80% of the population depends directly on the land for their survival. The people here are farmers and herders, many of whom rely on trees for fuel wood to heat their homes and cook their food. In this region, it is estimated that about half of the land is degraded from human activity, soil erosion, and rising temperatures (*The Economist*, 2016).

The idea of a green belt of trees gained the support of many organizations and the African Union, which is a consortium of 54 countries that seeks unity and solidarity among African countries. So, the Great Green Wall Initiative concept gained in popularity, and its goals and programs have evolved.

Of course, tree planting is part of the initiative, but many other goals have grown in importance: reducing the impacts of climate change, increasing food

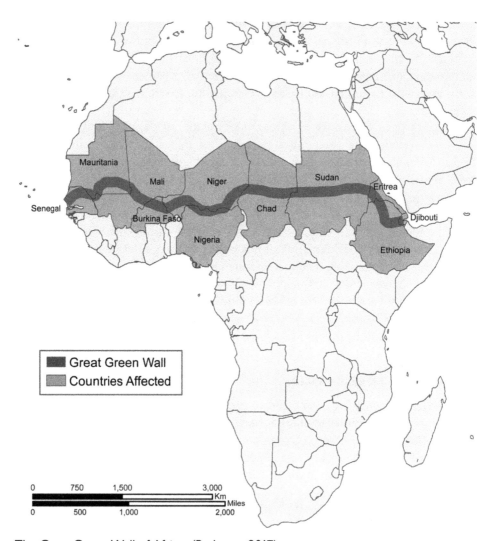

The Great Green Wall of Africa. (Bathgate, 2017)

supplies, and providing economic growth. Stated goals include restoration of 50 million hectares of land, creating 350,000 new jobs, improving food security for 20 million people, and sequestering 250 million tons of carbon (Monks, 2016).

So each country has looked within and determined its own goals; then the bigger overarching project works to promote shared goals (*The Economist*, 2016). Individual countries may focus on community development, education, farmer training, business development, water management, agro-forestry, and other goals that are locally specific. Together these actions combine to create the collective Great Green Wall Initiative, which could provide resiliency to the region.

The initiatives are based on integrated landscape approaches that allow each country to address land degradation, climate change adaptation and mitigation, biodiversity, and forestry within its local context. Time and again, examples of low-cost, grassroots initiatives have been very successful.

Specific country action plans have been developed and shared among partner countries. Here are a few:

Niger is one of the poorest countries in the world, but farmers in this central African country have worked to restore huge areas of desert land with very little money. Niger's farmers are regenerating the land using innovative methods such as digging "half moon" pits to store water, stimulating tree roots to revive them, and carefully caring for plants during drought time. These methods have helped restore about 200 million trees and 5 million hectares (19,000 square miles) of farmland. It is estimated that this will feed over 2 million people. In Niger, a key factor is that local people have control of the project, and this empowered small land owners to protect the trees and restore the land. This is called farmer-managed natural regeneration, or FMNR, and many organizations promote this approach over a large national government controlling the projects. These FMNR practices have now spread to Malawi and Ethiopia (FAO, 2018).

In **Mali**, farmers also use FMNR and have fully embraced agro-forestry, or the integration of trees into their cropping. They note the many benefits of planting trees: shade, livestock fodder, drought protection, firewood, and even the return of small wildlife. Farmers claim their crops have increased by 50% because of the trees, and livestock numbers have increased due to better grazing land. By nurturing native species of trees that sprout naturally, farmers create a resilient landscape at a low cost. A big issue here was the need to change old colonial land and resource rights that made it illegal for farmers to use the trees on their land. Historically, only the colonial rulers were allowed to cut trees and sell timber. Once these archaic laws were changed so that farmers could manage the trees on their land, the use of agro-forestry increased with great success. FMNR has spread from farmer to farmer and village to village, because when people see the benefits, they adopt the agro-forestry practices themselves (Hertsgaard, 2011).

The West African nation of **Senegal** has been actively involved in the Great Green Wall Initiative since 2008 and is now seeing the results of their work. The Senegal National Agency for the Great Green Wall works with local people who have planted 2 million seedling trees along a 545-kilometer (340-mile) strip of land that is their portion of the bigger African Green Wall. During the spring seasons, about 400 people work in tree nurseries, growing the seedlings until they are ready for planting in August. Six indigenous tree species were specially selected by scientists and local people because of their economic uses and hardiness. Then nearly 1,000 people work to plant about 2 million trees during the rainy season, before the long, dry season comes. The trees must be fenced off from animals for six years, allowing them to grow high enough that the grazing and wild animals cannot forage them down. A huge (7.5 hectares, 20 acre) gardening project has also been initiated and relies on an aquifer for irrigation; it employs 250 women who learn how to grow numerous fruits and vegetables for their communities. Overall, Senegal has planted more than 27,000 hectares (66,700 acres) of trees and restored more than 4 million hectares (9.9 million acres) of land as their part of the Great Green Wall Initiative (Powers, 2014).

WHO ARE THE KEY PLAYERS?

Local people, with assistance from national governments, nongovernmental organizations, scientists, and educators, are the foundation of the Great Green Wall.

Efforts are increasingly widespread: over 25 outside organizations and agencies are working to support the Great Green Wall Initiative. Together, many partners, including governments, the United Nations, the European Union, and other nonprofit groups, are working together to address the problems of desertification (African Union, 2018).

With large collaborative projects, it is common to have an "alphabet soup" of organizations involved, meaning that there are many groups with long names that are better known by their acronyms. The Green Wall certainly reflects this trend. The Great Green Wall Initiative for the Sahara and Sahel Initiative (GGWSSI) has over $4 billion of funding pledged from many diverse partners. The Community of Sahel-Saharan States (CEN-SAD) and the African Union Commission have partnered in regional efforts. The Sahel and West Africa Program (SAWAP) is the World Bank and Global Environmental Facility's contribution to the Great Green Wall Initiative, with $1 billion in distinct projects in 12 countries. The Food and Agricultural Organization (FAO) of the UN used funds from the European Development Fund (EDF) to develop the Action Against Desertification program to restore degraded lands in their regions to address the environmental, social, and economic impacts of desertification (FAO, 2018).

The Building Resilience through Innovation Communication and Knowledge Services (BRICKS) is a project that focuses on monitoring the ecological conditions and building technical knowledge in the region. This six-year project is funded by the World Bank and began in 2013.

The Local Environmental Coalition for a Green Union (commonly known in French as "Front Local Environnemental pour une Union Verte," or FLEUVE) is an initiative of the Global Mechanism (GM) of the United Nations Convention to Combat Desertification funded by the FAO and the European Union. A key goal of FLEUVE is to develop and build the capacity of local nongovernmental organizations so they can implement landscape-level projects for sustainable land management. Small investments are being made with many local groups to help people increase economic opportunities in local communities.

LOOKING AHEAD

Desertification and land degradation along the southern edge of the Sahara Desert have been occurring for at least 50 years and at a more rapid rate in recent decades. This is due to both human activity and environmental conditions. Population pressure, resource use, and climate change in the region have led to significant problems with land degradation, both in terms of vast land areas and affected populations. Specifically in sub-Saharan Africa, it is estimated that 500 million people live on land that is severely degrading and undergoing desertification (Schleeter, 2013).

Given the diversity and vastness of this initiative, the Great Green Wall is becoming more of a mosaic of greening initiatives tailored to local conditions, rather than a vast line of forest, which may not be possible or even helpful in some areas. Unity and collaboration seem to be key aspects of this initiative, which crosses many political, cultural, and ecological boundaries.

As we look ahead, there is reason for optimism. According to scientists and partners from international organizations, the collaborative efforts have been very positive: "Here, we saw political leaders, heads of state, and ministers in different countries wanting to work on common environmental issues and wanting to tackle land degradation issues together" (Schleeter, 2013.)

See also: Section 1: Agriculture, Climate Change, Deforestation; *Section 2:* Drought, Global North and Global South, Human Modification of Ecosystems; *Section 3:* Sustainable Development

FURTHER READING

African Union. 2018. "Great Green Wall for the Sahara and the Sahel Initiative." Food and Agriculture Organization. http://www.fao.org/docrep/016/ap603e/ap603e.pdf.

The Economist. 2016. "What Is Africa's 'Great Green Wall'?" The Economist Explains. August 31. http://www.economist.com/blogs/economist-explains/2016/09/economist-explains.

FAO. 2018. "Expanding Africa's Great Green Wall." Food and Agriculture Organization. http://www.fao.org/in-action/action-against-desertification/en/.

GGWSSI. 2018. "Great Green Wall for the Sahara and the Sahel Initiative." Online Virtual Reality Film. http://www.greatgreenwall.org/.

Hertsgaard, Mark. 2011. "Farmers Beat Back Drought and Climate Change with Trees." *Scientific American*. January 28. https://www.scientificamerican.com/article/farmers-in-sahel-beat-back-drought-and-climate-change-with-trees/.

Laing, Aislinn. 2016. "The 'Great Green Wall': Thousands of Miles Long Could Be Built across Africa to Stop the Spread of the Sahara." *The Telegraph*. May 3. http://www.telegraph.co.uk/news/2016/05/03/great-green-wall-thousands-of-mile-long-could-be-built-across-af/.

LPAA. 2018. "Great Green Wall: 'Growing a World Wonder.'" Lima-Paris Action Agenda. http://newsroom.unfccc.int/lpaa/resilience/great-green-wall-growing-a-world-wonder-restoring-the-productivity-and-vitality-of-the-sahel-region/.

Monks, Kieron. 2016. "Can the Great Green Wall Change Direction?" CNN. September 22. http://www.cnn.com/2016/09/22/africa/great-green-wall-sahara.

Powers, Diana S. 2014. "Holding Back the Sahara: Senegal Helps Plant a Great Green Wall to Fend Off the Desert." *New York Times*. November 18. http://www.nytimes.com/2014/11/19/business/energy-environment/senegal-great-green-wall-sahara-desert.html.

Schleeter, Ryan. 2013. "The Great Green Wall: Sahel-Sahara Project Aims to Combat Land Degradation." National Geographic Society. November 4. http://nationalgeographic.org/news/great-green-wall/.

UN. 2018. "World Day to Combat Desertification: 17 June." United Nations. http://www.un.org/en/events/desertificationday/background.shtml.

UNCCD. 2018. "United Nations Convention to Combat Desertification." UNCCD Secretariat. http://www2.unccd.int/.

WRI. 2005. "Millennium Ecosystem Assessment." World Resources Institute. http://www.millenniumassessment.org/documents/document.355.aspx.pdf.

Key Concepts

Alternative Agriculture

Alternative agriculture is the broad term used to describe several forms of agricultural production that are seen as distinct from the majority of industrial agriculture carried out in more developed countries. This includes several specific types of farming, each with a geographical influence: biodynamic, fair trade, local food, organic, and permaculture.

Biodynamic farming is based on the work of Dr. Rudolf Steiner, who developed these methods to address declining soil fertility in the 1920s. Today, biodynamic methods are practiced around the world, with some farms receiving certification by an organization called the Demeter Association. Biodynamic farmers seek balanced farm ecosystems in a given geographical location that work with nature and do not demand outside inputs. In fact, "biodynamic farming is much more than a method, it is a belief system" (Biodynamic Association, 2018). Biodynamic farming uses nine "preparations" made from herbs, minerals, and animal manures that are applied in very small amounts (similar to homeopathic mixtures for humans) that are thought to promote plant growth. Many activities on a biodynamic farm follow the astronomical calendar, so cultivating, planting, applying preparations, and harvesting often occur in certain moon phases (Biodynamic Association, 2018).

Fair trade is a certification and labeling system that signifies farmers receive fair wages for producing their crops. It was developed because many farmers in developing countries had only a few low-paying options for selling their crops for export. In the 1970s fair trade coffee was first imported to the Netherlands from small farms in Guatemala. Now fair trade coffee and chocolate are well established in mainstream supermarkets in developed countries, which is a direct example of geographic linkages among places. As food and farming have become global, some crops are produced in developing countries and only sold for export, rather than being consumed locally. For example, crops like coffee and chocolate are produced in tropical countries exclusively for export to North America and Europe. Large corporations have been able to set the prices, and farmers had no choice but to sell their crops cheaply. There is inequality in this system: cocoa farmers cannot even afford to buy a chocolate bar, for example. Fair trade is set up to guarantee that farmers earn a living wage for their work. "Fair Trade Certified products were made with respect to people and [the] planet" (Fair Trade USA, 2018). Indeed, this organization is the largest fair trade certifier in the United States, and they follow "rigorous social, environmental and economic standards work to promote safe, healthy working conditions [and] protect the environment." When consumers choose to purchase products with a fair trade label, this "can improve an entire community's day-to-day lives" because farmers earn a reasonable wage for their crops (Fair Trade USA, 2018).

Local food is based on geography: decreasing the distance food travels from farm to table and building relationships between farmers and consumers. There is no set distance that defines whether food is "local"—some say it is 100 miles, but others say it includes food produced within one state. Instead of miles, it is more common to consider local food activities: farmers' markets, community gardens, and community-supported agriculture (CSA), for example. Farmers' markets, if they are producer based (meaning that only farmers who grow the food can sell it), allow consumers to come and meet the person who grew the food they buy. Community gardens are common spaces, often organized by a park district or community organization, where people come together to grow crops and share in the harvest. CSAs are a type of farming where consumers pay up-front to be members of the farm and then they receive a basket of food each week during the growing season. These are all examples of how local food promotes a place where consumers and farmers can interact and get to know each other. This is accomplished through farmers' direct-to-consumer sales, which is called direct marketing (USDA-ERS, 2017).

Organic agriculture is a method of production that is certified by outside agencies that are accredited by the U.S. Department of Agriculture. As part of the 1990 U.S. Farm Bill, the Organic Foods Production Act (OFPA) established uniform national standards for the production, handling, and processing of foods labeled "organic" (USDA-AMS, 2018). The law does not address social issues such as farm size or ownership; rather, it specifically states that no synthetic chemical inputs may be used in production. These standards provide environmental benefits of lower pesticide usage, which protect watersheds, but do not indicate that the farm is small or locally owned. Indeed, there are vast fields (26,000 acres) of organic carrots owned and managed by one company in California, for example, which are shipped across the country (Eddy, 2012). Benefits of organic production methods include higher antioxidant levels and lower chemical residues on foods (Barański et al., 2014).

Permaculture is a term created by joining the concepts of permanent and agriculture. Bill Mollison is founder of permaculture and wrote numerous books on the topic, including *Introduction to Permaculture* (1991). Permaculture principles imitate nature through closed-loop systems without chemical inputs, based on a philosophy of working with nature. A key aspect of permaculture is planning and mapping a farm or garden (rural or urban) with multiple layers of production that have different zones of crop intensity types radiating out from the farmhouse. So closest to the house are crops that are picked most often and require the most attention, then in the zones farther away from the house, the management is less intensive, with crops such as food forests, for example. Mollison's books provide numerous sketches that depict his design concepts and the geography of a farm landscape. Seeking balance with nature, permaculture is a design system in which human settlement is integrated into food production (Mollison, 1991).

See also: Section 1: Agriculture, Case Study 2; *Section 3:* Case Study 14, Composting

FURTHER READING

Barański, M., D. Srednicka-Tober, N. Volakakis, C. Seal, R. Sanderson, G. B. Stewart, . . . C. Leifert. 2014. "Higher Antioxidant and Lower Cadmium Concentrations and Lower Incidence of Pesticide Residues in Organically Grown Crops: A Systematic Literature Review and Meta-Analyses." *British Journal of Nutrition* 112(5): 794–811.
Biodynamic Association. 2018. "What Is Biodynamics?" https://www.biodynamics.com/what-is-biodynamics.
Eddy, David. 2012. "Grimmway Farms' Experiences in Carrots, Organic Production." Growing Produce. April 9. http://www.growingproduce.com/uncategorized/grimmway-farms-experiences-in-carrots-organic-production/.
Fair Trade USA. 2018. "What Is Fair Trade?" https://fairtradeusa.org/what-is-fair-trade.
Mollison, Bill. 1991. *Introduction to Permaculture*. Tyalgum, Australia: Tagari Publications.
USDA-AMS. 2018. "National Organic Program." U.S. Department of Agriculture-Agricultural Marketing Service. https://www.ams.usda.gov/about-ams/programs-offices/national-organic-program.
USDA-ERS. 2017. "Local Foods: Overview." U.S. Department of Agriculture-Economic Research Service. https://www.ers.usda.gov/topics/food-markets-prices/local-foods/.
WFTO. 2014. "History of Fair Trade." World Fair Trade Organization. http://wfto.com/about-us/history-wfto/history-fair-trade.

Assessments and "Footprints"

Environmental assessment allows comparisons among various activities in order to make smart decisions that promote environmental sustainability and global understanding. Citizens can be involved, both to gain personal information on their individual impacts and also to be informed when demanding improved environmental policies. There are numerous types of assessments, including the ecological footprint, water footprint, and CO_2 footprint. The term "footprint" provides a useful image, as it implies an impact (like a foot hitting the ground), an amount (such as how big the foot is), and that it could be temporary or changing (if people decide to change their environmental impacts).

ECOLOGICAL FOOTPRINT

The Ecological Footprint tool is a resource accounting tool that provides an estimate of how much ecologically healthy land and ocean would be needed to support a given number of people based on their use of resources. Thus, this geographic modeling considers people's demands for food, energy, forest products, and land for roads and buildings. Likewise, the amount of waste produced by society must be accounted for. The Ecological Footprint measures and models all this to provide an estimate of the number of global hectares (acres) that are needed to support this person or a group of people with this lifestyle. For example, if every person on Earth used resources and created waste at the level of the typical U.S. citizen, four planet Earths would be needed (EDN, 2018).

CARBON FOOTPRINT

A typical daily routine, like taking a hot shower, making coffee, or driving a car, creates greenhouse gas emissions. Adding up the emissions from all these activities makes a household's carbon footprint. Things like location, lifestyle, and personal choice greatly affect the footprint. And online calculators help people estimate their footprint through three areas: transportation, home energy, and waste (EPA, 2016). The Nature Conservancy provides an online carbon footprint tool that accounts for location, travel, home size and characteristics, food choices, and shopping activities. Thus there are several approaches, but the goal is the same: to inform people about how their activities affect CO_2 emissions (TNC, 2018).

WATER FOOTPRINT

Obvious water use is flushing the toilet or watering the grass, although people still do not fully realize how many gallons of water each of these activities uses. More subtle water use is found in many other areas of everyday life: water to produce their hamburger or salad, water used in the manufacturing of their cell phone, or water needed to make electricity for switching on the light in their office. The water footprint tool helps people learn about their obvious and subtle uses of water. Calculators gather information on direct water use (in the bathroom, kitchen, laundry, lawn, pool, and car washing) and indirect water use (through energy, gasoline, electricity, shopping, recycling, food, and diet). As an example, the water footprint would be higher for a family in a large house, using high levels of electricity, eating a lot of meat, and watering a large lawn (Water Calculator, 2017; WFN, 2018).

See also: Section 1: Case Study 1, Case Study 4, Biodiversity Loss, Climate Change; *Section 3*: Sustainable Development, Water Conservation

FURTHER READING

EDN. 2018. "Ecological Footprint Calculator." Earth Day Network. http://www.earthday.org/take-action/footprint-calculator/#ecofootprint.

EPA. 2016. "Carbon Footprint Calculator." U.S. Environmental Protection Agency. https://www3.epa.gov/carbon-footprint-calculator/.

TNC. 2018. "Carbon Footprint Calculator." The Nature Conservancy. http://www.nature.org/greenliving/carboncalculator/index.htm.

Water Calculator. 2017. "What's Your Water Footprint?" Grace Communications Foundation. http://www.watercalculator.org/.

WFN. 2018. "Personal Water Footprint Calculator." Water Footprint Network. http://waterfootprint.org/en/resources/interactive-tools/personal-water-footprint-calculator/.

Composting

Composting is the natural biological process through which various microorganisms, including bacteria and fungi, break down organic matter into simpler substances. Organic matter, or biowaste, includes food scraps (like apple cores and orange peels) and yard waste (like grass clippings and leaves) that comprise about 30% of what Americans throw way. When this gets into the landfill, it takes up space and releases methane, which is a potent greenhouse gas (EPA, 2017). In the European Union, this waste causes an estimated 3% of total greenhouse gas emissions, and policy makers note that "[u]nquestionably, landfilling is the worst waste management option for bio-waste" (EU, 2016). Biowaste is defined as "biodegradable garden and park waste, food and kitchen waste from households, restaurants, caterers and retail premises, and comparable waste from food processing plants" (EU, 2016).

A woman pours vegetable peels and organic waste into a compost bin. (Jurate Buiviene/Dreamstime.com)

Taking this organic matter (biowaste) out of the waste stream makes sense, especially because it can be composted and turned into something useful: humus, a nutrient-rich natural fertilizer. When added to soil, this rich compost helps hold water, provide nutrients for plants, and reduce the need for adding chemical fertilizers. Large-scale programs to take biowaste and make compost can create jobs and help engage the community.

Composting requires four ingredients, commonly called greens, browns, air, and water. Greens are fresher things like fruit peels, vegetable scraps, coffee grounds, and grass clippings; these provide nitrogen. Browns are dried leaves and twigs that provide carbon. A compost pile must be equal parts greens and browns, in layers,

Zero-Waste Communities

In urban areas, garbage trucks typically collect trash weekly and take it to a dump, otherwise known as a municipal solid waste landfill (MSWLF). But this is expensive and takes up land, so communities often seek to reduce this impact on the environment.

The U.S. Environmental Protection Agency (EPA) has an online tool for towns that seek to transform their waste stream: https://www.epa.gov/transforming-waste-tool. Over 100 policies and programs can be implemented to "reduce the amount of waste disposed in landfills and promote waste prevention and materials reuse across waste generation sectors." The tool carefully defines all the relevant terms and shows a detailed spreadsheet to help keep track of all the data. Examples of 100 cities across the United States show how a waste reduction program can be successfully implemented (EPA, 2017).

Zero Waste Europe is an organization that seeks to support local groups working to reduce waste streams. They "want to re-design our society so that all superfluous waste is eliminated and everything that is produced can be re-used, repaired, composted or recycled back into the system. Anything that can't be repaired, composted or recycled should be re-designed and replaced or banned from entering the market" (ZWE, 2017).

Indeed, cities, businesses, and even individual buildings can manage their resource use in order to reduce, reuse, and recycle. These goals save money and can be expanded to a waste management hierarchy that includes (from most to least preferred) prevention, reduction, recycling, recovery, and disposal (UNEP, 2011: 294). This hierarchy emphasizes the fact that disposal or throwing something "away" into a landfill should be a last-ditch action.

The term "net-zero waste building" was defined by a Presidential Executive Order in 2015 as "a building that is operated to reduce, reuse, recycle, compost, or recover solid waste streams (with the exception of hazardous and medical waste) thereby resulting in zero waste disposal" (*Federal Register*, 2015).

Taking this one step further, various organizations such as universities, hospitals, and even military sites are realizing the environmental and cost-saving benefits of net-zero waste.

The U.S. Army, for example, has been working toward waste reduction goals since 2010. The Army plans to reach net-zero waste on 5 installations by 2020, and 25 installations by 2030. "Every day, more recycling strategies are developed moving beyond metals, paper and cardboard to include mattresses, glass, plastics, batteries, computer printers and motor oil. The best strategy is to consider the waste stream when purchasing items, reduce the volume of packaging, reuse as much as possible, and recycle the rest. . . . A net zero waste strategy eliminates the need for landfills, protects human health, optimizes use of limited resources and keeps the environment clean" (Army, 2010).

Further Reading

Army. 2010. "Net Zero: A Force Multiplier." Office of the Assistant Secretary of the Army. http://www.asaie.army.mil/Public/IE/netzero_info.html.

EPA. 2017. "Managing and Transforming Waste Streams: A Tool for Communities." Environmental Protection Agency. https://www.epa.gov/transforming-waste-tool.

Federal Register. 2015. "Planning for Federal Sustainability in the next Decade." Executive Order 13693, Presidential Documents. 80(57), 25. https://www.gpo.gov/fdsys/pkg/FR-2015-03-25/pdf/2015-07016.pdf.

UNEP. 2011. "Towards a Green Economy: Pathways to Sustainable Development and Poverty Eradication." United Nations Environmental Programme. https://www.unep.org/greeneconomy/sites/unep.org.greeneconomy/files/field/image/green_economyreport_final_dec2011.pdf.

ZWE. 2017. "Empowering Our Communities to Redesign." Zero Waste Europe. https://www.zerowasteeurope.eu/about/.

with different sized bits of waste. Air and water must be available and at the right amounts to encourage the natural microbial process of breaking down the organic materials.

Allowable items for composting include what people generally assume will break down in the environment: cardboard, coffee grounds, eggshells, fruits, grass clippings, hair, hay, houseplants, leaves, newspapers, nutshells, paper, sawdust, straw, and vegetables. Things that cannot be composted include coal or charcoal ash, dairy products, eggs, fats, grease, meat or fish bones or scraps, pet feces, and yard trimmings treated with chemical pesticides. The reasons these materials must be excluded are because they do not break down, will attract rodents, or will kill the beneficial composting organisms (NRDC, 2016).

There are two geographical scales for composting: individuals composting at home or community-wide composting as part of a waste management system. Composting at home can be done by anybody, because there are several options. Composting at the municipal scale requires collection, contaminant separation, sizing and mixing, and biological decomposition.

At-home composting can be in the backyard or even indoors. Backyard composting requires somewhat moderate temperatures, a shady spot, and ideally about a 1×1 meter (3×3 foot) area. A compost pile can be started right on the ground, or with wire fencing on top of a wooden pallet, to provide air flow. People can make or buy bins, and there are sideways barrels that can be tumbled to aerate the compost, which makes it break down faster. For people with no outdoor space, an indoor compost bin can be made or purchased that uses red worms to help process the food scraps. Overall, it is fairly simple: "Food is going to rot, no matter what. All you have to do is help" (NRDC, 2016). Done properly, the organic waste does not smell and can be turned into useable, rich compost in about five weeks (EPA, 2017).

Municipal solid waste management of biowaste has two components: yard waste and food waste. Twenty states have banned yard waste from landfills in an effort to decrease the volume of materials going in and the greenhouse gases emitted. There are now more than 3,400 composting operations for yard trimmings in the United States. However, food waste has a much lower composting rate, with approximately 347 composting facilities across the country. About 180 communities have food scrap collection programs, and many schools, restaurants, supermarkets, and other businesses also separate their food waste for composting, but the infrastructure is truly inadequate (Platt and Goldstein, 2014). More cities could follow the lead of Seattle and Toronto, or countries like the Netherlands and Austria, which have curbside collection of food waste. Just as residents have a blue recycle bin for paper and a green one for cans and glass bottles, there could be a third one for food scraps (Earth Share, 2018). This would promote the development of a separated waste stream, where these biowastes could be efficiently turned into compost.

A recent report by the World Bank notes that composting should be more widespread, especially in developing countries. "Over 50% of an average developing country city's municipal solid waste stream could be readily composted." Composting could save money when integrated into municipal waste management programs (Hoornweg et al., 2000).

From an environmental geography perspective, composting is a place-based industry. So "advancing composting and compost use in the U.S. is a key sustainability strategy to create jobs, protect watersheds, reduce climate impacts, improve soil vitality, and build resilient local economies" (Platt and Goldstein, 2014).

See also: Section 1: Agriculture, Solid Waste; *Section 2:* Case Study 6; *Section 3:* Alternative Agriculture, Sustainable Development, Sustainable Diet

FURTHER READING

Earth Share. 2018. "The Future of Garbage: Curbside Compost." http://www.earthshare.org/2014/02/compost.html.
EPA. 2017. "Reducing the Impact of Wasted Food by Feeding the Soil and Composting." U.S. Environmental Protection Agency. March 20. https://www.epa.gov/sustainable-management-food/reducing-impact-wasted-food-feeding-soil-and-composting.
EU. 2016. "Biodegradable Waste." European Commission. June 9. http://ec.europa.eu/environment/waste/compost/index.htm.
Hoornweg, Daniel, Laura Thomas, and Lambert Otten. 2000. "Composting and Its Applicability in Developing Countries." Urban Waste Management Working Paper Series, The World Bank. http://documents.worldbank.org/curated/en/483421468740129529/Composting-and-its-applicability-in-developing-countries.
NRDC. 2016. "Composting Is Way Easier than You Think." Natural Resources Defense Council. June 16. https://www.nrdc.org/stories/composting-way-easier-you-think.
Platt, Brenda and Nora Goldstein. 2014. "State of Composting in the U.S." *BioCycle* 55(6): 19. https://www.biocycle.net/2014/07/16/state-of-composting-in-the-u-s/.

Earth Day

Before 1970 there were almost no national environmental regulations in the United States, so factories could legally dump chemicals in the water or spew toxins in the air. Glaring examples of environmental catastrophes include the Cuyahoga River in Cleveland, Ohio, which actually caught fire in 1969 due to the high levels of industrial waste. Several huge oil spills also drew concern from citizens, as did frequent air pollution warnings—often so severe that parents were told to keep their children indoors (History.com, 2018).

By the late 1960s, a very broad coalition of people came together to demand environmental protection. Indeed, the first Earth Day was April 22, 1970, and it marked "a rare political alignment, enlisting support from Republicans and Democrats, rich and poor, city slickers and farmers, tycoons and labor leaders" (EDN, 2018). In fact, over 22 million Americans—at thousands of schools, universities, and cities—took to the streets in peaceful demonstrations to voice their support for environmental protection (EPA, 2017).

The first Earth Day was a turning point in the establishment of environmental protection in the United States and the rest of the world. It all began because of a grassroots effort to draw attention to environmental disasters and the need to establish protection of our water, land, and air. Senator Gaylord Nelson, a Democrat from Wisconsin, initiated the idea of a "teach-in" (or a general educational forum)

A woman holds up a sign at the March for Science during Earth Day celebrations in Atlanta, Georgia on April 22, 2017. (Russ Ensley/Dreamstime.com)

to discuss environmental topics at many college campuses across the country. This notion quickly grew into a nationwide movement because so many Americans were concerned about environmental problems. Much of the organization was accomplished by young people, all volunteers. Think about how this was accomplished before social media—these students literally had to send postcards and flyers by snail mail and wait for written replies, but eventually Earth Day events were organized at numerous sites.

Congress took a break (called a recess) so that politicians could be in their home districts to hear people's concerns—and there were many. Acting in response to public demands, within a year "the first Earth Day had led to the creation of the United States Environmental Protection Agency and the passage of the Clean Air, Clean Water, and Endangered Species Acts" (EDN, 2018). In the following decade, additional new environmental protections were put in place through laws to address environmental impacts of large building projects, pesticides and toxic substances regulation, protection of some marine life, cleanup of toxic sites, and removal of lead from gasoline. It is indeed difficult to imagine the condition of the current natural world if these environmental protections had not been established (Howard, 2016).

The next major Earth Day was April 22, 1990, and it went global. Interest surrounding this event led to renewed environmental participation and green activities, which helped to initiate the globally significant 1992 United Nations Earth Summit in Rio de Janeiro. In 1995, Senator Nelson was awarded the Presidential Medal of Freedom for being the founder of Earth Day and due to his long-term work on environmental issues. From 2000 to the present, annual Earth Day festivities have been truly global, with more than 1 billion people in hundreds of countries taking part each April 22 (EDN, 2018).

Also in the present day, the World Bank uses the global Earth Day movement to underscore the linkages between environmental protection and sustainable development. Environmental sustainability is necessary to reduce poverty. In low-income countries, 75% of income comes from forests, for example. Likewise, good policies to address climate change will create jobs and economic growth. For example, solar jobs in the United States have grown 12 times faster than overall job growth, and China has 35% more people working in renewable energy than in the petroleum industry. Further, air pollution costs money and harms health, causing 1 out of every 10 deaths globally. In terms of global oceans, environmental protection of marine life would generate an estimated $83 billion in additional benefits for fisheries as fish stocks could recover from overexploitation. Finally, restoring degraded lands and forests, particularly in Africa and Latin America, would have huge benefits for the economy and climate. Over 2 billion hectares (4.9 million acres) could be reestablished as productive landscapes with benefits including better food production, cleaner water, increased forest resources, and reduced climate risks. In the spirit of Earth Day participation, the World Bank promotes the concept that environmental and economic sustainability are mutually beneficial (World Bank, 2017).

Currently Earth Day provides opportunities for people to get involved and learn more about the planet. One example is seen in a recent initiative by the U.S. space agency, NASA, that encourages people to "adopt" one of 64,000 individual pieces of Earth as seen from space and receive fascinating NASA Earth science data collected for that location (NASA, 2017). This promotes local-to-global environmental understanding.

Earth Day exemplifies the power of individual people coming together to express their concerns and demand political action. The outcome of this peaceful grassroots activism from 1970 and beyond is the establishment of meaningful new policies that have ultimately led to vital environmental protections.

See also: Section 1: Biodiversity Loss, Climate Change, Water Pollution; *Section 3:* Case Study 12, Green Consumerism, Recycling, Renewable Energy

FURTHER READING

EDN. 2018. "The World's Largest Environmental Movement." Earth Day Network. https://www.earthday.org.

EPA. 2017. "Earth Day." U.S. Environmental Protection Agency. December 21. https://www.epa.gov/earthday.

History.com. 2018. "Earth Day." http://www.history.com/topics/holidays/earth-day.

Howard, B. C. 2016. "46 Environmental Victories since the First Earth Day." April 26. http://news.nationalgeographic.com/2016/04/160422-earth-day-46-facts-environment/.

NASA. 2017. "NASA Celebrates Earth Day by Letting Us All #AdoptThePlanet." April 6. https://www.nasa.gov/feature/goddard/2017/nasa-celebrates-earth-day-by-letting-us-all-adopttheplanet.

World Bank. 2017. "Five Facts for Your Environmental and Climate Literacy." April 22. http://www.worldbank.org/en/news/feature/2017/04/22/earth-day-2017—five-facts-for-your-environmental-and-climate-literacy.

Electric Cars

Electric vehicles (EVs) include battery-electric, plug-in hybrid electric, and fuel cell electric passenger cars and light-duty vehicles. Globally, in 2016 there were 2 million such vehicles on the roads, which is a significant increase, in fact double, from the previous year. Still, when considering the entire stock of vehicles on the road, EVs only comprise 0.2% of the global total. Sales are highly concentrated, with 95% of EV sales occurring in just 10 countries: China, United States, Japan, Canada, Norway, Britain, France, Germany, the Netherlands, and Sweden. These are total numbers of cars, not per person, so it is not surprising that given China's large population and national policies encouraging EV sales, this country comprises about 40% of all EVs in the world (IEA, 2017).

The history of electric cars shows their varying popularity, as they outsold the other two types of cars (gasoline and steam) in 1900. These early electric cars were basically a battery-powered horseless carriage that did not smell, rumble, or vibrate like gasoline cars of the day. Once Henry Ford began mass production, however, the price of gasoline cars dropped to only $650 compared to $1,750 for an electric car. By the 1930s, gasoline cars won out and ruled the world up to the modern era. By the 1970s, a few new electric car models were built, but few sold. Finally, in the 1990s, EV manufacturing began to ramp up due to emissions regulations set in place in many countries and U.S. states (Thompson, 2017).

Of the major traditional car companies, GM, Nissan, and Toyota have put the most effort into developing EVs, and this shows in their sales. In the first half of 2017, for example, fully electric car sales were up 96% from the previous year and plug-in hybrid sales were up 42% among these large car companies (Shahan, 2017a). The effort to shift to EVs can be stimulated by public purchasing power. A partnership of 30 U.S. cities, led by Los Angeles, Seattle, San Francisco, and Portland, is uniting to make a mass purchase of over 110,000 EVs for their public fleets that include police cars, street sweepers, and trash haulers (IEA, 2017).

A key policy to push the increase in electric vehicles in the United States is zero emissions vehicles (ZEV) mandates that are in place in 10 states: California, Connecticut, Maine, Maryland, Massachusetts, New Jersey, New York, Oregon, Rhode Island, and Vermont. Under these policies, automakers' sales must include a certain percentage of ZEVs, which can include sales of hybrids and low-emission conventional vehicles. The goal of ZEV mandates is to ensure that car companies research, develop, and market EVs that emit fewer climate-warming gases than gasoline cars and do not create tailpipe pollution. Indeed, states with the ZEV mandate have significantly higher sales of electric cars (Shahan, 2017b).

France and the UK have banned sales of new gasoline and diesel cars starting in 2040. In Norway only zero emission vehicles can be sold after 2025. In India all vehicles sold after 2030 must be electric. Numerous countries, including Austria, China, Denmark, Germany, Ireland, Japan, the Netherlands, Portugal, Korea, and Spain, have official target dates for electric car sales (Petroff, 2017).

Many experts say that EV usage will surge in the next decade. Indeed, some early examples are Norway tallying EVs as 29% of all new cars purchased in 2016. According to current estimates based on data from car manufacturers, the EV market will transition from early hesitant purchases to mass market adoption very soon. Up to 20 million electric cars could be on the road by 2020 and up to 75 million by 2025 (IEA, 2017).

Another business estimate shows that by 2040, EVs will comprise 54% of all car sales globally. This would cause a huge change in the fossil fuel industry, as this estimated 530 million electric cars on the road would require 8 million fewer barrels of oil a day to run, but more power from the electric grid (NBC, 2017). These EVs would decrease oil dependence and reduce greenhouse gas emissions, which is good for mitigating climate change, but more electricity will need to be produced. So how climate friendly are EVs? The answer is complex and depends on how their electricity is produced. In 37 states, the sources of energy used in power plants was clean enough (from higher usage of renewable energy) that electric cars are more climate friendly than their gasoline-engine counterparts. In other words, in 13 states, the electricity is heavily reliant on fossil fuels like coal and natural gas, so electric cars still cause high overall emissions (Climate Central, 2018). Keep in mind that tailpipe emissions cause air pollution problems in highly urbanized areas, whereas power plants are typically not located in downtown areas and can have stricter emissions standards that are easier to monitor than individual vehicles.

See also: Section 1: Air Pollution, Case Study 1, Case Study 5, Climate Change; *Section 3:* Case Study 11, Green Consumerism, Renewable Energy

FURTHER READING

Climate Central. 2018. "Climate Friendly Cars." http://climatefriendlycars.climatecentral.org/.
IEA. 2017. "Electric Vehicles Have Another Record Year, Reaching 2 Million Cars in 2016." International Energy Agency. June 7. https://www.iea.org/newsroom/news/2017/june/electric-vehicles-have-another-record-year-reaching-2-million-cars-in-2016.html.
NBC. 2017. "The Electric Car Revolution May Come Sooner than We Thought." NBC News. July 7. https://www.nbcnews.com/mach/tech/electric-car-revolution-may-come-sooner-we-thought-ncna780516.
Petroff, Alanna. 2017. "These Countries Want to Ditch Gas and Diesel Cars." CNN Money. July 26. http://money.cnn.com/2017/07/26/autos/countries-that-are-banning-gas-cars-for-electric/index.html.
Shahan, Zachary. 2017a. "Big Auto's Fully Electric Car Sales Up 102% In USA." Clean Technica. July 5. https://cleantechnica.com/2017/07/05/big-autos-fully-electric-car-sales-102-usa/
Shahan, Zachary. 2017b. "US Electric Car Sales by State." Clean Technica. May 4. https://cleantechnica.com/2017/05/04/us-electric-car-sales-state-whos-1-ohio-california/.

Thompson, Cadie. 2017. "How the Electric Car became the Future of Transportation." Business Insider. July 2. http://www.businessinsider.com/electric-car-history-2017-5.

Environmental Justice

According to the U.S. Environmental Protection Agency (EPA): "Environmental justice is the fair treatment and meaningful involvement of all people regardless of race, color, national origin, or income, with respect to the development, implementation, and enforcement of environmental laws, regulations, and policies" (EPA, 2017). The concept of environmental justice developed out of environmental racism, which occurs when communities of racial minorities are disproportionately exposed to higher levels of pollution and/or do not have access to clean air, soil, and water.

At the international level, environmental racism refers to the relationships where the Global South is subjected to waste and pollution caused by the industrialized nations in the Global North. Within one country or region, environmental racism can be seen in differences between neighborhoods where racial minorities and low-income people are concentrated in areas that are heavily industrialized and polluted. Environmental decisions are often based on political power, and poorer minority areas typically have less political power than wealthier, majority-white areas, so their voices may not be heard.

The environmental justice movement began in the 1960s as several key events drew attention to the problem of environmental racism. First, in the early 1960s,

Demonstrators march in Warren County, North Carolina, to protest the siting of a dump for toxic wastes in their community, October 21, 1982. The event was a catalyzing moment in the environmental justice movement. (Bettmann/Getty Images)

César Chávez organized Latino farmworkers to fight for their rights, including protection from dangerous pesticides. Not just adult workers, but also their children, were exposed to these poisons as parents were given no protective gear and thus came home contaminated with toxins. Chávez drew attention to the struggle of farm laborers who had no workers' rights or legal protections; this was a very early appeal for environmental justice based on workers' health and safety.

Then in 1982, the North Carolina governor and state politicians decided that Warren County, a poor, rural, majority–African American area would be the site for a new hazardous waste landfill. More than 6,000 truckloads of toxic PCB-contaminated soils were to be transported to Warren County, and some soil was spread alongside the roads leading to the new dump site. Fearing contamination of the soil and water in their communities, thousands of residents protested, some even laid down in the road to try to stop the trucks. Hundreds of people were arrested, and this is the first time people were imprisoned for protesting a landfill. Despite mass peaceful protests, the Warren County citizens lost their fight and the toxic waste was dumped in their nearby landfill. But their concern, determination, and voice became a symbol to others across the country who faced similar fears about toxins in their communities.

In 1983 the U.S. General Accounting Office (GAO) published the report "Siting of Hazardous Waste Landfills and Their Correlation with Racial and Economic Status of Surrounding Communities," based on Region IV of the EPA. This region is composed of eight southern states (Alabama, Florida, Georgia, Kentucky, Mississippi, North Carolina, South Carolina, and Tennessee) and demographic data showed that about 20% of the population is African American. The GAO report, however, found that hazardous waste landfills in Region IV were overwhelmingly concentrated in communities that were majority African American (DOE, 2018).

In 1987, the United Church of Christ Commission for Racial Justice published the report "Toxic Wastes and Race." This report stated that 60% of African Americans and Hispanics lived in communities with at least one hazardous waste site. Then in 1990, the book *Dumping in Dixie: Race, Class and Environmental Quality* by Robert Bullard gave specific examples of environmental justice problems in the U.S. South. Case studies of industrial and waste sites depicted very real concerns among African American communities facing environmental degradation (Skelton and Miller, 2016).

In 1991, the First National People of Color Environmental Leadership Summit was held with representatives from hundreds of communities across the United States. They discussed common problems and sought unified solutions. Together, they wrote the "Principles of Environmental Justice" document, which laid out goals for developing the national environmental justice movement.

In 1994, President Bill Clinton issued Executive Order 12898 "Federal Actions to Address Environmental Justice in Minority Populations and Low-Income Populations." This made environmental justice part of the federal decision-making process by focusing attention on health and environmental conditions in minority and low-income areas. Federal agencies were required to make environmental justice a central part of their mission and to establish an environmental justice strategy.

> ### Professor Maathai and Kenya's Green Belt Movement
>
> Professor Wangari Muta Maathai was the founder of the Green Belt Movement (GBM) in Kenya. The mission of this environmental organization is to "strive for better environmental management, community empowerment, and livelihood improvement using tree-planting as an entry point" (Green Belt Movement, 2018).
>
> In the 1970s, she saw that many women in Kenya had to walk farther and farther to gather fuel wood and water, as forests were declining and streams were drying up. To address both environmental concerns and social inequity, Dr. Maathai started GBM to encourage groups of women to grow and plant trees to take back land from desertification.
>
> Since that time, the GBM has planted over 51 million trees, based on a watershed approach that focuses on critical habitats on the verge of desertification. By paying a small amount to the local women for their work, the GBM organization has enabled communities to restore their rainwater, improve soil health, and thus provide food and fuel wood well into the future. These activities empowered women, their families, and communities. Now many rural areas have healthier environments, and women are gaining an education and standing up for their democratic rights (Green Belt Movement, 2018). Today GBM has both a Kenyan organization and a broader Green Belt Movement International, which together have become one of the most prominent grassroots women's organizations in the world.
>
> Professor Maathai's hard work proved successful in advancing both environmental and social sustainability. Environmental benefits include tree planting, water conservation, and soil building. Having trees nearby helps reduce the daily task of carrying fuel wood for rural women. The GBM creates thousands of part-time jobs, and this is important for women as it gives them some independence. Women are equal participants in the Green Belts, and they are trained in planting and cultivating seedlings.
>
> Professor Maathai was a remarkable woman, with numerous accomplishments, which include being the first East African woman to earn a PhD, writing four books, serving in the Kenyan Parliament, acting as Assistant Minister for Environment and Natural Resources in Kenya, and inspiring generations of environmental and social action.
>
> In 2004, she won the Nobel Peace Prize for her work, the first African woman to do so. In her acceptance speech, she generously noted that "[a]lthough this prize comes to me, it acknowledges the work of countless individuals and groups across the globe. They work quietly and often without recognition to protect the environment, promote democracy, defend human rights and ensure equality between women and men. By so doing, they plant seeds of peace" (Maathai, 2004).
>
> **Further Reading**
>
> Green Belt Movement. 2018. "Who We Are." http://www.greenbeltmovement.org.
> Maathai, Wangari. 2004. "Nobel Lecture." Nobel Prize. December 10. http://www.nobelprize.org/nobel_prizes/peace/laureates/2004/maathai-lecture-text.html.

More recently, the U.S. Environmental Protection Agency developed an online environmental justice mapping tool that provides a nationally consistent dataset and methodology called EJSCREEN. Environmental geographers appreciate this combination of spatially specific demographic and environmental data. A user can click on EJSCREEN and choose a geographic area, then the tool provides demographic and environmental information for that area. The EJ Index provides a combination of data: demographics including age, income, education, and racial minorities; and 11 environmental indicators that assess water, air, and soil quality, including variables such as air toxics assessments, ozone and particulate matter levels, and proximity to waste facilities. Although all these indicators are from publicly available data, EJSCREEN joins them together and displays this information for a specific geographic region (EJSCREEN, 2017).

As noted on the EJSCREEN and in numerous research articles, environmental justice is still a significant concern today. Several recent environmental conflicts in the United States may be linked to social justice, including the Flint, Michigan water crisis in which a city of thousands of poor, minority residents drank contaminated city water piped into their homes for years before authorities took action. Another example is the Standing Rock, North Dakota protests in which thousands joined the Sioux people to fight the construction of a pipeline through their sacred lands. Finally, environmental racism may be seen in "cancer alley," the 85-mile area of southern Louisiana with 150+ petrochemical plants and refineries, where the majority poor and black population have high rates of cancer and other health problems. Indeed, even recent research shows that highly polluting industrial facilities and the siting of hazardous waste storage are overwhelmingly located in poor, nonwhite neighborhoods (Sherman, 2016).

The history of environmental justice illustrates a truly grassroots environmentalism that is based on protecting public health, keeping workers safe, and preserving landscapes where people live, sometimes in urban areas. So "joining environmentalism to movements for economic and racial justice wouldn't be new" (Purdy, 2017), but it could make environmental issues relevant to more people. If people realize how they are linked to the environment in their daily lives, they may promote environmental justice and participate in environmental issues.

See also: Section 1: Case Study 5, Superfund; *Section 2:* Case Study 7

FURTHER READING

DOE. 2018. Environmental Justice History." U.S. Department of Energy. September 9. https://energy.gov/lm/services/environmental-justice/environmental-justice-history.

EJSCREEN. 2017. "EJSCREEN: Environmental Justice Screening and Mapping Tool." U.S. Environmental Protection Agency. August 17. https://www.epa.gov/ejscreen.

EPA. 2017. "Environmental Justice." U.S. Environmental Protection Agency. November 16. https://www.epa.gov/environmentaljustice.

Purdy, Jedediah. 2016. "Environmentalism Was Once a Social-Justice Movement: It Can Be Again." *The Atlantic.* December 7. https://www.theatlantic.com/science/archive/2016/12/how-the-environmental-movement-can-recover-its-soul/509831/.

Sherman, Erik. 2016. "If You're a Minority and Poor, You're More Likely to Live Near a Toxic Waste Site." *Fortune.* February 4. http://fortune.com/2016/02/04/environmental-race-poverty-flint/.

Skelton, Renee and Vernice Miller. 2016. "Environmental Justice Movement." Natural Resources Defense Council. March 17. https://www.nrdc.org/stories/environmental-justice-movement.

Environmental Nongovernmental Organizations (ENGOs)

People joining together to form environmental nongovernmental organizations (ENGOs) have been very effective in influencing environmental policy and encouraging the establishment of natural areas.

In the broadest sense, a nongovernmental organization (NGO) is a legally defined group established by individuals without the representation of any government body. The term NGO was originally developed by the United Nations in 1945 to signify any group that is not a for-profit business or a part of government (UIA, 2017). Although NGO is most commonly used, the terms civil society organization and nonprofit organization refer to the same types of groups.

The concept of nonprofit means that any money raised by the group stays within the organization to accomplish their goals, rather than going to shareholders or executives, as would occur in a for-profit business. At the same time, an NGO is not part of a government structure, but can work closely with the government and even get some funding from the government, often in the form of grant money. In the United States, such groups are defined in Section 501(c) of the United States Internal Revenue Code (26 U.S.C. § 501(c)) as a tax-exempt, non-profit corporation or association, meaning that these charitable organizations do not pay taxes.

NGOs vary in size from a handful of people to many million. They can be focused on education, health, sports, or any number of issues. For example, one of the earliest groups was the Red Cross, which was established in 1863, initially to provide neutral help for soldiers injured in war. NGOs can be active within every geographic scale: local, regional, state, national, or international (UN, 2018).

Environmental geographers are specifically interested in ENGOs. Adding the "E" for environment simply means that the NGO has an environmental focus. Considered one of the earliest ENGOs, the Sierra Club was established in 1892, initially to protect wilderness and nature in the western United States, but eventually expanded to a nationwide organization. Global ENGOs include Greenpeace, Natural Resources Defense Council, Ocean Conservancy, The Nature Conservancy, World Wildlife Fund, and World Resources Institute. Each has a specific mission, whether focused on environmental protection, policy, or education.

ENGOs can act to mobilize, facilitate, or partner. Mobilizing means that an ENGO can assemble resources to provide assistance quickly to help species in immediate harm. Facilitating indicates that an ENGO can promote change through campaigns, education, and lobbying policy makers to create environmental regulations. ENGOs can be important partners, working alongside other groups in industry and government to tackle problems and effect ecological change.

Although ENGOs have operational goals—they seek to accomplish specific things—they must also be dedicated to campaigning, both for donations and pressing for policy change. These activities are crucial for any ENGO, as they depend on fund-raising to exist; however, it can be difficult for volunteers to spend time on public relations efforts or political activities when they are truly more motivated to be active outdoors. This can be a real challenge for ENGOs, as their members tend to be outdoorsy people who do not typically spend time in the political arena. But time and again, successful ENGOs have reached their goals through this policy process, either by setting up new environmental regulations or working with policy makers to establish parks and preserves to protect nature. Overall, ENGOs can have a significant influence on environmental policy, which leads to long-term ecological benefits.

ENGOs vary in their approach. Just consider the great variety of environmental issues and problems in the world—each requires a distinct method to accomplish the best possible outcome. Some groups must first try to alleviate human poverty before they can focus on ecological issues; others may need geographical technologies such as remote sensing, geographic information systems (GIS), and mapping in order to bring sustainable landscape management. Some ENGOs must learn from local people to initially understand the ecosystem before attempting to suggest protection policies, whereas others may need to train local people in order to have long-term ecologically beneficial management.

More recently, some ENGOs have specifically been working with businesses in an effort to make corporations more ecologically aware. For example, Ceres developed the Global Reporting Initiative to help companies report their sustainability information just as they report their profit data. The Environmental Defense Fund has worked directly with large corporations to transform their practices (one notable example was their collaboration with McDonald's to end their use of foam-plastic food boxes). GreenBlue is an ENGO that works with businesses to make manufactured goods more environmentally sustainable; its Sustainable Packaging Coalition has developed packaging guidelines, metrics, and tools to assess ecofriendly packaging.

See also: Section 2: Case Study 8; *Section 3*: Case Study 15, Earth Day, Environmental Policy

FURTHER READING

Charity Navigator. 2018. "What Do Our Ratings Mean?" September 8, https://www.charitynavigator.org/.
Herrera, Tilde. 2010. "10 Green NGOs Businesses Should Know About." GreenBiz. June 14. https://www.greenbiz.com/blog/2010/06/14/10-green-ngos-businesses-know-about.
UIA. 2017. "What Is a Non-Governmental Organization (NGO)?" Union of International Associations. September 6. http://www.uia.org/faq/yb2.
UN. 2018. "Civil Society." United Nations. September 8. http://www.un.org/en/sections/resources/civil-society.
WANGO. 2018. World Association of Non-Governmental Organizations. www.wango.org.

Environmental Policy

"Earth provides enough to satisfy every man's need, but not every man's greed" said Mahatma Gandhi, the nonviolent resistance leader in India in the early 1900s. "We need to build stronger economies and economic opportunities for citizens while at the same time protecting the environment," stated Justin Trudeau, prime minister of Canada, speaking nearly a century later (Schumacher, 1973; Smith, 2017).

These statements indicate the historically complex, dynamic, and varying influences at play in environmental policy, which can be implemented at the local, state, national, or global level. Broadly defined, "an environmental policy is a statement about an organisation's environmental position and values . . . [its] overall environmental performance intentions and direction" (ISO, 2018).

Environmental policy must be informed and implemented through knowledge and expertise on numerous themes, including ecology, economics, geography, history, international relations, national politics, and local culture. Stakeholders include citizens, environmental nongovernmental organizations (ENGOs), government, businesses, and the planet itself. These are often at odds with one another, and each group may try to place their own interests above others. Of course, the planet has no voice, other than ENGOs who may promote nature for its own sake.

Why do people and companies follow environmental policies? Because the policies are set up to encourage or force compliance in several ways: 1) economic (also called market-based) incentives are things like taxes or tax exemptions, tradable permits, and fees that must be paid if a policy is not followed; 2) grants and incentives may be offered through policies to encourage voluntary compliance with environmental goals; 3) governmental involvement, such as specific agreements between the government and private companies to press voluntary environmental actions; or 4) governments can lead by example, such as by purchasing energy-efficient and green products.

Environmental laws are closely associated with social values, because regulations would not be developed or followed if people did not push for them and accept them. Thus, other issues are tied to environmental policy, such as poverty, inequality, gender, education, and development. Most environmental issues are tied to social concerns: climate change affects poor people more; biodiversity affects human health and medicines; and deforestation is aggravated by increased biofuel usage.

Here are just a few of the potential conflicts present within environmental policy implementation: preservation of nature versus development of resources; conflicts between science, uncertainty, and propaganda; and economic solutions such as regulatory taxes versus voluntary compliance.

In many countries, environmental laws have been a scapegoat—they are blamed—for economic problems, especially unemployment. A specific example is seen in the United States beginning in the 1980s and continuing to the present. Although conservative U.S. President Ronald Reagan stated that "I believe in a sound, strong environmental policy that protects the health of our people and a wise stewardship of our nation's natural resources," he repeatedly blamed "job-killing regulations" for the country's economic woes (Reagan, 1983). Research data, however, show the opposite because environmental regulations force change and innovation that create new jobs. Initially, this shift may be scary because jobs transition to new sectors of the economy. A realistic historical example would be after new regulations banned lead additives in gasoline, the factory that makes these additives closed, but new jobs were created in a factory that makes catalytic converters that control vehicle emissions. Thus some workers benefit from regulations, while others lose (Semuels, 2017). The economy constantly goes through cycles and updates, seen, for example, in recent declines in manufacturing because of increased automation. The economy has shifted toward information technology, which has experienced significant job growth.

Another important aspect of environmental policy is the difference between law and regulation. In the North American context, a law is a formal written statute, voted on and passed by either the U.S. Congress or Canadian Parliament, for example. Regulations, on the other hand, are the rules developed by administrative agencies, such as the U.S. Environmental Protection Agency or Environment Canada, that determine how laws are implemented and enforced. So the U.S. Clean Air Act of 1972, states in U.S. Code, Title 33, Chapter 26, Subchapter I, § 1251 (3): "it is the national policy that the discharge of toxic pollutants in toxic amounts be prohibited," but the U.S. EPA must define what the terms "discharge," "toxic," "pollutants," "amounts," and "prohibited" actually mean and how such rules will be enforced. Typically, there is a rule-making process during which industry, environmental, and other interest groups make comments to encourage specific approaches. The EPA balances economic realities with acceptable risk levels to determine the actual guidelines for water regulations, like which chemicals in which amounts are allowed.

Other examples include Australia, with a similar regulatory situation, except that energy is merged into the same agency, so their Department of Environment and Energy "designs and implements Australian Government policy and programs to protect and conserve the environment, water and heritage, promote climate action, and provide adequate, reliable and affordable energy." The European Union is an example of national policies merged into regional agreements among the 27 member countries. Here treaties are the basis of "primary laws" that are binding agreements agreed upon by all members and implemented by each country's environmental agency (also called ministry). The EU notes that it "has some of the world's highest environmental standards. Environment policy helps green the EU economy, protect nature, and safeguard the health and quality of life of people living in the EU" (EU, 2018).

At the international level, since 1972 the United Nations Environment Programme (UNEP) has been active in coordinating and assisting countries in their implementation of environmental policies. Their mission "is to provide leadership and encourage partnership in caring for the environment by inspiring, informing, and enabling nations and peoples to improve their quality of life without compromising that of future generations." The UNEP helps all other UN agencies with issues related to the environment and particularly focuses on helping developing nations with environmental policy goals. They work to assess national, regional, and international environmental data and trends, which are used by agencies to promote sustainable environmental management. The UNEP works on seven key environmental topics: chemicals and waste, climate change, disasters and conflicts, ecosystem management, environmental governance, environment data sharing, and resource efficiency. UNEP has been instrumental in the collaborative efforts to write guidelines and treaties on issues that cross political boundaries like international air and water pollution, chemicals, and waste. They have a document repository or online library, where many international environmental reports and documents can be accessed.

See also: Section 1: Climate Change Policies; *Section 3*: Case Study 13, Environmental Nongovernmental Organizations (ENGOs), Green Political Party

FURTHER READING

Australia Environmental Agency. 2018. http://www.environment.gov.au/.
Environment Canada. 2018. "Environment and Natural Resources." www.canada.ca/en/services/environment.html.
EU. 2018. "Regulations, Directives, and Other Acts." European Union. https://europa.eu/european-union/eu-law/legal-acts_en.
ISO. 2018. "ISO 14001: Environmental Policy." International Organization for Standardization. http://www.environmentalpolicy.com.au/.
Reagan, Ronald. 1983. "Radio Address to the Nation on Environmental and Natural Resources Management." June 11. https://www.reaganlibrary.archives.gov/archives/speeches/1983/61183a.htm.
Schumacher, Ernst Friedrich. 1973. *Small Is Beautiful: A Study of Economics as If People Mattered.* London: Blond & Briggs.
Semuels, Alana. 2017. "Do Regulations Really Kill Jobs?" *The Atlantic.* January 19. https://www.theatlantic.com/business/archive/2017/01/regulations-jobs/513563/.
Smith, Joanna. 2017. "Trudeau to Talk Balancing Oil with the Environment." March 9. https://www.thestar.com/news/canada/2017/03/09/trudeau-to-talk-balancing-oil-with-the-environment-at-energy-summit-in-houston.html.
UNEP. 2018. "What We Do." United Nations Environment Programme. http://www.unenvironment.org.
U.S. EPA. 2018. www.epa.gov.

Green Buildings

According to the World Green Building Council: "a 'green' building is a building that, in its design, construction or operation, reduces or eliminates negative impacts, and can create positive impacts on our climate and natural environment" (World GBC, 2018). In addition, green buildings reduce natural resource use and improve people's quality of life. Green buildings can be a new construction, or existing buildings may be retrofitted.

Green buildings are specifically planned and designed to reduce the overall impact of built structures on the environment and human health by reducing waste and pollution, efficiently using water and energy, and protecting workers' health to promote employee productivity. Numerous examples exist. Stepping into a green building, what would be obvious? From top to bottom, here are some things that may be present: solar panels, rooftop garden, improved insulation, natural lighting, water conservation, smart temperature controlled by sensors, ecofriendly building materials, high-efficiency appliances, and rainwater harvesting.

Thinking broadly, some green building concepts, such as passive solar design, have been used for centuries, as people built homes to gain solar radiation in the cool season and be shaded in the heat of summer. For example, many Native American tribes built their villages so that all homes got solar heat in the winter. In the modern era, builders and architects began the green building movement as part of the environmental movement of the 1960s and 1970s. This was motived by general concern for the environment, demand for friendlier construction materials, and need for energy efficiency. This was pushed by rising energy costs as oil prices rose from the 1970s on, which encouraged new research into energy efficiency and renewable energy.

Businesses concerned with their environmental impact can use sustainable materials and methods in the construction of new office buildings. (Ian Jeffery/iStockphoto.com)

Early rating systems for buildings came from the UK and Canada, which began in the late 1980s. An early nonresidential building rating system was called the British Research Establishment Environmental Assessment Methodology (BREEAM). Canadians developed a system called Building Environmental Performance Assessment Criteria (BEPAC).

The green building movement really became organized in the 1990s. The city of Austin, Texas, introduced the first local green building program in 1992, and U.S. President Bill Clinton set up the "Greening of the White House" program in 1993. One of the first U.S. standards was introduced in 1998, when the U.S. Green Building Council (USGBC) developed Leadership in Energy and Environmental Design (LEED).

Any building can be "green," including a single home, apartment complex, school, hospital, shopping mall, or office block. Typical green building categories include sustainable sites, energy efficiency, water efficiency, materials and resource use, indoor environmental quality, emissions, project/environmental management, and operations/maintenance. But there are many different ways to quantify how green a structure actually is, so several organizations have developed green building standards, which provide meaningful comparisons. Four examples of green building standards are:

1. International Code Council's 2012 International Green Construction Code (www.iccsafe.org): "The IgCC is the first model code to include sustainability measures for the entire construction project and its site—from design through construction, certificate of occupancy and beyond. The new code is expected

to make buildings more efficient, reduce waste, and have a positive impact on health, safety and community welfare." This code is adaptable, enforceable, and consensus-based with a focus on baseline green requirements, economic benefits, and professional development, and technical support.

2. American Society of Heating, Refrigeration, and Air-Conditioning Engineers' (www.ashrae.org) ANSI/ASHRAE/USGBC/IES Standard 189.1-2011, Standard for the Design of High-Performance Green Buildings Except Low-Rise Residential Buildings (ASHRAE 189.1): "Standard 189.1 provides total building sustainability guidance for designing, building, and operating high-performance green buildings. From site location to energy use to recycling, this standard sets the foundation for green buildings by addressing site sustainability, water use efficiency, energy efficiency, indoor environmental quality (IEQ), and the building's impact on the atmosphere, materials and resources."

3. Green Building Initiative's ANSI/GBI 01-2010: Green Building Assessment Protocol for Commercial Buildings (Green Globes) has certified 1,283 buildings totaling 21 million square meters (228 million square feet) in the United States. (www.thegbi.org) "Green Globes identifies opportunities and provides effective tools to achieve success. A nationally recognized green rating assessment, guidance and certification program, Green Globes works with you to realize sustainability goals for new construction projects, existing buildings and interiors."

4. U.S. Green Building Council's Leadership in Energy and Environmental Design (LEED) has certified over 93,000 projects amounting to over 576 million square meters (6.2 billion square feet) in 162 countries worldwide with four levels of achievement: Platinum, Gold, Silver, and Certified, depending on point totals. LEED version 4 (https://new.usgbc.org/leed-v4) certifies five types of projects: homes, neighborhood developments, building operations and maintenance, interior design and construction, and building design and construction. Highly detailed point systems are used for both required and optional activities that contribute to green buildings: energy, health and human experience, innovation, global/regional/local, location, materials, regional impacts, sustainable sites, transportation, waste, and water. The LEED website notes that "buildings are alive, and because of that, there's an expanded focus on metering and monitoring, which encourages building owners to track energy, water and ventilation rates."

Concerns about standards and certification have emerged, however, particularly related to the widespread use of the LEED system, which has become the largest certification program in the world (*USA Today*, 2013). First, the process of certification and verification costs a lot of money, so some builders or owners simply cannot afford the seal that indicates they have a LEED-certified green building. Second, the complex point system has been criticized because a builder can put effort into "easy" actions, like installing a bike rack (that may not actually be used due to a location along a busy highway), but not really tackle challenges like reducing high water use in a desert region. Some of these concerns have been addressed in the latest version of LEED. The stakes are high because hundreds of agencies at the city, state, and national levels now require LEED certification for newly constructed public buildings, and many towns give big tax breaks to businesses that build LEED-certified buildings.

Green building certification has become a notable label in current society; the question remains if it truly advances the intended goals of promoting environmental and human health. Overall, these goals can be achieved without a certification if city planners, architects, builders, and environmental geographers work to implement green design. Indeed, all buildings could simply be built as green buildings in the future.

See also: Section 3: Assessments and "Footprints," Sustainable Cities, Water Conservation

FURTHER READING

EPA. 2018. "Green Building Standards." U.S. Environmental Protection Agency. January 23. https://www.epa.gov/smartgrowth/green-building-standards.
GBI. 2018. "Comparing Green Globes & LEED." Green Globes. Green Building Initiative. https://www.thegbi.org/files/training_resources/Comparing_Green_Globes_LEED.pdf.
NRCAN. 2018. "Energy Efficiency in Buildings." Natural Resources Canada. http://www.nrcan.gc.ca/energy/efficiency/eefb/buildings/13556.
USA Today, 2013. "In U.S. Building Industry, Is It Too Easy to Be Green?" June 13. https://www.usatoday.com/story/news/nation/2012/10/24/green-building-leed-certification/1650517/.
World GBC. 2018. "About Green Building." World Green Building Council. http://www.worldgbc.org/what-green-building.

Green Consumerism

In industrialized, consumer-driven societies, the choices people make when they spend money shopping are seen as a way to engage in pro-environmental, sustainable behavior. Green consumerism is the purchasing actions that people take with the intention of promoting positive environmental outcomes.

In 1990, *The Green Consumer* was one of the first books to address this topic. It notes that "the marketplace is not a democracy; you don't need a majority opinion to make change. Indeed, it takes only a small portion of shoppers—as few as one person in 10—changing buying habits for companies to stand up and take notice" (Elkington et al., 1990: 9). Even today, consumers tend to believe that purchasing greener products is a method of "voting" with their wallet that can lead to major shifts in manufacturing and production.

In fact, now manufacturers and retailers realize that they must pay attention to the environmental concerns expressed by consumers and advocacy groups. Many companies have responded to this consumer demand by creating more environmentally friendly products through greener manufacturing processes (Haws et al., 2014).

Consumer actions like installing solar panels on the roof are a "big green gesture" that cost a lot of money, so it is helpful to know that there are also smaller, less expensive actions that can still help the environment. These include 1) buying

Fleece Jackets from Recycled Bottles!

All plastics are made from petroleum. For every ton (1,000 kilograms or 2,000 pounds) of plastic recycled, about 3.8 barrels (169 gallons) of oil are saved (Recycle Everywhere, 2017). But recycling plastics can be confusing because there are thousands of different types and community recycling programs only take specific kinds. That is because each type is made from a different chemical and petroleum content so they cannot be mixed for recycling.

The most common plastics accepted for recycling are two types of plastic beverage containers. The thinner, single-use soda and water containers are called PET (polyethylene terephthalate) and labeled #1. And the thicker bottles and containers for detergent and other liquids are called HDPE (high-density polyethylene) and are labeled #2. Plastic shopping bags are also usually #2, but cannot be recycled with #2 bottles (Maine.gov, 2016). Clear plastic is preferred by the recycling companies, as it is more versatile and gets a higher price. Many recycling facilities only take #1 and #2 bottles. Even if a facility takes water bottles, they rarely accept the harder plastic bottle lids. And when the price of oil goes down, some recycling facilities refuse to take plastics at all because they cannot sell them for a profit.

So what actually happens to a plastic water bottle that is placed in a recycle bin? Here's the process: 1) A truck empties the bin and takes the empty plastic bottles to a materials recovery facility to be sorted by type and color. 2) Old bottles are chopped up and sorted again through a flotation process. 3) The pieces are then melted and squeezed into strands that are dyed and spun into fiber or chopped into pellets. 4) This plastic is used to make new items (Recycle Everywhere, 2017).

What items can be made from recycled plastic? Well, #1 bottles can be used to make carpet, backpacks, polar fleece, or sleeping bag insulation. Recycled #2 milk jugs can become plastic "lumber" for park benches or play sets. The #2 detergent bottles are used to make hard plastic buckets, Frisbees, or stadium seats (Maine.gov, 2016).

Much of the recycled plastic in the United States is gathered and crushed into 1,000-pound bales. For many years, most of U.S. recycled plastic was exported to China, where it was used for plastics in new consumer items, but now China's importing laws are changing. People say that American companies should increase their processing and use of recycled plastics, and there are certainly some examples of this occurring (Staube, 2017).

Outdoor companies are making fleece jackets made from recycled plastic bottles—it takes about 25 recycled plastic water bottles to make one jacket! On the other hand, until consumers demand more recycled-content items, companies find that new plastic is easier and cheaper to source. The amount of clothing made from recycled fibers is "miniscule" compared to that made of new plastic (CNBC, 2017). Consumers could demand that these big companies help "close the loop" and make more products out of recycled plastic.

Further Reading

CNBC. 2017. "Almost No Plastic Bottles Get Recycled into New Bottles." April 24. http://www.cnbc.com/2017/04/24/almost-no-plastic-bottles-get-recycled-into-new-bottles.html.

Maine.gov. 2016. "What Do Your Recyclables Become?" Main Department of Environmental Protection. https://www1.maine.gov/dep/waste/recycle/whatrecyclablesbecome.html.

Recycle Everywhere. 2017. "What They Become." Canadian Beverage Container Recycling Association. http://www.recycleeverywhere.ca/recycling-info/what-they-become/.

Staube, Christopher. 2017. "How National Sword Is Upending Exports." Plastics Recycling Update. May 24. https://resource-recycling.com/plastics/2017/05/24/national-sword-upending-exports/.

environmentally friendly cleaning supplies to avoid adding toxins to the water supply; 2) stopping the purchase of bottled water—to reduce single-use plastic waste; 3) buying products made from post-consumer recycled content to help "close the loop" and encourage companies to manufacture such products; 4) unplugging appliances and devices when not in use because vampire power increases electricity use and related emissions; and 5) eating local food to reduce the carbon footprint of food distribution and transportation (Pisani, 2010).

According to surveys, 70% of consumers say they consider the environmental impact of their purchasing and 85% of consumers say they buy green products, but marketing data show that only 8% actually do. This is likely due to ethical reasoning, according to an ethics expert, who notes that "buying green products presents people with a social dilemma: they have to be willing to pay premium prices—not for their own direct benefit, but for the greater good." Thus the difference between survey claims and actual behavior occurs because "while people love to voice their idealism to survey companies, the cold facts are they almost always put their self-interest first" (Entine, 2011). Consumers buy green products if they are as effective and of the same quality as cheaper nongreen items. And then only if there is proof to convince them that the higher price actually signifies a measureable environmental value.

What began as a somewhat hippie, free-spirited consumer movement in the 1970s to buy natural products and "save the Earth" has now shifted. These same consumers, a bit older, are quite skeptical of green consumerism. The majority of seniors are indifferent to green purchasing, and 25% actually believe that "it makes no difference." Still others, holding true to their environmentalist roots, say that green consumerism is an oxymoron (two words that collide as opposites) because the only real solution to environmental problems is to reduce or eliminate consumerism (Sachdeva et al., 2015).

Corporations realize the value of green consumerism among younger people who are still interested and older consumers who now need convincing has led to more scrutiny of environmental marketing claims. Now companies must provide specific information to differentiate themselves, as random, vague ecofriendly statements on a label no longer satisfy consumers. Companies must be transparent, specific, and creative by focusing on innovative green technology and honest corporate sustainability commitments to enhance their brand identity. Specifically, they need documented green credentials, which can be verified by consumers through organizations such as Cradle to Cradle manufacturing, the Forest Stewardship Council, Green Seal, or the U.S. FTC Green Guides, among others (*Scientific American*, 2016).

At the same time, green marketing may be a reputational risk for some big companies because so many activists and consumers will claim such efforts are just greenwashing, which is "when environmentally imperfect companies make green claims." Ironically, this has actually led many companies to covert environmentalism, or "burying mentions of their green deeds in their websites or corporate responsibility reports, rather than tout them on products or advertisements and risk the wrath of critics" (Makower, 2010).

In fact, business experts say that green consumerism now means much more than individual people voting with their wallet and deciding to buy environmentally friendly products. Instead it goes to the corporate level, where companies are shifting to sustainable production methods that lower their long-term costs. Companies realize that greening their supply chain can increase profits. Consumers demanding transparency and real sustainability actions further encourage these efforts (Entine, 2011).

There are three main critiques of green consumerism. First, it shifts responsibility of environmental degradation onto the consumer instead of business or government. Indeed, corporations could simply implement innovative green technologies across their manufacturing process, and government regulations could simply require that all manufacturing follow "green" methods. Second, green consumerism ignores or underemphasizes that the basic cause of environmental problems is overconsumption or consumerism. The bottom line is: if you don't need it, don't buy it. Third, green consumers must be diligent and well informed to avoid spending money on products that are actually part of a profitable greenwashing effort to cover up a company's true negative environmental activities.

See also: Section 3: Earth Day, Electric Cars, Green Buildings, Renewable Energy

FURTHER READING

Denniss, Richard. 2017. "To Cure Affluenza, We Have to Be Satisfied with the Stuff We Already Own." *The Guardian*. October 29. https://www.theguardian.com/business/2017/oct/30/to-cure-affluenza-we-have-to-be-satisfied-with-the-stuff-we-already-own.

Elkington, John, Julia Hailes, and Joel Makower. 1990. *The Green Consumer*. London: Gale Group.

Entine, Jon. 2011. "Eco Marketing: What Price Green Consumerism?" Ethical Corporation. September 1. http://www.ethicalcorp.com/environment/eco-marketing-what-price-green-consumerism.

Haws, Kelly L., Karen Page Winterich, and Rebecca Walker Naylor. 2014. "Seeing the World through GREEN-Tinted Glasses: Green Consumption Values and Responses to Environmentally Friendly Products." *Journal of Consumer Psychology* 24(3): 336–354.

Makower, Joel. 2010. "The Green Consumer, 1990–2010." Green Biz. March 29. https://www.greenbiz.com/blog/2010/03/29/green-consumer-1990-2010.

Pisani, Joseph. 2010. "Green Your Routine: Eight Easy Ways to Be a Green Consumer." CNBC.com. August 2010. https://www.cnbc.com/id/27416057.

Sachdeva, Sonya, Jennifer Jordan, and Nina Mazar. 2015. "Green Consumerism: Moral Motivations to a Sustainable Future." *Current Opinion in Psychology* 6: 60–65.

Scientific American. 2016. "Green . . ." Earth Talk. April 18. https://www.scientificamerican.com/article/are-green-labels-legitimate-or-just-greenwashing/.

Green Political Party

The Green Party is both a broad term signifying any political group that promotes environmental issues and a specific term meaning a defined political party with a specific platform. This is somewhat confusing, but is due to varying viewpoints and strong, independent ideas within the green movement through time. There is both an overarching history to the green movement, but also distinct variations among green parties in specific countries.

Overall, the green political movement began in the 1960s, but actual green political groups and candidates were created later, in the 1970s. Although these pro-environmental parties were forming, it took longer for them to win elections. By the early 1980s, Belgium had Green Party candidates earning over 5% of the vote and winning some local elections. In Italy, the Greens won national parliamentary elections by the late 1980s. British voters also have favored some green candidates, often with the important 5% threshold that grants them representation in parliament.

Indeed, a key point is that in a parliamentary system, found in Canada, Australia, New Zealand, and most European countries, multiple political parties are elected and have power at the national level. Multiple political parties must work together and compromise to get laws enacted and to get enough support within parliament to select the prime minister. This is different from the U.S. system, for example, where it is a winner-take-all approach, so only the two major parties get seats in the U.S. Congress and the president is elected separately. In a parliamentary system, smaller parties, such as the Greens, often earn seats in the national legislative body and thus influence politics.

Germany is known as the leader of Green Party influence, in part because its parliamentary political system allows for multiple parties (any party over a 5%

Democratic Green Party candidate Frank Habineza has his finger inked at a polling station in Kigali, Rwanda after casting his vote in the Rwandan Presidential election on August 4, 2017. (Stringer/Anadolu Agency/Getty Images)

threshold of votes) to have a voice. Die Grünen (The Greens) was established in 1979 and came on the scene in the early 1980s. In 1983, they earned a million votes that comprised 5.6% of the total and granted the party 28 out of 497 seats in the national German parliament (Die Grünen, 2018).

Whereas most political parties are specific to each individual country, green parties have broader common interests, so these groups work together. By 1984, the European Coordination of Green Parties was formed with representation from Belgium, France, Germany, the Netherlands, Sweden, Switzerland, and the UK. This grew as several key issues led to the need for additional cooperation (namely, the fall of the communist Eastern European political system and several environmental tragedies, such as the Chernobyl nuclear disaster). By 1993, the European Federation of Green Parties became the overarching group. At the same time, the Federation of Green Parties of the Americas was set up for collaboration among greens in North and South America.

Within the United States, the Association of State Green Parties (ASGP) was formed in the 1990s to help existing state parties develop. The ASGP became the Green Party of the U.S. (GPUS), but a significant divide grew within the movement, which also occurred in the German Green movement at the time: fundamentalists versus pragmatists. The fundamentalists want the Green Party to be a different type of political party, specifically a membership group that is active in demonstrations, protests, and activism to promote environmental and social issues. The pragmatists want the Greens to be focused on elections and be structured more like the major political parties in the United States. In the end, the Green Party of the U.S. evolved more pragmatically to compete in local, state, and national elections without accepting outside, corporate funding. The main four guiding pillars of the Green Party of the U.S. are ecology, social justice, democracy, and peace (Ballotopia, 2017; Green Party US, 2018).

This divide within the green parties exemplifies just how challenging it is to address complex environmental issues because some people believe that drastic change is the only way to truly have an impact, whereas others want to work within the existing political and social system. Further, many Green Party issues tend to have a local focus, with an estimated 200+ local politicians recognized as Green Party members, and thus attempts to unify so many unique groups is challenging. Finally, the open-minded, big-thinking, alternatively driven members of Green parties may have a difficult time coming together in a unified political party. But the Greens have had many local, state, and even national political candidates in U.S. elections since the 1990s. In each presidential election, for example, there is a Green Party ticket for president and vice president. In the 2000 presidential election, their candidate, Ralph Nader, took 2.7% of the vote, the most ever for the Green Party in the United States, but many argued this took away votes from the known environmentalist Al Gore, who ran as a Democrat (Urlaub, 2016).

Despite differing visions of their preferred political structure early on, there are now commonalities among green parties worldwide. The Global Greens Charter was adopted in 2001 by the green parties from 72 countries meeting in Canberra, Australia. This charter defines the common principles of ecological wisdom, social justice, participatory democracy, nonviolence, sustainability, and respect for diversity. These principles were upheld in the 2012 Global Greens Congress in Dakar, Senegal, when delegates signed an action plan with goals to address urgent problems in the world (Global Greens, 2018).

See also: Section 3: Case Study 13, Environmental Nongovernmental Organizations (ENGOs), Green Consumerism

FURTHER READING

Ballotopia. 2018. "Green Party." https://ballotpedia.org/Green_Party.
Die Grünen. 2018. "Partei: Wer Wir Sind." http://www.gruene.de.
Global Greens. 2018. "Global Greens Charter." https://www.globalgreens.org/globalcharter.
Green Party US. 2018. "Everything We Do Is Based on Our Four Pillars." http://www.gp.org/.
Urlaub, Per. 2016. "In Europe, the Green Party Is a Force. In the U.S., It's Irrelevant. Here's Why: Governing Requires Compromise." *Washington Post*. October 19. www.washingtonpost.com/posteverything/wp/2016/10/19/in-europe-the-green-party-is-a-force-in-the-u-s-its-irrelevant-heres-why.

Green Technology

Technology is the application of knowledge for practical purposes. Green technology (greentech) specifically draws on environmental science to develop methods and materials that are useful in everyday life to mitigate or reverse the negative impacts of human activity on the environment. Greentech is an exciting area, as these innovations are constantly being developed to simultaneously meet people's demands and the needs of the planet. Because new technology is not typically profitable right away, private companies can be hesitant to tackle such innovations. Often government programs are needed to stimulate cutting-edge, innovative greentech.

Greentech innovation requires inventions based on scientific discovery, which some countries recognize and thus provide support through public funding of research and development in a broad range of scientific fields. Countries such as Canada, Estonia, Finland, Italy, Japan, Mexico, and New Zealand allocate relatively high percentages of their overall public research and development budgets to energy and the environment (OECD, 2018). This will foster greentech innovations in various sectors. Although there are thousands of examples of greentech, a few applications include energy and transportation.

In terms of renewable energy, we typically think of wind and solar. But ocean energy also holds great promise as a greentech answer to energy demand. Over 70% of the Earth's surface is covered by oceans. This massive area collects thermal radiation from the sun and produces mechanical energy from tides and waves. Several types of new energy technologies are being developed based on ocean energy. Marine and hydrokinetic technologies capture energy from oceans and rivers—including waves, tides, ocean currents, free-flowing rivers, streams, and ocean thermal gradients—to generate electricity. These technologies are at a very early stage of development and "represent an emerging industry with hundreds of potentially viable technologies" (EERE, 2018). For example, tidal energy could be gathered from the oceans because there are two high tides and two low tides over a 24-hour period; a "fence" could collect tidal energy and change it into electricity by forcing water through ocean turbines, which would trigger a generator. Another greentech innovation could be wave energy. Anybody who has been to an ocean beach knows the sound of the waves crashing on shore. This energy could be harnessed and converted into electricity by offshore or onshore systems. Wave energy technologies are working to take energy directly from surface waves or from pressure changes that occur below the surface of the ocean. There is a huge potential for tidal and wave energy, but installing the systems is very expensive, and some people worry about harming marine wildlife. If such problems can be overcome, these innovative greentech systems would have low maintenance costs and long-term benefits because their fuel, ocean energy, is free.

Another area of greentech innovation is transportation, which can be approached from three directions, according to the U.S. Environmental Protection Agency: increasing the efficiency of vehicle technology, changing how we travel and transport goods, and using lower-carbon fuels. We need all three to help achieve our societal goals on reducing our climate-changing greenhouse gas emissions (EPA, 2017).

For example, truck manufacturers are developing new fuel cell technologies that would allow vehicles to run on oxygen and hydrogen, so their only emissions are water and heat. Currently, the production of hydrogen demands high fossil fuel use, which is not sustainable. But this could be powered by renewable energies, which would make hydrogen fuel cell vehicles a clean green alternative for truck transportation. European companies are taking the lead in this technology, with one UK firm trying to bring a hydrogen fuel cell–powered vehicle to the general public by 2020. Also related to transportation innovation are lithium-air (or lithium-oxygen) batteries, which have up to 15 times the specific energy of our current lithium-ion batteries using just oxygen, a renewable resource with no pollution. This technology can produce high energy output in proportion to their weight, which is a great innovation for green transportation. Lithium-air storage units have been in development for decades, but scientists hit a few road blocks, notably rapid loss of battery power and unpredictable short-circuiting. Recent advancements dealt with the first problem, so with a bit more greentech innovation, scientists could solve the second problem and develop long-range electric cars.

These and many other exciting greentech innovations are being developed each day by scientists, engineers, and even at-home inventors, all of which can stimulate collaboration and a shift toward more greentech deployment and usage. For example, each year *Scientific American* publishes a list of the "Top 10 Emerging Technologies" and magazines like *Popular Science* also announce their top new innovations in science and technology in a "Best of What's New" issue.

The greentech innovations noted here are actually interrelated! For example, if lithium-air batteries are developed, the use of electric cars will expand, and these could be recharged by electricity from tidal power. Government funding plays a significant role in encouraging such innovations, because these activities may not be immediately profitable and thus a risky endeavor for private companies to tackle alone. Science and greentech are closely associated, and these innovations clearly represent a bright future.

See also: Section 1: Case Study 4, Technology: Innovation and Consequences; *Section 3:* Case Study 11, Electric Cars, Renewable Energy, Sustainable Cities

FURTHER READING

Calcuttawala, Zainab. 2017. "5 Green Energy Innovations that Could Change the World." Business Insider. May 10. http://www.businessinsider.com/green-energy-innovations-change-the-world-2017-5.

Doerr, John. 2009. "The Green Road to Prosperity: Federal Investment in Entrepreneurs Will Create Jobs, Boost the Economy and Raise Energy Security." *Scientific American.* March 1. https://www.scientificamerican.com/article/the-green-road-to-prosperity/.

EERE. 2018. "Renewables." Office of Energy Efficiency and Renewable Energy. U.S. Department of Energy. April 18. https://energy.gov/eere/renewables.

EPA. 2017. "Green Vehicle Guide." U.S. Environmental Protection Agency. November 16. www.epa.gov/greenvehicles/routes-lower-greenhouse-gas-emissions-transportation-future.

OECD. 2018. "Green Technology and Innovation." Organization for Economic Cooperation and Development. https://www.oecd.org/sti/outlook/e-outlook/stipolicyprofiles/newchallenges/greentechnologyandinnovation.htm.

Popular Science. 2018. "Best of What's New." https://www.popsci.com/best-of-whats-new-2017.

Scientific American. 2018. "Top 10 Emerging Technologies." https://www.scientificamerican.com/article/10-emerging-technologies-to-watch/.

Recycling

Consumer choices matter when it comes to recycling. First, people can choose products that are easily recyclable. Second, they can participate in a recycling program (curbside pickup or dropping off at a recycling center). Third, shoppers can help "close the loop" by buying products with a high recycled content. If consumer demand for these recycled products remains low, then companies will not purchase recycled materials, and the entire cycle is broken.

There are several key terms to help inform consumers:

- *Recyclable product* means an item can be collected, processed, or manufactured into a new product after it has been used, but this item does not necessarily contain any recycled material, and each community accepts different recyclables.
- *Recycled content* products are manufactured with recycled materials, often from waste recovered during manufacturing (like fabric that falls to the factory floor after useful pieces are cut out).
- *Post-consumer content* indicates that an item was made from recycled materials that were collected from consumers in a recycling program. This is the term that really indicates that there is a "closed loop" from manufacturing, to consumer use, to recovery of a material.

Recycling rates of municipal solid waste (trash) in the United States vary significantly by type of material: paper, aluminum cans, glass, and plastics, for example.

- *Paper* comprises the largest component of municipal waste at about 30% of all trash. Of this, about 65% is recycled. This recycled paper is used to make new paper items, thus saving trees and fresh water needed in paper production. Most recycle programs, both dropoff and curbside pickup, accept paper and brown (nonwaxed) cardboard. Consumers can seek out products made from recycled paper, as this helps build a strong demand for it.
- *Aluminum* soda and beer cans are recycled at a rate of 54%, so about half still go to landfills. This wastes both mineral and energy resources, as it requires 95% less energy and water to create a can from recycled aluminum than from new raw materials. As an example, recycling one aluminum can would save the equivalent energy needed to power a TV for three hours. Another comparative statistic: the amount of aluminum cans that Americans send to the landfill every three months is enough to build the entire U.S. commercial airline fleet (Recycle Across America, 2018).
- *Glass* only has a 26% recycling rate, meaning that nearly three-quarters of all glass containers (11.5 million tons in the United States annually) end up in landfills (EPA, 2016). This is particularly illogical, as glass food containers can be recycled and used again and again. Historically, it was typical that people took glass milk jugs and glass soda pop bottles back to the distributor to be sanitized and refilled. But this refilling option is no longer available in most places. So recycling is second best, as making new glass from recycled glass is cheaper than consuming raw materials.
- *Plastics* are hugely popular in industry and among American consumers, who generate 33 million tons, or 13% of the waste stream. Only 9% of this is recycled. The numbers 1 to 7 are printed on the bottom of many plastic containers in an attempt to help consumers figure out if their local recycling program accepts each type. The typical numbers and the type of plastic resin are: #1: PET (polyethylene terephthalate), #2: HDPE (high-density polyethylene), #3: Vinyl, #4: LDPE (low-density polyethylene), #5: PP (polypropylene), #6: PS (polystyrene), and #7: Other (mixed plastics) (EPA, 2018). The #1 plastic water bottles are most accepted in recycling programs because the market for this type of plastic is more established. For example, many companies now make items from this recycled plastic: recycling five #1 plastic bottles is enough fiber to fill one ski jacket. Because Americans throw away 2.5 million plastic bottles every hour, there is clearly opportunity to expand the use of these fibers in manufacturing.

Recycling levels in the United States have not increased in the last 20 years despite millions of dollars spent on recycling awareness and education. The recycling industry is constantly battling "contamination," which is when recycled materials are incorrectly mixed together so the recycling facility cannot sell the materials to be refabricated and used to make new products. For example, if plastics of varying

Zero-Waste Home

Garbage or trash is called municipal solid waste (MSW) by environmental managers, and it consists of all the things people use and throw away: clothing, bottles, food packaging, newspapers, food scraps, batteries, paint, appliances, etc. In the United States, over 167 million tons of garbage, which is equal to 3 pounds per person, per day, is thrown away each year, most of which goes into dumps or landfills (EPA, 2016). A municipal solid waste landfill (MSWLF) is a defined area of land where waste is deposited below, at, or above ground level and is managed to receive waste (UNITR, 2013). There are over 3,000 landfills in the United States today.

When seeking to reduce the waste that ends up in landfills, people talk about the 3Rs: reduce, reuse, and recycle. Reduce the material items ("things") each person buys. Reuse things already owned rather than buying new items. Recycle materials so they may be used to make another item.

This leads to the notion of zero waste, which takes a shift in lifestyle. "Creating sustainable lifestyles means rethinking our ways of living, how we buy, what we consume and how we organize our daily lives" (UNEP, 2018).

Downsizing an apartment or home is one first step; then a family can organize the essential items that are truly needed. In addition, by reducing the amount of horizontal surfaces, like shelves, people are forced to give up extra stuff they don't really need. To specifically address the issue of trash, buying natural-fiber products, like cotton and linen, allow them to be composted. Buying things like vinegar and baking soda in bulk and then making your own simple cleaning supplies greatly reduces the amount of waste from such products. Buying food in bulk greatly reduces the plastic wrappings, which go straight into the trashcan.

According to the author of the book *Zero Waste Home*, people can move toward achieving this goal by following the 5R's: refuse, reduce, reuse, recycle, and rot. Refuse what you do not need. Reduce what you do need. Reuse by using reusable containers. Recycle what you cannot refuse, reduce, or reuse. And rot (compost) the rest (Johnson, 2013).

It might be tricky to completely change lifestyles and create only a shoebox full of trash in an entire year (as many dedicated families are able to do). Perhaps there is a moderate, "doable" alternative, whereby each person rethinks their consumption and reduces the waste they produce. Zero waste is the ultimate goal, but even if each person reduced their garbage by 20% to 25%, this would make a significant, positive difference on the environment.

Further Reading

EPA. 2016. "Advancing Sustainable Materials Management: 2014 Fact Sheet." United States Environmental Protection Agency. https://www.epa.gov/sites/production/files/2016-11/documents/2014_smmfactsheet_508.pdf.

Johnson, Béa. 2013. *Zero Waste Home: The Ultimate Guide to Simplifying Your Life by Reducing Your Waste.* New York: Scribner.

UNEP. 2018. "Sustainable Lifestyles: Consume Differently." United Nations Environmental Programme. http://web.unep.org/resourceefficiency/what-we-do/sustainable-lifestyles-consume-differently.

UNITR. 2013. "Guidelines for National Waste Management Strategies." United Nations Institute for Training and Research. http://cwm.unitar.org/national-profiles/publications/cw/wm/UNEP_UNITAR_NWMS_English.pdf.

numbers are dumped into one bin, this prohibits all of them from being melted down into new fiber because the various types of petrochemicals in the plastics cannot be combined. This contamination is due to confusion about separating recycled materials, which leads to apathy and skepticism about recycling. Recycling levels could increase if standardized labels were displayed on recycling bins throughout society. One estimate is that if recycling levels reach 75% in the United States, 1.5 million new jobs would be created and the ecological equivalent would be like removing 50 million cars from the road each year (Recycle Across America, 2018).

Leading the world, Germany recycles and composts 65% of its municipal solid waste. South Korea and most Western European countries recycle about half of theirs. The United States, Poland, Hungary, Canada, and several other central European countries have recycling rates at about 30%. At the low end of the spectrum are New Zealand (0%), Turkey (1%), Chile (1%), and Mexico (5%) (McCarthy, 2016). Most countries in the Global South do not have formal waste recycling programs, but do in fact recycle at a high rate out of economic need and lack of resources. An estimated 15 million people rely on salvaging recyclable materials from the waste stream. Overall, the recycling industry employs the second largest number of people globally; only agriculture employs more people (Planet Aid, 2015).

In the United States alone, recycling is a $200 billion a year industry that generates up to 10 times more jobs than do landfills. In addition to economic benefits, the ecological gains include reduced exploitation of natural resources like minerals, timber, and water and decreased dependence on fossil fuels, which produce greenhouse gas emissions linked to climate change.

See also: Section 1: Case Study 4; *Section 3*: Composting, Green Technology, Renewable Energy

FURTHER READING

EPA. 2016. "Advancing Sustainable Materials Management: 2014 Fact Sheet." U.S. Environmental Protection Agency. https://www.epa.gov/sites/production/files/2016-11/documents/2014_smmfactsheet_508.pdf.
EPA. 2018. "Reduce, Reuse, Recycle." U.S. Environmental Protection Agency. https://www.epa.gov/recycle.
McCarthy, Niall. 2016. "The Countries Winning the Recycling Race." *Forbes*. March 4. https://www.forbes.com/sites/niallmccarthy/2016/03/04/the-countries-winning-the-recycling-race-infographic.
Planet Aid. 2015. "Recycling Rates Around the World." September 2. http://www.planetaid.org/blog/recycling-rates-around-the-world.
Recycle Across America. 2018. "Did You Know? Get the Facts Here." http://www.recycleacrossamerica.org/recycling-facts.

Renewable Energy

Globally, over 9.8 million people are employed in renewable energy, and these jobs are increasing rapidly. This renewable energy employment is concentrated in China, Brazil, United States, India, Japan, and Germany, with China alone accounting for

3.64 million workers and a 3% per year jobs growth rate. In fact, jobs in solar and wind energy have doubled in just the last four years, and it is expected that "the number of people working in the renewables sector could reach 24 million by 2030, more than offsetting fossil-fuel job losses and becoming a major economic driver around the world" (IRENA, 2017).

Renewable energy resources are constantly replenished and will never run out. This in contrast to nonrenewable energy sources, such as coal, oil, and natural gas; the supply of these fossil fuels is finite and they will eventually become too expensive to obtain. Currently, the world depends on nonrenewable energy sources, but production of renewable energy is increasing. These regenerative energy resources include solar energy, wind energy, biomass, and geothermal energy.

In the United States, about 15% of electricity produced at power plants ("utility-scale facilities") is from renewable sources, composed of the following specific types: hydropower 6.5%, wind 5.6%, solar 0.9%, geothermal 0.4%, and future innovations in ocean water energy. Keep in mind, this does not include the rooftop solar panels that are located on people's homes or the small solar arrays that communities, schools, and businesses build for their own electricity production because these are separate from the utility power plants (EIA, 2018).

Water in motion creates energy, which can be harnessed and turned into electricity. There are both large- and small-scale hydropower technologies, and hydropower is the most common and least expensive source of renewable electricity in the United States. Nearly all states employ some hydropower, and a handful of states (Oregon, Idaho, and Washington, for example) get the majority of their electricity

Wind turbines and a collection substation where power is consolidated and then sent out to the electricity grid. (iStockphoto)

from hydropower. Because this is a relatively cheap source of energy, citizens in these states tend to have lower bills than in other states. Historically, hydropower has been used to grind grain into flour since the Greek era, 2,000 years ago. By the late 1800s, technology was developed to use hydropower to provide street lights. In present times, technology improved so that hydroelectricity became more efficient and could be transmitted longer distances, with very consistent energy production. Although hydroelectric dams can negatively affect wildlife and cost high sums up-front, the benefits of hydropower are high availability in specific regions and the absence of greenhouse gas emissions.

Wind energy takes energy in the Earth's winds and, through various technologies, uses it to make electricity, charge batteries, pump water, or even grind grain. In fact, wind energy has been used for hundreds of years—think about historical windmills in Holland or the plains of North America (NREL, 2018). The modern term is wind turbine, and they can be much larger and more efficient than in the past. Turbines are mounted on a tower to obtain the most possible energy and at 30 meters (100 feet) or higher, they get less turbulent, faster winds. Most turbines have two or three blades that look like propellers. These turn around a rotor that is connected to a shaft, which spins a generator to create electricity. Wind turbines can be built on land or offshore in the ocean or large lakes. Wind farms are groups of large wind turbines (producing several megawatts of power) that provide bulk power to the electrical grid. Individual small turbines that produce less than 100 kilowatts can be used for homes or water pumping. These can generate power in remote locations that are not connected to the electric grid. In 2016, in the United States, 102,000 workers were employed at wind farms across the nation, which was an increase of 32% from the previous year (Energy.gov, 2017).

Solar energy has great potential because the sun is a powerful source of energy. Every hour, the Earth receives more energy from the sun than is used by all people on the planet for a whole year (NREL, 2018). Two main technologies can convert this solar radiation to usable energy for people's daily lives: PV and CSP. PV is photovoltaic and it is what most people are familiar with as rooftop solar panels. When the sun shines onto the panel, its cells absorb photons from the sunlight, creating an electric field across the PV layers, causing electricity to flow. Concentrating solar power (CSP) is the technology used in very large solar power plants. Mirrors are used to reflect and concentrate sunlight onto receivers that collect solar energy and convert it into heat that is used to produce electricity. In addition, there are two commonly used options for homes and businesses: solar water heating and passive solar design. Solar water heaters use the sun's heat to create hot water for a home, either through directly heating the water or transferring heat within the collector to a well-insulated hot water tank. Passive solar design is simply taking advantage of a home's site, climate, and materials to minimize energy use (NREL, 2018). For example, having large, south-facing windows can warm a house during the winter, and with careful planning of the roofline, these windows can be shaded from sunlight during the summer months. Designing a passive solar home reduces heating and cooling needs by working with the geography of the site. In 2016, in the United States, 374,000 individuals worked, in whole or in part, solar energy jobs, which was an increase of 25% from the prior year (Energy.gov, 2017).

Geothermal energy is heat from within the Earth in the form of hot water or steam that is reached by drilling. Some geothermal sources are nearer the surface and some are very deep. There are a variety of geothermal resources and applications from small to large scale. Most people have heard of geothermal heating and cooling of homes. This is a system of pipes buried underground near a building. Liquid (usually water) circulates through the loops of pipes to gain or lose heat within the ground. In the winter, the geothermal system gathers heat from the warmer ground, while in the summer it moves heat from the warmer inside air into the cooler ground. This form of energy production is possible due to constant shallow-ground temperatures. Larger-scale technologies are used in geothermal power plants where hot water located a few miles below the surface causes steam, which rotates a turbine that activates a generator, which produces electricity (EERE, 2018).

See also: Section 1: Climate Change, Energy; *Section 3:* Case Study 11, Case Study 13, Sustainable Development

FURTHER READING

EERE. 2018. "News." Office of Energy Efficiency and Renewable Energy. U.S. Department of Energy. April 18. https://energy.gov/eere/office-energy-efficiency-renewable-energy.

EIA. 2018. "What Is U.S. Electricity Generation by Energy Source?" U.S. Energy Information and Administration. https://www.eia.gov/tools/faqs/faq.php?id=427&t=3.

Energy.gov. 2017. "2017 U.S. Energy and Employment Report." U.S. Department of Energy. https://energy.gov/downloads/2017-us-energy-and-employment-report.

IRENA. 2017. "Renewable Energy Employs 9.8 Million People Worldwide, New IRENA Report Finds." International Renewal Energy Agency. May 24. http://www.irena.org/newsroom/pressreleases/2017/May/Renewable-Energy-Employs-98-million-People-Worldwide-New-IRENA-Report-Finds.

NREL. 2018. "Learning about Renewable Energy." National Renewable Energy Laboratory. https://www.nrel.gov/workingwithus/learning.html.

Sustainable Cities

By the year 2050, about two-thirds of the world's population will live in urban areas. This represents a doubling of the urban population, from 3.6 billion to over 6 billion, in just 40 years. Demands for environmental resources such as water, electricity, and waste removal will need to keep up with this immense growth. The concentrated populations in urban areas create both benefits and challenges to sustainability initiatives. For example, public transit is more viable with larger populations in a given geographic area, but densely populated areas increase water usage, so demand may exceed supply.

Sustainable cities, also known as green cities, are urban areas that have implemented environmentally friendly policies and adopted sustainable practices. People often have more influence on their local politicians, so achieving sustainability initiatives in your home town may be more feasible than trying to influence national or global environmental policies. Many criteria are included in the goals of a green city. In general, "sustainable cities have cleaner air, less traffic congestion and fewer greenhouse gas emissions" (Bloomberg and de Lille, 2016). Indeed, cities are currently at the forefront of climate mitigation efforts. For example, over 7,000 urban areas have committed to developing climate action plans as part of the Global Covenant of Mayors for Climate and Energy.

What cities are actually "green" and how do we know? One model attempts to rank U.S. cities with populations over 100,000 based on criteria from the U.S. Census Bureau and the National Geographic Society's Green Guide. The four categories are: 1) electricity (renewable sources for utilities and incentives for citizens to install small-scale renewable systems); 2) transportation (public transit, carpool programs, and air quality); 3) green living (green space, sustainable certified buildings); and 4) recycling (comprehensive city programs and citizens' views of environmental issues). Scoring at the top of this U.S. green cities list are Portland, Oregon; San Francisco, California; Boston, Massachusetts; Oakland, California; and Eugene, Oregon (Svoboda, 2016).

Another global assessment of green cities is based on comprehensive data regarding the three pillars of sustainability: social (quality of life, including health, education, income equality, crime, work-life balance, and living costs); environmental (ecological concerns, including energy consumption, renewable energy use, green space, recycling and composting rates, greenhouse gas emissions, drinking water, sanitation, and air pollution; and economic (business perspective, including transport infrastructure, ease of doing business, tourism, GDP per capita, global economic networks, connectivity, and employment rates). Zurich, Singapore, Stockholm, Vienna, London, Frankfurt, Seoul, Hamburg, Prague, and Munich were the Top 10 for 2016 (Batten, 2016).

Key issues include funding and political will to implement sustainability policies. Often green cities must make massive investments in the construction of green infrastructure, such as public transit systems. There must be commitment among citizens and local businesses to approve higher taxes in order to build now for a green future; likewise, public-private collaborations are often useful in such construction projects.

Cities in less developed countries need help to gain green technology transfer and to gain access to well-constructed housing, water and sanitation, and electricity. Indeed, the United Nations notes that building sustainable cities requires investment in renewable energy sources, efficient use of water and electricity, design and construction of compact cities, retrofitting of buildings and an increase in green space/parks, reliable and affordable public transportation, and improved waste and recycling systems (UN, 2013).

In addition to developing countries and developed Western countries in Europe and North America, other cities provide examples of changing goals to promote green actions. Dubai is a city of 2.8 million in the oil-rich nation of the United Arab Emirates (UAE) that is known for its modern architecture and a massive ecological footprint due to heavy fossil fuel usage. This city provides an example of extremes: a desert city with an indoor ski slope that demands extremely high amounts of electricity. But ruler Mohammed bin Rashid Al Maktoum announced that his city will get 75% of its energy from renewable sources and have the world's smallest urban ecological footprint by 2050. Many innovative green technologies have recently been introduced here; for example, Dubai built several giant solar power plants in the desert that will produce clean and cheap electricity. Innovative approaches are necessary, with impending freshwater shortages, so Dubai now burns natural gas to generate electricity, and the leftover heat is used to distill seawater, removing the salt. This massive facility now produces 10 gigawatts of electricity and 1.9 billion liters (500 million gallons) of desalinated water each day (Kunzig, 2017). Still, this is an example of a fossil fuel–based economy, which is unsustainable in the long term.

Urban areas have several opportunities to build a green economy that is based on renewable energy, clean technology, reduced consumerism, and acknowledging the value to ecosystem services. For example, solar energy can be promoted by changing city laws that require homeowners go through a complicated permitting process in order to install solar panels on their homes. Cities can encourage energy efficiency and remove the inflexible governmental obstacles to green building. Finally, green city design is a key component, as urban areas must build bike- and pedestrian-friendly transportation routes (EDN, 2018). Such transportation options are related to smart growth, which is a form of urban development based on a mix of land-uses and buildings, diverse housing, and transportation options within neighborhoods. Community involvement is at the core of smart growth so that an urban area is developed to provide services within walking or biking distance of home: people can live, work, go to school, and shop within their neighborhood. This drastically reduces dependence on fossil fuel transportation, thus mitigating climate change emissions and creating the basis for a sustainable city. By making green infrastructure investments, sustainable cities can make the transition to a cleaner, safer, and climate-resilient future.

See also: Section 1: Case Study 1, Urbanization; *Section 2:* Case Study 6; *Section 3*: Sustainable Development

FURTHER READING

Batten, John, 2016. "Sustainable Cities Index 2016: Putting People at the Heart of City Sustainability." Arcadia. https://www.arcadis.com/en/global/our-perspectives/sustainable-cities-index-2016/.

Bloomberg, Michael and Patricia de Lille. 2016. "Green Cities: Why Invest in Sustainable Cities?" CNN. December 1. http://www.cnn.com/2016/12/01/opinions/sustainable-cities-opinion/index.html.

Coalition for Urban Transitions. 2018. http://www.coalitionforurbantransitions.org/home/about.

EDN. 2018. "Green Cities." Earth Day Network. https://www.earthday.org/campaigns/green-cities/.

Kunzig, Robert. 2017. "The World's Most Improbable Green City." *National Geographic*. April 4. http://www.nationalgeographic.com/environment/urban-expeditions/green-buildings/dubai-ecological-footprint-sustainable-urban-city/.

NLC. 2016. "Sustainable City Institute." National League of Cities. http://www.nlc.org/program-initiative/sustainable-cities-institute.

Svoboda, Elizabeth. 2016. "America's Top 50 Green Cities." *Popular Science*. February 8. https://www.popsci.com/environment/article/2008-02/americas-50-greenest-cities.

UN. 2013. "World Economic and Social Survey: Towards Sustainable Cities." United Nations, Sustainable Development Challenges. http://www.un.org/en/development/desa/policy/wess/wess_current/wess2013/Chapter3.pdf.

Sustainable Development

The term sustainable development came to be recognized after the 1987 publication of the United Nations report "Our Common Future." This document noted that economic development and industrialization were not enough. Instead, the world should address sustainable development, which is "development that meets the needs of the present without compromising the ability of future generations to meet their own needs" (UN, 1987). In the United States, President Bill Clinton's Council on Sustainable Development wrote in 1996 that it was "an evolving process that improves the economy, the environment, and society for the benefit of current and future generations" (EPA, 2015). The three pillars of sustainability—social, environmental, and economic—must all be addressed in order to truly reach long-term sustainable development goals.

In 1992, 172 countries participated in the United Nations Conference on Environmental and Development in Rio de Janeiro, Brazil. This meeting is also known as the Earth Summit or the Rio Summit, and here lengthy negotiations produced a document called the Earth Charter, which outlined goals for a sustainable and peaceful world in the following century. The action plan for these goals was a document called Agenda 21, as an acknowledgment to the actions needed to address sustainable development into the 21st century. Public participation in policy making and an integration of environmental and social concerns were key points.

In 2000, the Millennium Declaration was established by leaders from 189 countries. This document linked social welfare principles, such as freedom and equality, with the respect for nature. Thus the Millennium Declaration, as noted by the United Nations, stresses international human rights and humanitarian laws, as well as sustainable development goals, again stressing social and natural resources that must be protected for future generations.

Addressing "The Future We Want," the United Nations held an international meeting in 2012, again in Rio de Janeiro, which specifically noted sustainability goals. These included the integration of economic, social, and environmental concerns to eradicate poverty; shift to a green economy of sustainable production and consumption; and promote freedom, peace, security, and human rights. Overall, the leaders noted that transparent, democratic institutions must advance these goals with public participation that includes citizens, nongovernmental organizations, and businesses. There is some concern, however, that the term "sustainable development" is so general that "it can mean all things to all people" (UNECE, 2018).

The concepts of sustainability flourished over the next decades until most nations and international organizations had stated goals related to these concepts. For example, in the United States, the EPA notes that within the three pillars of sustainability the following components must be acknowledged in their research and decision

Auroville, City of Peace

Are there any examples of people intentionally living together in peace and harmony? Yes, Auroville, India, was established in 1968, but it has not been completely successful, partly due to a lack of environmental sustainability.

"Auroville wants to be a universal town where men and women of all countries are able to live in peace and progressive harmony above all creeds, all politics and all nationalities. The purpose of Auroville is to realise human unity" (Auroville.org, 2018).

Auroville is a city in southeastern India that was founded by a French woman, Mirra Alfassa, known as "The Mother," with people from 124 nations in attendance. Spiritual leaders seeking peace joined with regional lawmakers in India to provide the land for this experiment. Since it began from barren land, the architects could start from scratch and attempt to build a livable city.

At the center of town is the Matrimandir, a dome structure with a solar power plant that is surrounded by carefully tended gardens. Four zones radiate out from this glowing dome: Residential, Industrial, Cultural (& Educational), and International. Around the edge is a natural area for environmental research, farms, and other housing.

The Auroville Charter is based on four unifying principles: 1) Auroville belongs to humanity as a whole; 2) Auroville will be the place of an unending education and constant progress; 3) Auroville wants to be the bridge between the past and the future; and 4) Auroville is a site of actual human unity. According to the city's website, "The Charter thus forms an omnipresent referent that silently guides the people who choose to live and work for Auroville."

In the modern day, Auroville officially has a population of 2,487, but people say it is much larger, up to 10,000, and there is a severe housing shortage. Many here are international people seeking a unique way of life that is outside mainstream society. The town, however, is not separated from the surrounding regions, so many people arrive from local areas and work service jobs (as maids or laborers) in Auroville.

Although Auroville's goal is for people to live together in harmony, this is not always the case. There is violence, some bureaucracy and regulations seem illogical, corruption occurs, thefts happen, and people are not all equal (Crowell, 2015). Some people from surrounding areas come into the city to work as servants, for example, which does not fit with the charter concepts of unity.

Part of the issue is that a community cannot be fully insulated from the outside world: Auroville's citizens claim that much of the violence is caused by people from outside. But environmental concerns are also very real, because water, fuel, and food are in short supply and must be purchased from outside.

Auroville is a fascinating experiment in how people can try to build a harmonious community, but still face challenges, the most significant of which may be the lack of sustainable natural resource availability.

Further Reading

Auroville.org. 2018. "Auroville: The City of Dawn." http://www.auroville.org/.
City of Dawn. 2012. Film by New Momentum for Humanity Unity. 55 minutes.
Crowell, Maddy. 2015. "Trouble in Utopia." Slate. July 24. http://www.slate.com/articles/news_and_politics/roads/2015/07/auroville_india_s_famed_utopian_community_struggles_with_crime_and_corruption.html.

making: social criteria (human health, environmental justice, participation, education, and communities); environmental variables (ecosystem services, green engineering, air quality, water quality, and resource integrity); and economic indicators (jobs, incentives, supply and demand, natural resource accounting, costs, and prices) (EPA, 2016).

According to the European Union, sustainable development is "at the heart of" most European policies because they recognize that social, economic, and environmental dimensions should be tackled together. Although initiated decades ago, most recently sustainable development has been mainstreamed into EU policies through the 2010 EU Sustainable Development Strategy. Indeed, the EU viewpoint is that "[a] life of dignity for all within the planet's limits and reconciling economic efficiency, social inclusion and environmental responsibility is at the essence of sustainable development" (EU, 2017).

At the global scale, world leaders voted in 2016 to adopt 17 sustainable development goals in the "2030 Agenda for Sustainable Development." These goals apply to all the world's citizens, as countries will voluntarily mobilize to "end all forms of poverty, fight inequalities and tackle climate change, while ensuring that no one is left behind" (UN, 2018). The United Nations outlines these sustainable development goals, which also follow the three pillars of sustainability. Social goals will address poverty, hunger and food security, health, education, gender equality and women's empowerment, and water and sanitation. Economic goals include sustainable energy, economic growth, infrastructure and industrialization, inequality, cities, and sustainable consumption and production. Finally, environmental goals focus on climate change, oceans, and biodiversity. Overall, the UN notes that peace, justice, and strong institutions must be built through partnerships across the globe. Countries have the primary responsibility to review their progress in implementing the goals, but this all adds up to a united global effort to promote sustainable development.

Indeed, each nation has the responsibility to work toward the global sustainability goals. As noted by former U.S. President Barack Obama: "We commit ourselves to new Sustainable Development Goals . . . we recognize that our most basic bond—our common humanity—compels us to act . . . we reaffirm that supporting development is not charity, but is instead one of the smartest investments we can make in our own future" (USAID, 2016).

See also: Section 2: Case Study 8; *Section 3*: Case Study 15, Green Consumerism, Green Technology, Sustainable Cities

FURTHER READING

EPA. 2015. "Sustainability Primer." U.S. Environmental Protection Agency. https://www.epa.gov/sites/production/files/2015-05/documents/sustainability_primer_v7.pdf.

EPA. 2016. "Sustainability." U.S. Environmental Protection Agency. https://www.epa.gov/sustainability.

EU 2017. "Sustainable Development." European Union. http://ec.europa.eu/environment/sustainable-development/.

UN. 1987. "Our Common Future:" Report of the World Commission on Environment and Development." United Nations Report. http://www.un-documents.net/our-common-future.pdf.
UN. 2018. "Sustainable Development Goals: 17 Goals to Transform Our World." United Nations. http://www.un.org/sustainabledevelopment/.
UN Library. 2018. "UN Documentation: Environment." United Nations. http://research.un.org/en/docs/environment/conferences.
UNECE. 2018. "Sustainable Development: Concept and Action." United Nations Economic Commission for Europe. http://www.unece.org/oes/nutshell/2004-2005/focus_sustainable_development.html.
USAID. 2016. "Sustainable Development Goals." United States Agency for International Development. https://www.usaid.gov/GlobalGoals.

Sustainable Diet

Does taking a shower or eating a hamburger use more water? By far, the burger (450 gallons) consumes more water than the shower (25 gallons). This is one example of the impact that food—people's dietary choices—has on the environment.

"Sustainable diets are those diets with low environmental impacts which contribute to food and nutrition security and to healthy life for present and future generations. Sustainable diets are protective and respectful of biodiversity and ecosystems, culturally acceptable, accessible, economically fair and affordable; nutritionally adequate, safe and healthy; while optimizing natural and human resources" (FAO, 2010). Indeed, environmental and human health go hand in hand, and both are in crisis. In the United States, for example, nearly 40% of adults and 19% of youth are obese, which is the highest rate ever (Larned, 2017).

Agriculture uses or degrades many natural resources: soil and water, of course, but also significant amounts of fossil fuels—directly for running farm machinery, but also indirectly because of farm chemicals like synthetic chemical fertilizers and pesticides made from fossil fuels. In addition to using natural resources, agriculture affects water supplies and greenhouse gas emissions, among other environmental resources.

Think about the water needed to produce various food, directly from watering the plant, and indirectly as plants are fed to animals. A whole head of lettuce makes a big salad and growing it requires 12 gallons of water. But animal products demand much more water, mostly because of all the feed it takes to make one pound of flesh; it is an inefficient process of converting plants into animal protein. So it takes 53 gallons of water to make one egg, 468 gallons to produce a pound of chicken, 880 gallons to make one gallon of milk, and—incredibly—1,800 gallons of water to produce one pound of beef (Kristof, 2015).

In addition to water use, livestock's contribution to the emission of GHG is due to several activities. First, fossil fuels are burned to produce mineral fertilizers used in feed production. Methane emissions are caused by the breakdown of fertilizers and from animal manure. Land-use changes and land degradation are caused by livestock feed production and livestock grazing. Finally, fossil fuel use in livestock feed and animal production, transportation, and processing causes increased GHG emissions. Geographically, it is worth noting that many fast food restaurants actually source meat products from far-away countries, which further increases GHG emissions. Overall, livestock cause significant contributions to total global anthropogenic emissions of methane (40%), carbon dioxide (9%), nitrous oxide (65%) and ammonia (64%) (FAO, 2006; IATP, 2009).

A review of numerous research articles on the topic of dietary impacts shows that GHG emissions could be reduced over 70% and water use by 50% by shifting the typical Western diet to more environmentally sustainable diets (Aleksandrowicz et al., 2016).

The European Union published a report calling for a shift to more sustainable diets that will benefit the environment and improve human health. This outlines six principles: eat more plants; eat a variety of foods; waste less food; moderate your meat consumption (both red and white); buy food that meets a credible certified standard; and eat fewer foods that are high in fat, salt, and sugar. The report outlines how countries can promote healthier and more sustainable diets:

1. Policies should revise national dietary guidelines to reflect sustainability goals, strengthen green procurement among government agencies, and support food education.
2. Update national agricultural policies so that environmental, economic, and social values influence food production and consumption.
3. Take action to prevent obesity through healthy, sustainable diets as a public health goal.
4. Better tailor economic policies to promote healthy, sustainable diets.
5. Build local to global synergies to achieve sustainable food consumption.
6. Collaborate to build supportive "best practices" between countries and other stakeholders.
7. Encourage the food and agricultural industry to participate in this shift to sustainable diets (Alarcon and Gerritsen, 2014).

Geographically, there are two very different situations for agriculture in the Global North and South. Big money is invested in agriculture in North America and Europe, where CEOs and shareholders in corporations gain profits from large-scale, industrial food production, which has led to overproduction of calories and protein in the last 50 to 70 years. In these areas, there is relatively productive land and access to water, technology, big machinery, government subsidies, and capital (money). This system is based on thinking of short-term economic gains, but causes environmental degradation and health problems, like increasing obesity, type 2 diabetes, and heart disease.

On the other hand, small-scale farmers, who comprise the majority of food production in the world, grow food on vulnerable soils with little technology. Many of them are located in the Global South, and they are constrained by land degradation, water shortages, climate change, and policies that have promoted globalized industrial agricultural production rather than long-term ecosystem and social health. In fact, 25% of the world's farmland is "highly degraded" and much of this is in the Global South, where local farmers must provide food for the growing populations, often facing health concerns, in the face of environmental change (FAO, 2011).

Globally, sustainable diets are those that are appropriate for local environmental conditions, support farmers, and build regional economies. Environmental geographers study these various components and understand how to promote policies to build local food systems based on sustainable diets.

See also: Section 1: Animal Agriculture, Case Study 2; *Section 2:* Case Study 6; *Section 3:* Alternative Agriculture, Case Study 14

FURTHER READING

Alarcon, Brigitte and Erik Gerritsen. 2014. "On Our Plate Today: Healthy, Sustainable Food Choices." LiveWell for LIFE. WWF and Friends of Europe—Les amis de l'Europe. http://livewellforlife.eu/wp-content/uploads/2014/12/LiveWell-for-LIFE_Rec-Report_English_Final.pdf.

Aleksandrowicz, Lukasz, Rosemary Green, Edward J. M. Joy, Pete Smith, and Andy Haines. 2016. "The Impacts of Dietary Change on Greenhouse Gas Emissions, Land Use, Water Use, and Health: A Systematic Review." *PLoS One* 11(11): e0165797.

FAO. 2006. "Livestock's Long Shadow." Food and Agriculture Organization. http://www.fao.org/docrep/010/a0701e/a0701e00.HTM.

FAO. 2010. "Sustainable Diets and Biodiversity: Directions and Solutions for Policy, Research and Action." Food and Agriculture Organization. http://www.fao.org/docrep/016/i3004e/i3004e00.htm.

FAO. 2011. "The State of Land and Water Resources for Food and Agriculture." Food and Agricultural Organization. https://www.globalpolicy.org/images/pdfs/SOLAW_EX_SUMM_WEB_EN.pdf.

Hamblin, James. 2017. "If Everyone Ate Beans Instead of Beef." *The Atlantic.* August 2. https://www.theatlantic.com/health/archive/2017/08/if-everyone-ate-beans-instead-of-beef/535536/.

IATP. 2009. "Agriculture and Climate—the Critical Connection." Institute for Agriculture and Trade Policy. https://www.globalpolicy.org/images/pdfs/SocEcon/2009/Hunger/Web_JimK_Cop15.pdf.

Kristof, Nicholas. 2015. "Our Water-Guzzling Food Factory." *New York Times.* May 30. https://www.nytimes.com/2015/05/31/opinion/sunday/nicholas-kristof-our-water-guzzling-food-factory.html.

Larned, Victoria. 2017. "Obesity among All U.S. Adults Reaches All-Time High." CNN. October 13. http://www.cnn.com/2017/10/13/health/adult-obesity-increase-study/index.html.

Water Conservation

Of all precipitation that falls on Earth each year, 39% goes to surface runoff (flowing into rivers and lakes) and groundwater (filling aquifers), which are called renewable freshwater resources. People install infrastructure to withdraw water from

A residential rain barrel collects water for homeowners to use on their property. (Denise P. Lett/Dreamstime.com)

these rivers and aquifers. Across the planet, 844 million people do not have access to safe water, so water conservation can have a positive impact in many regions (water.org, 2018). In the United States, the population has doubled in the past 50 years, but water usage has tripled. This has led to increasing concern, as 40 states anticipate water shortages by 2024 (EPA, 2017).

Globally, the average ratios of water use withdrawals by country are 59% agriculture (mostly irrigation), 23% municipal (also called domestic), and 18% industry (FAO, 2016). In each water-use sector, there are opportunities for conservation.

Municipal water goes to homes and offices and is highly dependent on individuals' water-use patterns. Of course, there are basic things that each person can do to conserve water at home, such as taking shorter showers and turning off the water while brushing their teeth. Some people even set up systems to reuse gray water, which has been used but is not polluted, for other uses. An example of this would be to flush your toilet with water that fills in a splash bucket set at the back of your shower. People can reduce water use by watering their yard less and planting more water-efficient plants, which is called xeriscaping. In addition to these household actions, people oftentimes do not realize that their individual decisions affect agricultural and industrial water consumption.

In fact, people make decisions every day that affect their water use through diet, transportation, and consumerism. Dietary use of water relates to the amount of water it takes to produce various types of food, with meat and dairy demanding a large amount of water. It takes about 460 gallons of water to produce a quarter-pound hamburger and 500 gallons for a serving of chicken, but only about 200 gallons for a pound of bread and 13 gallons for an orange. Thus different kinds of diets have different levels of water consumption (USGS, 2016).

In terms of transportation choices, people opting to take public transit or drive a high-fuel-efficiency car save water because the petroleum industry is highly water intensive. Petroleum production, transportation, and processing use a lot of water for drilling fluids, pressure maintenance, and construction. In fact, it takes about 13 gallons of water to produce 1 gallon of gas. Jet fuel is more highly refined; thus airplane flights use huge amounts of water. It is estimated that to produce the fuel burned by flying 3,000 miles across North America (New York to Los Angeles, for example) is about as much water as 24,000 water-efficient toilet flushes at 1.6 gallons each (NGS, 2018).

Speaking of toilets, government regulations have influenced domestic water use, as various fixtures can be regulated. For example, since 1994, all toilets sold in the United States must meet efficiency standards and use less than 1.6 gallons per flush, which is significantly lower than older toilets, which commonly used up to 7 gallons. Shower heads must also spray less than a specific gallon-per-minute water amount.

Industrial water use is more complex and has different methods of regulation. For example, since 1996, the U.S. Safe Drinking Water amendments have required the EPA to develop guidelines to help utilities develop water conservation plans. Still U.S. industries withdraw a significant—248 cubic kilometers (66 trillion gallons)—amount of water each year, of which 219 cubic kilometers (5.8 trillion gallons) goes to cooling electric power plants and 7.35 cubic kilometers (200 billion gallons) for mining (FAO, 2016).

Industry and manufacturing use water in numerous ways during the manufacturing process, for example, to change or maintain temperature, clean equipment, or halt drying between stages in an assembly line. Likewise, because industrial facilities are mostly self-contained, there are obvious ways they can tally and reduce their water use, such as optimizing processes (efficiency techniques) and reusing water onsite. But more important, the industries are dependent on consumers, so if people buy fewer items, the industries will use less water.

Consumers can conserve water simply by recycling their trash. For example, Americans throw away 62 billion kilograms (136 billion pounds) of paper each year. This is a water-intensive product that could be reused or recycled. A person saves about 13.3 liters (3.5 gallons) of water by recycling one pound of paper, which is about the amount of a typical newspaper (NGS, 2018).

Water conservation can be as simple as reducing consumption of manufactured goods, because if fewer items are made, less water is used by industries. Nearly every material item is produced in a factory that uses water, so each item a person buys is part of their water usage. Plastics are omnipresent, from shampoo bottles to toys to cell phones. Producing plastic is water intensive, as it takes about 24 gallons of water to make one pound of plastic. In fact, the amount of water required to make a plastic water bottle is about six or seven times what is actually inside the bottle. It takes about 12,760 liters (3,371 gallons) of water to manufacture one smart phone (Friends of the Earth, 2015). Cotton production is highly water intensive. It takes an estimated 650 gallons of water to grow just one pound of cotton (USGS,

2016). This means that it takes a lot of water to make one new cotton t-shirt, an item so common among young people that it is almost considered disposable in industrialized societies.

Overall, water conservation addresses water shortages by improving both the quality and quantity of water resources. Most fresh water that is withdrawn is "used" and then returned to the environment at a later time as wastewater. This water may be polluted or diminished in quality, and thus require water treatment before it is safe for use or reintroduction into the environment. Water conservation reduces withdrawals, thus protecting water resources and promoting future water resource availability.

See also: Section 1: Case Study 2, Water Scarcity; *Section 3:* Assessments and "Footprints," Recycling

FURTHER READING

EPA. 2017. "Water Conservation at EPA." U.S. Environmental Protection Agency. March 7. www.epa.gov/greeningepa/water-conservation-epa.

FAO. 2016. "Water Uses." Food and Agriculture Organization. http://www.fao.org/nr/water/aquastat/water_use/.

Friends of the Earth. 2015. "Mind Your Step: The Land and Water Footprints of Everyday Products." https://www.foe.co.uk/sites/default/files/downloads/mind-your-step-report-76803.pdf.

NGS. 2018. "Water Conservation Tips." National Geographic. http://www.nationalgeographic.com/environment/freshwater/water-conservation-tips/.

USGS. 2016. "Choose How Much Water It Takes to Make/Grow." U.S. Geological Survey Water Science School. https://water.usgs.gov/edu/activity-watercontent.html.

Water.org. 2018. "The Water Crisis." https://water.org/our-impact/water-crisis/.

Glossary

Adaptation
the process of changing to better fit a situation; a feature or behavior that helps one survive in a changing environment.

Agrarian
related to farming and agricultural production; rural areas and landscapes.

Anthropocene
a proposed new epoch in geologic time that denotes significant, detectable human impact on the Earth's ecosystems.

Anthropogenic
something that is the result of the influence of human beings on nature; changes in nature that are caused by people.

Biodiversity
also called biological diversity; the variety of life on Earth; includes variation and diversity within species, between species, and of ecosystems.

Circular Economy
no-waste system in which natural resources in manufacturing are returned to the Earth without degradation; consumer goods designed to be repaired, not replaced, and natural resources reused repeatedly to make other items. Opposite of the current linear economy in which items are made, used, and disposed of.

Citizen Science
the general public participating in the collection or analysis of environmental data to assist scientists.

Climate Change, Anthropogenic
alteration of the Earth's global climate system due to human activities, including industrialization, burning fossil fuels, and land-use change, that lead to higher levels of atmospheric greenhouse gases; causes global warming, sea level rise, changes in weather and storms, and other environmental change.

Colonialism
countries acquiring political control over another region and exploiting its people and natural resources.

Columbian Exchange
transfer of animals, plants, humans, cultures, ideas, and technologies between Europe and its colonies after voyages by Columbus and other explorers in the late 1400s.

Conservation
protecting and using natural resources to ensure the highest economic and social benefits over the long term.

Consumerism
social and economic system that encourages consumption of ever-increasing amounts of goods and services; exemplified as an obsession with shopping.

Desertification
process by which fertile land becomes desert; typically caused by drought, deforestation, poor agricultural techniques, or climate change.

Earth Day
annual event held on April 22 to demonstrate support for environmental protection; began in 1970.

Ecocentric
approach or philosophy that has a nature-centered system of values.

Ecological Footprint
measurement of the land area required to sustain a population with a given level of consumption; the amount of nature needed to uphold a specific standard of living.

Ecosystem Services
environmental benefits people obtain from ecosystems.

Ecotourism
vacationing in a place to experience nature; if done appropriately, sustainable travel that supports the local environment and minimizes ecological impacts.

Electronic Waste
also called e-waste; electronic products such as cell phones, computers, and televisions that are discarded with increasing frequency; can leak lead, mercury, arsenic, and other toxins into landfills.

Emergency Management
organizations that deal with humanitarian responses to hazards, including natural disasters; based on preparedness, response, and recovery.

Endangered Species
a plant or animal species that experts have categorized as very likely to become extinct.

Environmental Justice
equal protection from environmental hazards; inclusion of all people, regardless of race, gender, culture, income, or class, in environmental policy development, implementation, and enforcement.

Environmental Management
planning, developing, implementing, and maintaining policies and actions that protect the environment.

Environmental Nongovernmental Organization (ENGO)
also called a civil society organization; a not-for-profit organization working on environmental issues.

Environmental Policy
laws and regulations regarding environmental issues.

Environmental Protection Agency (EPA)
established in 1970; the mission of this U.S. government agency is to protect human health and the environment by creating and enforcing regulations based on laws passed by the U.S. Congress.

Environmentalist
a person who cares about the natural environment and seeks to protect it.

Extinction
dying out or termination of a species on Earth; rate accelerated by human activities.

Fair Trade
a movement to help farmers and producers in developing countries earn a fair wage for their crops and products sold in wealthier countries.

Feedback Loop
a situation where part of the output becomes a new input in a system; can enhance change already underway, particularly in an ecological system.

Food and Agriculture Organization of the United Nations (FAO)
established in 1945 with headquarters in Rome, Italy; the mission of this international agency is to reduce hunger by helping developing countries modernize and improve agricultural practices.

Food Justice
when all people have equal rights to grow, sell, and eat healthy food that is affordable, nutritious, and healthy for the land, workers, and animals.

Food Security
when all people have social, economic, and physical access to enough nutritious food to meet their dietary needs.

Fossil Fuel
nonrenewable energy resources that formed over hundreds of millions of years deep under the Earth's surface from decaying plants and animals; includes oil, coal, and natural gas; burning of these releases greenhouse gases (carbon dioxide, methane, nitrous oxide) and other pollutants into the air.

Fracking
full name is hydraulic fracturing; injection of chemical liquids into the ground to force open existing cracks in the rock so that oil or gas can be extracted.

Geographic Information System (GIS)
a computer system used to gather, manage, analyze, and present geographic data; allows for layers of geographic information to be displayed on one map.

Geographic Scale
map scale is the ratio of distance on a map to the corresponding distance on the Earth's surface; more broadly, geographers often discuss phenomenon at the local, regional, national, and global scales.

Global North
similar to more developed countries or First World countries; nations that have more advanced economies based on high levels of industrialization and urbanization.

Global South
similar to less developed countries or Third World countries; nations that have economies based on agriculture, with low levels of industrialization and urbanization.

Global Warming
rise in average temperatures of Earth's oceans and air; one example of climate change.

Grassroots
local, community participation, or involvement that is not typically associated with government or existing policy.

Green Building
also called green construction or sustainable building; environmentally responsible and resource-efficient methods of a building's construction, renovation, operation, and maintenance.

Green Space
undeveloped land that has grass, trees, or other vegetation; can include parks and community gardens.

Green Technology
also called greentech, clean technology, or environmental technology; use of science and technology to make products and practices more environmentally friendly.

Greenhouse Gas (GHG) Emissions
release of gases that trap heat in the lower atmosphere making Earth warmer; human actions such as burning of fossil fuels and deforestation have caused almost all of the observed increase in atmospheric greenhouse gases in the last 150 years.

Gross Domestic Product (GDP)
a common measure of a country's economy; the total value of all goods and services produced in a specified place and time.

Hazardous Waste
waste that poses a real or potential threat to the environment or human health.

Industrialization
process of shifting a society's economic structure from agriculture to manufacturing; related changes include greater urbanization, intensified natural resource consumption, and increased fossil fuel use.

Intergovernmental Panel on Climate Change (IPCC)
established in 1988; this international group of experts is organized through the United Nations to provide objective, science-based information on climate change and its impacts.

Land-Use Change
modification of natural landscapes and wilderness due to human actions such as agriculture, urbanization, and deforestation; affects ecosystems, biodiversity, carbon storage, and climate change.

Least Developed Countries
defined by the United Nations as low-income countries with severe economic, political, and environmental barriers to sustainable development; 47 countries currently listed, which means they have access to specific development assistance.

Life Cycle Assessment (LCA)
a tool used to evaluate potential environmental impacts of a product, process, or activity through all its stages: from extracting raw materials, to manufacturing, use, and final disposal.

Lobbyist
someone hired by a business to persuade policy makers to make laws that support that business; a person who tries to influence legislation on behalf of a special interest; the number of registered lobbyists in the United States is 10,000.

Mitigation
reducing the frequency or severity of exposure to risks; efforts to reduce the impact of a natural disaster event.

Model
a representation of an environmental system or process using data, statistics, and geographical variables; can be used to inform environmental management or predict how changes in policies may alter ecological conditions.

Multinational Corporation (MNC)
a large company that does business in more than one country, so lacks geographic loyalty; some MNCs have political influence.

National Aeronautics and Space Administration (NASA)
established in 1958; this U.S. government agency is responsible for aerospace research, flight technologies, scientific programs, and the space program.

National Oceanic and Atmospheric Administration (NOAA)
established in 1970; this U.S. government agency within the Department of Commerce focuses on scientific research and education related to oceans and the

atmosphere; responsible for daily weather forecasts, severe storm warnings, coastal restoration, and supporting marine commerce; NOAA affects over one-third of the U.S. economy.

National Weather Service (NWS)
established in 1870; this U.S. government agency within NOAA delivers weather, water, and climate data; provides forecasts and warnings help protect people and property.

Natural Disaster
a naturally occurring environmental event that has severe negative consequences for humans, including property damage, injuries, or fatalities; includes droughts, earthquakes, floods, hurricanes, tornadoes, tsunamis, wildfires, and volcanic eruptions; can be rapid or slow onset; *example:* an earthquake.

Natural Hazards
natural environmental events that cause some risk to humans; includes droughts, earthquakes, floods, hurricanes, tornadoes, tsunamis, wildfires, and volcanic eruptions; can be rapid or slow onset; *example:* living near a fault line.

Natural Resource
something found in nature, such as minerals, fresh water, forests, or animals that is valuable or useful to humans.

Overpopulation
when the numbers of a certain species exceed the environment's carrying capacity in a given place, causing degradation of the environment, depletion of resources, and eventually population decline.

Pollution
substances or energy introduced into the environment that have negative consequences such as harming human health or impairing ecosystems.

Precautionary Principle
a cautious strategy for dealing with risk; if the full effects of an activity or product are unknown, then the activity or product should not be allowed until it is proven to be safe; the European Union follows this approach.

Preservation
protecting, maintaining, or keeping ecosystems in their natural state well into the future; protecting the environment by limiting or prohibiting human use.

Protected Area
a clearly defined geographical place that is recognized and managed to achieve long-term preservation or conservation of ecosystems and natural processes.

Public Participation
process used by government and organizations to include affected individuals and communities in decision making and policies; seeks collaborative, acceptable decisions.

Renewable Energy
energy gathered from infinite sources that naturally replenished on a human timescale, such as wind, sun, waves, and geothermal.

Smart Growth
an urban planning approach based on public engagement that encourages mixed uses, diverse housing and public transportation options, and development within existing neighborhoods to reduce urban sprawl.

Social Vulnerability
the social, economic, and other demographic variables that make a person or a group of people more susceptible to risk, such as natural disasters.

Spatial Distribution
the geographic arrangement of a phenomenon across the Earth's surface; often mapped to provide an important tool in environmental geography.

Stakeholder
a person or group of people who have an interest or concern in a given situation or activity.

Sustainability
a complex concept involving the endurance of a system into the future; typically involves balancing environmental, social, and economic factors for the long term.

Sustainable Development
balancing social, economic, and environmental concerns in development that allow people to meet their needs today and into the future.

3 Pillars of Sustainability
the economic, social, and environmental factors that must be balanced to make a system or situation sustainable over the long term.

3Rs: Reduce, Reuse, Recycle
key terms when discussing waste: *reduce* the amount of trash people create, *reuse* items so they do not get thrown away, and *recycle* things to make new objects.

United Nations (UN)
established in 1948 with headquarters in New York City; the mission of this international organization is to maintain international peace, security, and cooperation among all nations on Earth; currently 193 member states.

United Nations Environmental Programme (UNEP)
established in 1972 with headquarters in Nairobi, Kenya; this United Nations agency provides leadership and encourages partnership on environmental activities around the globe.

Vegan
a person who eats no animal products; may also avoid using anything made from animals.

World Health Organization (WHO)
established in 1948 with headquarters in Geneva, Switzerland; this United Nations agency addresses international public health issues.

Zero Waste
planning and manufacturing products to be reused and recycled so that no trash is sent to landfills.

Bibliography

INTERNATIONAL AND GOVERNMENTAL ENVIRONMENTAL AGENCIES

International

Food and Agriculture Organization of the United Nations (FAO) www.fao.org
Intergovernmental Panel on Climate Change (IPCC) www.ipcc.ch
International Federation of Red Cross and Red Crescent Societies (IFRC) www.ifrc.org
International Union for Conservation of Nature (IUCN) https://www.iucn.org/
United Nations (UN) www.un.org
United Nations Convention to Combat Desertification (UNCCD) www2.unccd.int
United Nations Development Programme (UNDP) http://www.undp.org/
United Nations Educational, Scientific and Cultural Organization (UNESCO) https://en.unesco.org
United Nations Environment Programme (UNEP) www.unenvironment.org
United Nations Framework Convention on Climate Change www.cop23.unfccc.int
United Nations Population Fund (UNFPA) www.unfpa.org
World Bank http://www.worldbank.org/
World Meteorological Organization www.wmo.int

United States

Environmental Protection Agency (EPA) www.epa.gov
National Aeronautics and Space Administration (NASA) www.nasa.gov
National Oceanic and Atmospheric Administration (NOAA) www.noaa.gov
National Parks Service (NPS) www.nps.gov
National Weather Service (NWS) www.weather.gov
Peace Corps www.peacecorps.gov
U.S. Fish and Wildlife Service (FWS) www.fws.gov
U.S. Forest Service (USFS) www.fs.fed.us
U.S. Geological Survey (USGS) www.usgs.gov

Examples from Other Nations

Australian Department of the Environment and Energy www.environment.gov.au
Environment and Climate Change Canada www.ec.gc.ca
Environmental Protection Authority of New Zealand www.epa.govt.nz
European Environment Agency www.eea.europa.eu
Ministry of Environment (Japan) www.env.go.jp/en
Ministry of Environment, Forest and Climate Change (India) www.envfor.nic.in
Ministry of Environmental Protection (People's Republic of China)
    www.english.mep.gov.cn
Secretariat of Environment and Natural Resources (Mexico) www.gob.mx/semarnat

ENVIRONMENTAL GEOGRAPHY MAGAZINES AND ONLINE SOURCES

Alternatives Journal www.alternativesjournal.ca/
Audubon www.audubon.org
E&E News www.eenews.net
E/The Environmental Magazine www.emagazine.com
Earth www.earthmagazine.org
Earth First! Newswire www.earthfirstjournal.org
Earth Times www.earthtimes.org
Ecologist www.theecologist.org
Environment News Service www.ens-newswire.com
Environmental News Network www.enn.com
ESA Observing the Earth www.esa.int/ESA
Freecycle www.freecycle.org
Geographical www.geographical.co.uk
Google Earth www.google.com/earth
Green Living www.greenlivingonline.com
Grist www.grist.org
Group on Earth Observations (GOE) www.earthobservations.org
How Stuff Works www.howstuffworks.com
LiveScience www.livescience.com
Living on Earth www.loe.org
Massive www.massivesci.com
Mother Earth News www.motherearthnews.com
Mother Nature Network www.mnn.com
National Aeronautics and Space Administration www.nasa.gov
National Geographic www.nationalgeographic.com
Natural History http://www.naturalhistorymag.com
Nature www.nature.com
New Scientist www.newscientist.com
Organic Life www.rodalesorganiclife.com
Orion Magazine www.orionmagazine.org
Our Planet (UN) www.ourplanet.com

Planet Ark (Australia) www.planetark.org
Planet Earth www.nerc.ac.uk/planetearth/
Popular Science www.popsci.com/
Real Climate www.realclimate.org/
Rocky Mountain Institute www.rmi.org
Science www.sciencemag.org/
ScienceDaily www.sciencedaily.com
ScienceDirect www.sciencedirect.com
Scientific American www.scientificamerican.com
Smithsonian www.smithsonianmag.com
Space www.space.com/about
Treehugger www.treehugger.com
Yale E360 www.e360.yale.edu
Yes! www.yesmagazine.org/

ENVIRONMENTAL NONGOVERNMENTAL ORGANIZATIONS (ENGOs)

International

350.org www.350.org
Amazon Watch www.amazonwatch.org
CERES www.ceres.org
Climate Action Network www.climatenetwork.org
Conservation International www.conservation.org
Cousteau Society www.cousteau.org
Earth Charter Initiative http://earthcharter.org
Earth Day Network www.earthday.org
EarthJustice Legal Defense Fund www.earthjustice.org
Earthwatch Institute www.earthwatch.org
Environmental Defense Fund www.edf.org
Forest Stewardship Council www.ic.fsc.org
Friends of the Earth www.foei.org
Greenpeace www.greenpeace.org
International Union for Conservation of Nature www.iucn.org
National Geographic Society www.nationalgeographic.org
Natural Resources Defense Council www.nrdc.org
The Nature Conservancy www.nature.org
Ocean Conservancy www.oceanconservancy.org
Oxfam www.oxfam.org
People for the Ethical Treatment of Animals www.peta.org
Population Connection www.populationconnection.org
Rainforest Action Network www.ran.org
Resources for the Future www.rff.org
Slow Food www.slowfood.com
Union of Concerned Scientists www.ucsusa.org

Wildlife Conservation Society www.wcs.org
World Resources Institute www.worldwatch.org
World Wide Fund for Nature (World Wildlife Fund in U.S. & Canada) www.wwf.panda.org
Worldwatch Institute www.worldwatch.org

Examples of North American Organizations

American Farmland Trust www.farmland.org
American Rivers www.americanrivers.org
American Society for the Prevention of Cruelty to Animals (SPCA) www.aspca.org
Amigos de la Tierra México: Otros Mundos Chiapas www.otrosmundoschiapas.org
Canadian Council on Ecological Areas www.ccea.org
Canadian Environmental Network (RCEN) www.rcen.ca
Canadian Parks and Wilderness Society www.cpaws.org
Defenders of Wildlife www.defenders.org
Ducks Unlimited www.ducks.org
Environmental Defense Canada www.environmentaldefence.ca
Humane Society of the United States www.humanesociety.org
League of Conservation Voters www.lcv.org
National Audubon Society www.audubon.org
National Farm to School Network www.farmtoschool.org
National Parks Conservation Association www.npca.org
National Wildlife Federation www.nwf.org
Nature Canada www.naturecanada.ca
Pronatura México www.pronatura.org.mx
Sierra Club www.sierraclub.org
Trust for Public Land www.tpl.org
Wilderness Society www.wilderness.org

Selected Books Related to Environmental Geography

This list includes a selection of the outstanding books that are of relevant to environmental geography. Many of these authors are experts on environmental topics and have written other excellent books on related topics. So explore the library and keep reading!

Abbey, Edward. 1968. *Desert Solitaire: A Season in the Wilderness.* New York: McGraw-Hill.
Adams, Ansel. 1974. *Ansel Adams: Images, 1923–1974.* New York: New York Graphic Society.
Berry, Wendell. 1977. *The Unsettling of America: Culture and Agriculture.* San Francisco: Sierra Club Books.
Carson, Rachel. 1962. *Silent Spring.* Boston: Houghton Mifflin.
Colborn, Theo, Dianne Dumanoski, and John Peterson Myers. 1997. *Our Stolen Future: Are We Threatening Our Fertility, Intelligence, and Survival?* New York: Plume.
Commoner, Barry. 1971. *The Closing Circle: Nature, Man, and Technology.* New York: Knopf.
Diamond, Jared. 2004. *Collapse: How Societies Choose to Fail or Succeed.* New York: Viking.
Dr. Seuss. 1971. *The Lorax.* New York: Random House.
Friedman, Thomas. 2008. *Hot, Flat, and Crowded.* New York: Farrar, Straus and Giroux.
Goodall, Jane. 1971. *In the Shadow of Man.* Boston: Houghton Mifflin.
Gore, Al. 1992. *An Inconvenient Truth.* Emmaus, PA: Rodale Press.
Hawken, Paul, ed. 2017. *Drawdown: The Most Comprehensive Plan Ever Proposed to Reverse Global Warming.* New York: Penguin.
Heberlein, Thomas A. 2012. *Navigating Environmental Attitudes.* New York: Oxford.
Howard, Sir Albert. 1947. *The Soil and Health: A Study of Organic Agriculture.* Greenwich, CT: Devin-Adair Co.
Jones, Van. 2008. *The Green Collar Economy: How One Solution Can Fix Our Two Biggest Problems.* New York: HarperCollins.

Kimmerer, Robin Wall. 2013. *Braiding Sweetgrass: Indigenous Wisdom, Scientific Knowledge and the Teaching of Plants*. Minneapolis, MN: Milkweed Editions.

Kingsolver, Barbara. 2007. *Animal, Vegetable, Miracle: A Year of Food Life*. New York: HarperCollins.

Klein, Naomi. 2014. *This Changes Everything: Capitalism vs. The Climate*. New York: Simon & Schuster.

Kolbert, Elizabeth. 2014. *The Sixth Extinction: An Unnatural History*. New York: Henry Holt and Company.

Lappé, Anna. 2010. *Diet for a Hot World*. New York: Bloomsbury.

Lappé, Frances Moore. 1971. *Diet for a Small Planet*. New York: Ballantine Books.

Leonard, Annie. 2010. *The Story of Stuff: The Impact of Overconsumption on the Planet, Our Communities, and Our Health—And a Vision for Change*. New York: Simon Schuster.

Leopold, Aldo. *A Sand County Almanac and Sketches Here and There*. 1949. New York: Oxford University Press.

Mann, Michael E. 2013. *The Hockey Stick and Climate Wars*. New York: Columbia University Press.

Marsh, George Perkins. 1864. *Man and Nature: Or, Physical Geography as Modified by Human Action*. New York: C. Scribner.

McDonough, William and Michael Braungart. 2002. *Cradle to Cradle: Remaking the Way We Make Things*. New York: North Point Press.

McKibben, Bill. 2010. *The End of Nature*. New York: Random House.

McKibben, Bill, ed. 2012. *The Global Warming Reader*. New York: Penguin.

McPhee, John. 1989. *The Control of Nature*. New York: Farrar, Straus and Giroux.

Mollison, Bill. 1991. *Permaculture: A Designers' Manual*. Tyalgum, Australia: Tagari Publications.

Muir, John. 1912. *The Yosemite*. New York: The Century Co.

Nash, Roderick. 1967. *Wilderness and the American Mind*. New Haven, CT: Yale University.

Outwater, Alice. 1996. *Water: A Natural History*. New York: Basic Books.

Pollan, Michael. 2006. *The Omnivore's Dilemma: A Natural History of Four Meals*. New York: Penguin.

Pretty, Jules and Z. Pervez Bharucha. 2018. *Sustainable Intensification of Agriculture: Greening the World's Food Economy*. London: Routledge.

Reisner, Marc. 1986. *Cadillac Desert*. New York: Viking.

Schumacher, Ernst F. 1973. *Small Is Beautiful: Economics as if People Mattered*. London: Blond and Briggs.

Shiva, Vandana. 2000. *Stolen Harvest: The Hijacking of the Global Food Supply*. Cambridge, MA: South End Press.

Steingraber, Sandra. 1997. *Living Downstream: An Ecologist Looks at Cancer and the Environment*. Boston: Addison-Wesley Publishing.

Thoreau, Henry David. 1854. *Walden, or, Life in the Woods.* Boston: Ticknor and Fields.

Williams, Terry Tempest. 1991. *Refuge: An Unnatural History of Family and Place.* New York: Pantheon Books.

Wilson, Edward O. 1992. *The Diversity of Life.* Cambridge, MA: Harvard University Press.

Index

Actions: human actions and natural disasters, 102–104, 174–175; importance of taking action, xii; promoting sustainability, 202–205
Adaptation: definition, 289; key concept, 137–139; reducing vulnerability, 104
Adoption of Innovation Model, 88
"Affluenza," 178
Afghanistan, local efforts to promote conservation, 106, 121–126
Africa: climate refugees, 119; GGWSSI, 206, 232–237; highest losses from drought, 144; small-scale solar power, 140; weather-related crises, 119
African Union, idea of tree planting, 233
Agenda 21, 192, 279
Agent Orange, 49
Agrarian, definition, 289
Agriculture: alternative, 239–241; Animal agriculture, 5–7, 52–54; contribution to hypoxia in Gulf of Mexico, 27–31; impact on land use change, 4; key concept, 47–50; organic farming in Denmark, 206, 226–232
Agrobiodiversity loss, 55
AGW (Anthropogenic Global Warming). See Anthropocene/Anthropogenic ("human new") era
Air pollution: Bangladesh leather industry, 77; from burning coal and biomass, 67; China, 209, 210; key concept, 50–52; negative impact of cars, 16–17; Nigeria, 41; smog, 9, 17, 50, 51; warnings spurred Earth Day, 246
Alaska, coastal villages considering move inland, 116
Algalita Marine Research Foundation, 36
Aluminum, recycling, 271

Amazon, introduction of diseases by tourists, 152
Animal agriculture, 5–7, 52–54
Annan, Kofi, call for Millennium Ecosystem Assessment, 149
Anthropocene/Anthropogenic ("human new") era: Anthropogenic Global Warming (AGW), 110–112, 164, 172, 232, 283; definitions, 289–290; key concept, 139–142
Arbor Day Foundation, 62
Artisanal and small-scale mining (ASM), 79
Asia: hardest hit by weather-related crises, 119; South Asia Tsunami, 2004, 106, 131–136, 148
Asia Infrastructure Investment Bank, 213
Assessing environmental sustainability, 200–202, 241–243, 277
Atrazine, 9, 99
Austin, Texas, first local green building program, 260
Australia: Department of Environment and Energy, 258; Great Barrier Reef, 10–11, 22–27
Automobiles. See Cars; Electric vehicles (EVs)

Ban Ki-moon, 92, 109
Bangladesh, 9; impacts of leather tanning, 77
Behavior: agricultural subsidies, 30–31; consumer choices, 267; human vs. other species, 162; new technology can lead to worse environmental impacts, 88–89; reducing dependence on automobile, 18, 20; related to water use, 94; structural or behavioral adaptations, 137

Benefits: biodiversity, 56; focus on long-term, 154; forest, 61, 62; good waste management/recycling, 76, 83, 273; hydropower, 275; industrial agriculture, 49; nature's, 1, 2, 139, 147–150, 176–177; organic production methods, 240. *See also* Ecotourism

Benefits and costs: fossil fuels, 138; industrial development, 209; sharing in Costa Rica, 224–225; urban parks and green cities, 179, 276; vulnerability to natural disasters, 171–172

Benefits and risks: of GMO crops, 72–74; technological innovation, 86

Better Life Index, 198

Bicycles, car and bike sharing programs, 20

Biodegradable plastics, 37

Biodiversity: Costa Rica, 220–225; definitions, 4, 289; E. O. Wilson on protecting, 5; protection best way to address declining animal habitat, 160–161; UN Convention on Biological Diversity, 160–161, 225. *See also* Wildlife

Biodiversity loss: key concept, 54–57; primary drivers, 161

Biodynamic farming, 239

Biowaste (organic matter), 243–246

Birds: "Bird Cast" migration forecasts, 219; Feeder Watch, 218; Great Backyard Bird Count, 217; NestWatch, 218; observations, 205, 215–220; population decline, 57

Black Sea, recovery from hypoxia, 30–31

Brandt Commission/Brandt Line, 156

BRICS (Brazil, Russia, India, China), New Development Bank, 213

Brownfield sites, 76

Brundtland, G. H./Brundtland Commission, 192, 199

Buildings: codes to limit seismic risk, 146–147; green buildings, 259–262; siting to reduce disasters, 174

Bullard, Robert, *Dumping in Dixie*, 252

Butterfly Effect, xi–xii

Bycatch, 81

CAFOs (Confined feeding operations), 53, 93

Canada: BEPAC green building rating system, 260; national park management, 182; water quality guidelines, 92

"Cancer alley" in Southern Louisiana, 254

Cap and trade principle, 172–173

Carbon dioxide (CO_2): atmospheric increases, xix, 58, 111–112; Carbon footprint calculation, 242, Country emissions, xix, 110. *See also* Greenhouse gas (GHG) emissions

Carbon monoxide (CO), 51

Carbon neutrality vs. using no fossil fuels, 224

Carrying capacity, 166–167, 191

Cars: cars rule, 10, 15–21; electric vehicles, 249–251; environmental impacts, 137–138

Carson, Rachel, 100

Case studies. *See specific topics by name as indicated in the table of contents*

Case study format, xxii

Catlin Survey of Great Barrier Reef, 25

Central Park, New York, 176–177

CERCLA. *See* Superfund

Certification/rating systems for buildings, 259–262

Chaos Theory, definition, xii

Characteristic waste, 76–77

China: China's bold steps toward renewable energy, 205, 209–215; earthquake, 1556, 102; electric car mandate, 250; most REEs mined only in, 79

Chromium, hexavalent, 77

Circular economy: definition, 289; vs. linear economy, 201–202

CITES (Convention on International Trade in Endangered Species . . .), 64, 160

Cities, sustainable, 276–279

Citizen science: bird observations, 205, 215–220; definition, 289

Civil society organizations, 255

Clean Air Act, 17, 52, 53, 258

Clean Water Act (CWA): definition, 296; guidelines for drinking water, 93; impacts of climate change, 58; regulation of CAFOs, 53

Climate change: climate system changed by human activities, 164; global warming, definition, 292; impact on birds, 216–217; increased wildfires, 189; key concept, 57–60; recurring theme, xx–xxii; temperature rise in Norway, 55; as wicked environmental problem, xv–xvi. *See also* Anthropocene/Anthropogenic ("human new") era; Greenhouse gas (GHG) emissions

Climate change policy: controversy regarding need, xx–xxi, 24–26, 109–111; and facts, 105, 109–115; key concept, 59–60; may create jobs and economic growth, 248; mitigation and adaptation, 104, 170, 172

Climate refugees, 115–121

Clinton, Bill, 192, 252, 260, 279

Club of Rome, 192

Coal: in China, 209, 210, 212; in U.S., 166, 203. *See also* Fossil fuels

Coastal Community Preparedness Goals (NOAA), 130

Collaboration in Great Green Wall project, 236–237

Colonialism, 194, 290

Columbian exchange, definition, 290

Community gardens/community-supported agriculture, 240

Community-based natural resource management (CBNRM), 124

Complexity: of interactions between Earth and humans, xi–xii; of preparing for large disasters, 135; wicked vs. tame environmental problems, xv–xvi

Composting, 243–246

Comprehensive Environmental Response, Compensation, and Liability Act (CERCLA). *See* Superfund

Confined feeding operations (CAFOs), 53, 93

Consequences. *See* Unintended consequences

Conservation: definition, 290; groups' education programs regarding snow leopard, 124; vs. preservation, 2–4. *See also* ENGOs (environmental NGOs)

Conservation International, 203

Consumerism: definition, 290; education for smarter choices, 202; expansion of consumer class, 166–167; green consumerism, 262–265; happiness defined by amount of "stuff," 178; water usage, 285–286

Contamination of recycled materials, 271, 273

Convention of Parties (COP). *See* UN Framework Convention on Climate Change (UNFCCC)

Convention on International Trade in Endangered Species of Wild Fauna and Flora (CITES), 64, 160

Coral Reef Ocean Acidification Monitoring Portfolio (CROAMP), 25

Corals. *See* Great Barrier Reef

Corn ethanol, 66

Cornell Lab of Ornithology, 216–219

Corruption: fossil fuel exploitation and government, 39–40, 43–44; illegal profits from ecotourism, 152

Costa Rica, peace with nature, 205–206, 220–226

Cost/benefit analysis: indirect costs related to changing major economic drivers, 138; prevention vs. disaster losses, 172

Crime, environmental, 157, 160

Crop varieties commercially extinct, 55

Cruelty of livestock production, 6–7

Cultural degradation, 152

Cultural services, 148

Culture/Worldview, 204–205

Cuyahoga River, 9, 246

Dead zones, 11, 27–32

Deforestation, 4–5, 61–63, 112, 162, 257

Demeter Association, 239

Denmark, achievements in organic agriculture, 206, 226–232

Desertification: definition, 290; GGWSSI, 206, 236–237; Kenya's Green Belt Movement, 253; UN Convention to Combat Desertification, 232–233, 236. *See also* Drought; Land degradation

Development: alternative ways to measure, 196–198; stages of economic, 194–196; term underscores inequalities, 156. *See also* Jobs; Sustainable development

Disaster preparedness: geohazard risk reduction, 146, 175; hurricanes, 130; Indian Ocean countries, 135; tornadoes, 184; tsunami education, 134–135; wildfire, 188–189

Disasters and Conflicts Programme (UNEP), 103

Dominant social paradigm (DSP), 204

Drought, 143–145, 233. *See also* Desertification

Dubai, ecological footprint, 277–278

Earth Charter, 279

Earth Day, 246–249, 290

Earthquakes: Indian Ocean earthquake and tsunami, 131–136; key concept, 145–147; Ring of Fire, 186–187; rise where fracking increased, 71; Shaanxi, China, 102; Tohoku, Japan, earthquake-tsunami-nuclear sequence, 146

Ecocentric, definition, 290

Ecological footprint, 166, 191, 241–242, 290

Economic growth: countries with well-being and growth, 197–198; de-emphasizing in favor of well-being, 200; G20 Summit ideas, 212–213

Ecosystems: definition of ecosystem diversity, 56; human modification of, 161–165; management integral to disaster risk reduction, 103–104, 134; services, definitions, 56, 290; services, key concept, 147–150

Ecotourism: Costa Rica, 221; definition, 290; key concept, 150–153; Wakhan Corridor, 124–125

EEZs (Exclusive economic zones), 80–82

EJSCREEN, 253–254

Electric vehicles (EVs), 249–251, 269

Electricity: from renewable sources in China, 211; small-scale solar power in Africa, 140; sources determine climate friendliness of EVs, 250

Electronic waste/E-waste, 67–70, 290

Emerald ash borer, xiii–xiv

Emergency management, 290. *See also* Disaster preparedness

Endangered species: definition, 291; Endangered Species Act (ESA), 64; key concept, 63–65; snow leopard of Afghanistan, 123–124, 125; wildlife populations reduced by human actions, 121. *See also* Biodiversity loss

Energy: changing patterns, 65, 138; efficiency for sustainable cities, 278; key concept, 65–67; nearly half consumed by four countries, 65. *See also* Electricity; Fracking; Renewable energy; *specific types of energy by name*

ENGOs (environmental NGOs), 202–204, 255–256, 291

Environmental agreements, global/international, 163–164

Environmental assessment, 200–202, 241–242, 277

Environmental degradation: capitalism built on, 204; climate change example, 191; ecosystems, 149, 162; mining impacts, 78–79; overpopulation threats, 166; short-term economic gains cause, 283; tourism impacts, 151–152, 182; urbanization, 91. *See also* Environmental impacts; Habitat; Land degradation; Plastics; Pollution

Environmental geography: complexity, xi–xii; definition, xi; geography ideas, xii–xiv; information, knowledge, and concern, xv–xviii; key themes, xx–xxi, 141; local to global, xviii–xix; maps and models, xiv–xv; overview and use of this book, xxi–xxii; taking action, xii

Environmental impacts: of cars, 16–17, 138; of humans on biodiversity, 56–57; of Nigeria's oil industry, 39–45. *See also* Environmental degradation; Land degradation

Environmental justice, 39–45, 251–254, 291

Environmental management, definition, 291

Environmental NGOs (ENGOs). *See* ENGOs (environmental NGOs)

Environmental policy: Costa Rica, 220–225; definition, 291; ENGO influence on, 255–256; international variations, 9–10; key concept, 256–259; need for stakeholder support, xv–xviii, 257; public participation, 205, 215–216

Environmental systems and science, 98–99

Environmentalism: beginnings of, 1–4; "shades of green," 204–205

Environmentalist, definition, 291

EPA. *See* U.S. Environmental Protection Agency (EPA)

Ethanol, environmental impacts, 66

Ethics: claims of buying green products, 264; of companies claiming sustainability, 204–205; of pricing nature, 149

EU. *See* European Union (EU)

Europe: banning sales of gasoline and diesel cars, 249–250; green infrastructure for healthy ecosystems, 139; Green Party organizations, 266–267; lead in climate change mitigation policies, 172–173

European Environment Agency, 172
European Union (EU): 2010 Agenda for Sustainable Development," 281; Drinking Water Directive of 1998, 92–93; organic farming, 226–232; organic farming goals, 229–230; policies merged into regional agreements, 258; six principles for sustainable diets, 283; vs. U.S. waste disposal patterns, 83
Evacuation routes, 169–170
Exclusive economic zones (EEZs), 80–82
Externalities, economic, social, and environmental impacts, 30
Extinction: bird species, 164, 217; definition, 291; human impacts on mass, 139, 141; as result of environmental crime, 157; safety of food crops against, 55; snow leopard pushed to edge of, 106, 121; species, 4–5, 57, 58, 63–64, 87, 121, 166
ExxonMobil, xxi

Fair trade, 239–240, 291
FAO. *See* Food and Agriculture Organization of the United Nations (FAO)
Farmer-managed natural regeneration (FMNR), 235
Feedback loops: climate change, 116, 164; definition, 291; in environmental systems, 8
FEMA. *See* U.S. Federal Emergency Management Agency (FEMA)
Fertile Crescent, 101
Fertilizers, contribution to hypoxia in Gulf of Mexico, 27–31
First, Second, and Third World countries, 156
Fisheries: displaced by resorts, 152; economic impact of dead zones, 29; fish habitat provided by wetlands, 94–95; Illegal fishing, 81; overfishing, 57, 80–82, 134
5Rs: Refuse, reduce, reuse, recycle, and rot, 272
Flash floods, 154
Floods, key concept, 153–155
Food: Danish Ministry of Environment and Food, 228; diet and climate change, xix, 7; food miles, 48; GE/GM/GMO crops used in processed, 72; local food movement, 240; meat vs. plant based diets, 5–7, 191; separating out food waste, 245; sustainable diets, 282–284; water consumption producing, 285–286. *See also* Agriculture
Food and Agriculture Organization of the United Nations (FAO): Action Against Desertification program, 236; definition, 291
Food justice, definition, 291
Food security: definition, 291; as ecosystem service, 148; and organic farming, 230
"Footprints" and environmental assessments, 241–243
Forest Stewardship Council (FSC) principles, 61–62
Forests, 4–5, 61–63. *See also* Deforestation
Fossil fuels: 81% of world energy production from, 65; burning increases atmospheric CO_2, xix, 57, 109–113; definition, 292; imports and exports, 66–67; nonrenewable, 274; subsidies, 210, 212–213; technological adaptations rely on, 137; used in livestock production, 282–283. *See also* Coal; Transportation
Fracking (hydraulic fracturing): definition, 292; key concept, 70–71
Frontier mentality, 2
Fuel cell technologies, 269
Fuel efficiency of cars, 16
Fukushima accident. *See* Tohoku, Japan, earthquake-tsunami-nuclear sequence

G20 (Group of Twenty), 212–213
Gaia hypothesis, 98
Galapagos Islands, negative impacts of tourism, 150–153
Gandhi, Mahatma, 256
Garbage. *See* Great Pacific Garbage Patch; Waste
GE/GM/GMO crops: key concept, 72–74; prohibition in Europe, 229
Genetic diversity, 55. *See also* Biodiversity
Genetically modified (GMO)/genetically engineered (GE) crops. *See* GE/GM/GMO crops
Genuine Progress Indicator (GPI), 198
Geographic Information System (GIS): definition, 292; using to build environmental models, xiv

Geographic scale, xviii–xix, 292
Geography ideas, xii–xiv
Geothermal power: in China's five-year plan, 211; in Costa Rica, 224; technology, 276
Germany: as leader of Green Party influence, 266–267; recycling or composting of MSW, 273
Ghana, Agbogbloshie e-waste processing area, 69
Gibbs, Lois, 84–85
GIS. *See* Geographic Information System (GIS)
Glass, recycling, 271
Global environmental agreements, 163–164
Global Greens Charter/Congress, 267–268
Global North and Global South: access to safe drinking water, 94; contrasting urbanization rates, 90; definitions, 292; issues for sustainable diets, 283–284; key concept, 155–158; livestock production contrasts, 53; South subjected to North's waste and pollution, 251; variations in energy sources and usage, 67
Global Summit on Biodiversity, 2010 Future Policy Award won by Costa Rica, 223
Global warming. *See* Climate change
Governance, definition, 91
Grassroots initiatives: definition, 292; Earth Day, 246–247; successes of, 234
Great Backyard Bird Count, 217
Great Barrier Reef, 10–11, 22–27
Great Green Wall of the Sahara and the Sahel Initiative (GGWSSI), 206, 232–237
Great Pacific Garbage Patch, 11, 32–39
Green Belt Movement, 253
Green Building Assessment Protocol for Commercial Buildings (Green Globes), 261
Green buildings: certification/rating systems, 260–262; definition, 292
Green consumerism, 262–265
Green political parties, 265–268
Green space: definition, 292; historic stages of urban parks, 176; key concept, 176–180
Green technology (greentech): definition, 292; key concept, 268–270; negative, perverse or controversial consequences, 89; for sustainable cities, 277–278
Greenhouse gas (GHG) emissions: associated with agriculture, 48; CO_2 vs. methane, 71; contributions of waste, 82, 243, 245; definition, 292–293; Europe committed to cutting, 172–173; impacts of lifestyle decisions, xviii–xix; from increasing fossil fuel consumption, 67; one-third from transportation sector, 17; U.S. is second largest emitter, 60. *See also* Climate change
Greenpeace, 203
Greenwashing, 204–205, 264–265
Gross domestic product (GDP): of China higher than of U.S., 210; definition, 293; does not measure social well-being, 196–198
Group of Twenty (G20), 212–213
Gyres, 33–34

Habitat: and community form an ecosystem, 161; degradation's impact on species, 5, 43, 56–57, 81–82, 94–95, 218; destruction by livestock, 123; destruction by too many people, 179; and ecosystems conservation, 56; key concepts, 158–161; rainforests, 87; services, 148. *See also* Endangered species
Haiti, lives lost in Hurricane Matthew, 104
Hansen, James E., 110
Happiness: and sustainability, 178, 198–199; Sustainable Happiness Baseline Chart, 178; World Happiness Report, 227
Happy Planet Index, 199, 223
Hazardous waste: definition, 293; electronic/e-waste, 67–70, 290; industrial hazardous waste, 75–77; key concept, 74–78; storage in poor, nonwhite neighborhoods, 254
HAZUS software for disaster managers, 174
Health: agricultural chemical impacts, 47, 49; air pollution impacts, 16–17, 50–51, 209; benefits of green space and parks, 176–177; deaths caused by environmental problems, 77; e-waste processing impacts, 69; meat consumption increases risks, 6;

mining casualties, 79; water scarcity impacts, 94
"Heat islands," cities as, 18, 91
Hetch Hetchy Dam, 3–4
Home ignition zone (HIZ), 188–189
Household hazardous waste, 76
Humans: need for nature, 97–98; reliance on nature vs. control by, 100, 105, 132, 142; requirements for survival, 192
Hurricanes/cyclones/typhoons: Hurricane Katrina, 106, 126–131; Hurricane Matthew (2016), 104, 166–167; key concept, 167–170; name varies by region, 102
Hydraulic fracturing. *See* Fracking (hydraulic fracturing)
Hydrogen fuel cell-powered vehicles, 269
Hydrologic cycle, 99, 148
Hydropower: benefits and costs, 274–275; in China's five-year plan, 211; in Costa Rica, 224
Hyogo Framework for Action (HFA), 104
Hypotheses, testable, 99
Hypoxia. *See* Dead zones

Ijaw Youth Movement (IYM), 42
Illegal logging, 61
Incentives to follow environmental policies, 257
Income distribution, 194–195, 197–198
Independent Commission on International Development Issues (ICIDI), 156
India: Auroville, city of peace, 280; electric car mandate, 250
Indigenous peoples exploited by tourist industry, 152
Industrial Revolution: GHGs added since, 57; human alteration of Earth since, 164; significant increases in atmospheric CO_2 since, 112
Industrialization: of agriculture, 47–49; definition, 293; of fishing, 80–81; link to development, 192; and resource use, 1–2, 193–194
Inequality: cities demonstrate differences between rich and poor, 90–91; suggested by terms for development, 156–157. *See also* Social vulnerability
Infrastructure as adaptation strategy, 138–139

Insurance, earthquake, 146
Intergovernmental Panel on Climate Change (IPCC), xx, 59–60, 112–113, 139, 172, 293
International comparisons of urbanization difficult, 89–91
International Ecotourism Society, 125
International environmental agreements, 163–164. *See also specific UN conventions by name*
International Flood Initiative (IFI), 154
International Green Construction Code (IgCC), 260–261
International Institute for Sustainable Development (Canada), 203–204
International Organization for Standardization (ISO), 201
International Plant Protection Convention (IPPC), 159–160
International Union for Conservation of Nature (IUCN): invasive species control, 159; Snow leopard listed on Red List, 123; species threatened with extinction, 63–64; World Commission on Protected Areas (IUCN-WCPA), 180–181
"Internet of Things," xiv
Interstate highway system, 17–19
Invasive/exotic species: fire danger increased by, 189; impact on wildlife, 159–160
IPCC (Intergovernmental Panel on Climate Change). *See* Intergovernmental Panel on Climate Change (IPCC)
Island nations, sea level rise impacts, 115–121
Isle de Jean Charles, Louisiana, federal grants to resettle, 116
Ivory demand and elephant deaths, 157

Jobs: adaptation to change, 101, 138, 145; creation and destruction, 257–258; exploitation of locals, 152, 280; land restoration, 234, 253; loss in fishing industry, 81; renewable energy industry, xxi, 60, 205, 211–212, 248, 273–275; waste management/recycling, 76, 83, 243, 273
Justice: environmental justice, 39–45, 251–254, 291; food justice, definition, 291. *See also* Social vulnerability

Kenya, Green Belt Movement, 253
Key concepts. *See specific concepts by name as indicated in the table of contents*
Kiribati, response to climate change, 115, 117–118

Land degradation: Afghanistan, 122; Africa, 232–237, 253; benefits of restoration, 248; fracking impacts, 71; livestock production impacts, 53, 282–284. *See also* Soil
Land ethic, 98
Land use: change by humans, 4–5; definition of land-use change, 293; regional variations in forest, 61; zoning impacts, 18
Landfills, 37, 68, 83–84, 202, 243–245, 252, 271–273
Landscapes and geography, 100–101
Law vs. regulation/rule-making, 258
LCA. *See* Life cycle assessment (LCA)
LDC. *See* Least/Less Developed Countries (LDC)
Lead (Pb), 51
Leadership in Energy and Environmental Design (LEED), 261–262
Least/Less Developed Countries (LDC), 104, 156, 194–196, 277, 293
LEED green building certification, 261–262
Leopold, Aldo, 98
Life Cycle Assessment (LCA): definition, 293; environmental impacts of consumer products, 16–17, 201–202; food impacts on GHGs, 48
Lifestyle: consumption of resources, 166–167; environmental impact of typical American, xviii–xix, 191; individual decisions impact GHG emissions, xix; reducing waste, 272
Listed waste, 76
Lithium-air storage units, 269
Livestock, environmental impacts, 53, 123, 282–283
Lobbyist, definition, 293
Local food movement, 240
Local to global geographic linkages, xviii–xix
Location, absolute vs. relative, xii–xiii
Los Angeles, air pollution, 9, 50
Love Canal, 84–85
Lovelock, James, Gaia hypothesis, 98

Maathai, Wangari Muta, 253
Majority World, definition, 158
Mali, land restoration under GGWSSI, 235
Malthus, Thomas, 165
Maps and models, use in environmental geography, xiv–xv
Marine Debris Program (U.S. NOAA), 36
Marine life, 32–38, 81, 82. *See also* Fisheries; Oceans
Marshall Island, 34
MDC. *See* More Developed Countries (MDC)
Meadows, Donella, et al., *The Limits to Growth*, 165
Megacities, 90
Memory, environmental, xvii–xviii
Methane: gas leaks, xvii; generating electricity from decomposing biomass, 83; potency as greenhouse gas, 7, 71; released by biowaste in landfills, 243, 282–283
Mexico, banning of GMO corn seeds, 73
Mexico, Gulf of. *See* Dead zones
Midwest, contribution to hypoxia in Gulf of Mexico, 29–31
Millennium Declaration, 279
Millennium Ecosystem Assessment, 149
Minerals and mining: demand for electronics and e-waste, 68, 69; key concept, 78–80; rare earth elements (REEs), 79
Misinformation to create uncertainty, xviii, xxvi
Mississippi River, contribution to hypoxia in Gulf of Mexico, 27–31
Mitigation: addresses root causes of change, 138; definition, 293; of drought, 144–145; of hazards, 104; key concept, 170–173
Mixed waste, 77
MNC. *See* Multinational corporations (MNC), definition
Models, definition, xiv–xv, 293
Moment magnitude scale (MMS) to measure earth quakes, 131–132
Money. *See* Profit motive
Moore, Charles, 33, 36
More Developed Countries (MDC), 104, 194, 239
Motor vehicles per capita, international comparisons, 15

Mountains, opportunities and constraints, 101
Muir, John, 2–4, 97–98, 203
Multinational corporations (MNC), definition, 294
Municipal solid waste (MSW), 82–84, 245, 272, 273

NAAQs (national ambient air quality standards), 52
NASA. *See* U.S. National Aeronautics and Space Administration (NASA)
National Audubon Society, 216–217, 219
National Climate Assessment (NCA), 57–58, 171
National Geographic: explaining climate change impacts, 16, 58, 115; Green Guide, 277
National Institute of Building Sciences, 174
National People of Color Environmental Leadership Summit (1991), 252
National Weather Service (NWS), definition, 294
Natural disasters: increase of weather-related, 58; some partly caused by human actions, 102–104, 174–175. *See also* Disaster preparedness; *specific events by name or type*
Natural hazards: categories, 102; caused by volcanoes, 186; definition, 294; ecosystem services buffer against, 148; impact of sudden environmental events, 101–102; key concept, 173–176; mitigation, 170–173; to natural disasters, 102–104, 174
Natural resources: definition, 2, 294; necessary for a harmonious community, 280; pricing, 149; regional variations in consumption, 166–167, 191; stolen and traded illegally, 157; technology's impact on how people view, 87
Natural Resources Defense Council (NRDC), 203
Nature: Costa Rica's peace with, 205–206, 220–226; humans' need for nature, 97–98; reliance vs. perceived human control, 100, 105, 132, 142, 272
The Nature Conservancy (TNC), 62, 203, 242, 294
New Development Bank (BRICS), 213
New Orleans. *See* Hurricanes/cyclones/typhoons

Newspapers, environmental impacts of paper vs. digital, 201–202
Niger, land restoration under GGWSSI, 235
Nigeria, 11, 39–45
NIOSH (National Institute for Occupational safety and Health), 79
Nitrogen oxides (NO_x), 52
NOAA. *See* U.S. National Oceanic and Atmospheric Association (NOAA)
Non-governmental organizations (NGOs), 202–204, 255–256. *See also* Environmental NGOs (ENGOs)
Nonpoint vs. point-source pollution, 30, 93
Nonprofit organizations, 255
North Carolina, Warren County toxic waste protest, 251–252
North Pacific Gyre/North Pacific Trash Vortex. *See* Great Pacific Garbage Patch
Nuclear power: isotopes from weapons testing in geologic record, 141; radioactive waste, 77; Tohoku, Japan, earthquake-tsunami-nuclear sequence, 146

Obama, Barack, 69, 213, 281
Oceans: ocean acidification, 25–26, 58, 81; ocean energy, 268–269; opportunities and constraints, 101
Ogoniland, human rights abuses and environmental degradation, 41–43
Organic farming. *See* Agriculture
Organisation for Economic Cooperation and Development (OECD), 198
Organizations supporting AGW theory of climate change, 111
Our Common Future, 192, 199, 279
Outdoor Magazine, obituary for Great Barrier Reef, 25–26
Overfishing: destruction of coastal ecosystems, 134; fish species facing extinction in Europe, 57; key concept, 80–82
Overpopulation, 165–167; definition, 294
Ozone (O_3), 51, 52

Palm oil, 61
Paper, recycling, 271
Paris Agreement (UN Framework Convention on Climate Change, 2015), xx, 60, 113, 172–173
Parks, national, key concept, 180–183

Parks, urban, key concept, 176–180
Parks Canada, mission statement, 182
Parliamentary systems and green party successes, 266–267
Particulate matter (PM), 51–52
Partners, ENGOs as important, 255
Partnership for Environment and Disaster Risk Reduction (PEDRR), 103–104
Peer review, 99
Permaculture, 240–241
Pesticides: Carson's exposure of DDT damage, 100; Chávez and environmental justice, 252; hazardous waste, 74; organophosphates from war chemicals, 47, 49; prohibited in organic farming, 226, 228, 240
Pests, diffusion, xiii–xiv
Petroleum: compostable plastic alternatives, 37; oil spills and Earth Day, 246; water consumption, 286. *See also* Fossil fuels; Nigeria
Pimentel, David, on use of grain for livestock production, 6
Pinatubo, Mount (Philippines), eruption, 187
Pinchot, Gifford, 2
Pipeline systems, 65
Place: environmental concern affected by geography and time frame, xvi–xviii; place matters, xii–xiii
Planned obsolescence, 68
Plastics: diffusion of idea of refillable water bottles, xiii; recycling vs. new, 263, 271; trash in marine garbage patches, 32–39, 81; water consumption, 286
Poaching. *See* Crime, environmental
Point-source pollution, nonpoint pollution vs., 30, 93
Politics, green political parties, 265–268
"Polluter pays" approach, 85–86
Pollution: challenges of consumer lifestyle, 7–9; definition, 7, 294; environmental justice and, 251–254; key concept of water pollution, 92–94; polluters subject to penalty in Cost Rica, 224–225; worsened by urban industries, 91. *See also* Air pollution; Great Pacific Garbage Patch; Water pollution
Population, Human: distribution, 101, 166, 179, 210; growth increases flood damage, 154; human elimination of many factors that control, 165; key concept, 165–167; overpopulation, 294
The Population Bomb (Ehrlich), 165
Post-consumer content, definition, 270
Poverty. *See* Social vulnerability
Powell, John Wesley, 1
Precautionary principle, 8–9, 73, 225, 294
Preservation: vs. conservation, 2–4; definition, 294
President's Council on Sustainable Development (PCSD), 192, 279
Profit motive: vs. environmental protection, xviii, xxi, 44; food system based on short-term economic gain, 283; for GM/GE crops, 72–74; and greenwashing, 264–265; misinformation by fossil fuel industry about climate change, 109–110; short-term economic gains vs. laws of nature, 100
Protected areas: Costa Rica's parks and reserves, 222–223; definition, 295; key concept, 180–183; Marine Protected Areas, 82
Provisioning services, 148
Public Participation: in data collection and reporting, 216; definition, 295. *See also* Citizen science

Qualitative and quantitative methods, xiv–xv

Racism, environmental, 251–254. *See also* Environmental Justice; Social vulnerability
Rainforests, impact on species, 81–82
Rare earth elements (REEs), 79
Reagan, Ronald, blamed regulations for killing jobs, 257
Recycling: definitions, 270; fleece jackets from plastic bottles, 263; key concept, 270–273; materials recoverable from e-waste, 69; refillable water bottles, xiii; saves water, 286. *See also* Zero-waste
Red Cross: earthquake safety information, 147; one of earliest NGOs, 255
Red List of Endangered Species (IUCN), 63, 123
Refugees: displaced by Hurricane Katrina, 128; victims of sea level rise, 105, 115–121
Regulation services, 148

Remote sensing: for environmental models, xiv; for global comparisons, 90; location of wildfires, 189; shows rate of sea level rise, 115–116

Renewable energy: China, 205, 209–215, 248; Costa Rica, 223–224; definition, 295; Europe, 173; expansion, 65, 268–269; G20, 212; key concept, 273–276; pros and cons, 89; United States, 210, 248, 250, 274; "waste to energy," 83. *See also* Geothermal power; Hydropower; Solar power; Wind power

Research: citizen science, 205, 215–220, 289; public funding for energy and environment, 268.

Ring of Fire/Circum-Pacific Belt, earthquakes along, 186–187

Rio Earth Summit. *See* UN Conference on Environment and Development (1992)

Rivers, opportunities and constraints, 101

Roosevelt, Theodore, 2

Saffir-Simpson Hurricane Wind Scale, 127; 168–169

Sahel-Sahara region. *See* Great Green Wall of the Sahara and the Sahel Initiative (GGWSSI)

Saro-Wiwa, Ken, 41–42

Scale, geographic, xviii–xix, 292

Scapegoat, environmental laws as, 257

Science: citizen science, 205, 215–220; communicating to influence policy, xv; need for collaboration between natural and social, xii; scientific method, 98–99

Sea level rise, 58, 105, 115–121, 130. *See also* Climate change

Seeds: Doomsday Vault, 55; genetically engineered (GE), 72–73; saving, 47

Seismic codes, 146–147

Sendai Framework for Disaster Risk Reduction 2015–2030, 104, 146

Senegal, land restoration under GGWSSI, 235

Seychelles, call for all small island nations to unite, 119

Shaanxi, China, earthquake, 102

Shiva, Vandana, 73

Sierra Club, 203, 255

Silent Spring (Rachel Carson), 100

Silviculture, definition, 5

Smart growth, definitions, 20, 278, 295

Smog, 17, 50, 51

Snow leopard, 106, 123–124, 125

Social component of sustainability especially difficult to measure, 199–200

Social values, environmental laws associated with, 257

Social vulnerability: definition, 295; within disaster management, 174–175; and environmental degradation, xxi; growth of slums, 90–91; pollution and waste sites, 82–83, 251–254; poorer countries with no money for adaptation, 116, 144–145; poverty despite oil profits in Nigeria, 40; poverty increases disaster risk, 104, 128, 130, 139, 146, 155; Social Vulnerability Index (SoVI), 175. *See also* Justice

Social well-being: focus of Costa Rica's policies, 220–221; measuring development in terms of, 194–199

Soil: degradation by industrial agriculture, 49; degradation from corn ethanol production, 66; mining for fuel, 66

Solar power: Africa, 140; China top producer of, 211; jobs, 274; panels and arrays, 89, 110, 274; passive solar design, 259; prices dropping, 213; PV and CSP technologies, 275; solar and wind prices dropping, 213

Solving the E-Waste Problem (STEP) Initiative, 68–69

South Korea, electric car mandate, 250

Spatial distribution, definition, 295

Spatial thinking, space, place, and location, xii–xiii

Species diversity, definition, 56

St. Helens, Mount, eruption, 187

Stakeholders: definition, 295; need for support in setting policy, xv–xviii

Subsidies: agricultural, 6, 30–31, 66, 229–230, 283; of cars by U.S., 17–18; fertilizer, 31; inefficiencies of fossil fuel, 212–213; supporting environmental policy, 222

Sulfur dioxide (SO_2), 51–52

Superfund: CERCLA Brownfields Program, 76; Love Canal and creation of, 84–85; now footed by taxpayers, 85–86

Survival of humans requires limiting five areas, 192

Sustainability: actions to promote, 202–205; definitions, 192, 205; managing fisheries, 81–82; monetary profit vs., xxi, 44; in the real world, 201; role of population control, 165; understanding intersection between humans and nature and, xi–xii. *See also* 3 Pillars of sustainability

Sustainable development: 2010 and 2030 agendas for, 281; definition, 295; linkage with environmental protection, 248; managing tourism, 150–153; national and international activities, 192–193, 279–282; sustainable cities, 276–279

Sustainable Economic Development Assessment (SEDA), 197–198

Svalbard Global Seed Vault (Doomsday Vault), 55

Synthesis Report (IPCC), 113

Taupo supervolcano (New Zealand), 186

Technology: assembly line manufacturing, 87; fixes vs. changing human behavior, 18; human adaptations to, 137–138; innovation and consequences, 86–89. *See also* Green technology (greentech)

Tectonic plates, 145–146, 187

Thoreau, Henry David, 1

3 Pillars of sustainability, 152, 192–193, 199–200, 277, 279, 281, 289

3Rs: Reduce, reuse, recycle, 244, 272, 289

Throw-away society, 201–202

Tidal and wave energy, 268–269

Time-place concept, xvi–xviii

Tipping points, 8

Tohoku, Japan, earthquake-tsunami-nuclear sequence, 146

Tornadoes, key concept, 183–185

Tourism. *See* Ecotourism

Toxic waste. *See* Hazardous waste; Pollution

Transcendentalism, 1

Transportation: area of greentech innovation, 269; cars, 10, 15–21; electric vehicles, 137–138; green technology for, 269–270; new approaches, 20; in sustainable cities, 276–279

Trash. *See* Municipal solid waste (MSW)

Tree planting: Kenya's Green Belt Movement, 253; as part of GGWSSI, 233–234

Triple bottom line for sustainability, 199–200

Trudeau, Justin, on environmental protection, 256

Trump, Donald, 60, 71, 210

Tsunamis: South Asia, 2004, 106, 131–136; Tohoku, Japan, earthquake-tsunami-nuclear sequence, 146

UN Conference on Environment and Development (1992), 192, 247, 279

UN Convention on Biological Diversity, 160–161, 225

UN Convention to Combat Desertification (UNCCD), 232–233, 236

UN Educational, Scientific and Cultural Organization (UNESCO). *See* UNESCO

UN Environment Programme (UNEP): assessment of oil operation effects in Nigeria, 42–43; definition, 295; Disasters and Conflicts Programme, 103; environmental management assistance, 9, 258; Regional Seas Programme, 36; working to reduce hazard threats, 103

UN Framework Convention on Climate Change (UNFCCC), xx, 59–60, 112, 113

UN High Commissioner for Refugees (UNHCR), legal status of climate refugees, 118–119

UN International Strategy for Disaster Reduction (UNISDR), 103–104

UN Office of Disaster Risk Reduction, 102, 103

UN Secretary-General's Special Representative for Disaster Risk Reduction, 189

UN Universal Declaration of Human Rights, 44

Uncertainty, wicked vs. tame environmental problems, xv–xvi, xviii

UNESCO (United Nations Educational, Scientific and Cultural Organization): important to prevent and mitigate effects of natural disasters, 171–172; program on geohazard risk reduction, 146; World Heritage Sites, 23

Unintended consequences: of creating parks, 179; of human modification of the environment, 162; of technology and innovation, 86, 88–89, 137–138

United Church of Christ Commission for Racial Justice, 252
United Kingdom (UK): BREEAM green building rating system, 260; no sales of new gasoline and diesel cars from 2040, 249. *See also* Europe; European Union (EU)
United Nations (UN): classification of national economies, 194; definition, 295; working toward sustainable development, 279–281. *See also* UN organizations
United States: American dependence on automobile, 15–21; backing away from renewable energy, 210; costs of heat waves and major droughts, 144; vs. EU waste disposal patterns, 83; Green Party organizations, 267; national parks, 182; resources consumed in, 191, 242, 285
Universal waste, 76
Urbanization: geographic views, 89–91; key concept, 89–91; urban green space, 176–180
U.S. Army waste reduction goals, 244
U.S. Department of Agriculture: goal for AFO/CAFO operators, 53; organic food accreditation, 240
U.S. Environmental Protection Agency (EPA): definition, 291; establishment, 247; online tool for waste reduction, 244
U.S. Federal Emergency Management Agency (FEMA), 127–129, 170, 174
U.S. Geological Survey (USGS) Natural Hazards Mission Area, 102
U.S. Global Change Research Program (USCRP), 171
U.S. National Aeronautics and Space Administration (NASA): adaptation and mitigation, 172; definition, 294; evidence of rapid climate change, 112; program for adopting pieces of Earth, 248
U.S. National Flood Insurance Program (NFIP), 154–155
U.S. National Institute for Occupational Safety and Health (NIOSH), 79
U.S. National Oceanic and Atmospheric Administration (NOAA): Coastal Community Preparedness Goals, 130; definition, 294
U.S. National Oceanic and Atmospheric Association (NOAA): definition, 294; Marine Debris Program, 36; Ocean Acidification Program (OAP), 25; warnings for storms and natural hazards, 102
U.S. National Parks Service, 98
U.S. National Strategy for Electronics Stewardship, 69
U.S. Safe Drinking Water amendments, 286

Values, environmental policy and conflicting, xv–xvi
Vegan, definition, 295
Vesuvius, Mount, eruption, 102
Violence of multinational oil companies in Nigeria, 43–44
Volcanic Explosivity Index (VEI), 186
Volcanos, 102, 141, 185–188, 221, 224
Vulnerability. *See* Social vulnerability

Wakhan Corridor of Afghanistan, 121–126
Waltons, the richest family in America, 195
Warning systems: based on drought monitoring, 145; fewer available in Haiti, 104; Hurricane Katrina, 130; none available for South Asian tsunami, 131–135; often prohibitively expensive, 106; UNESCO helping in disaster preparation, 146
Waste: electronic waste, 67–70, 290; hazardous waste, 74–78; municipal solid waste (MSW), 82–84, 245, 272, 273; taking into account, 242; "waste to energy," 83. *See also* Oceans; Zero-waste
Wastewater, 287
Water: conservation, 284–287; desalinization by waste heat, 278; requirements for meat vs. food crops, 95, 282–283; Water footprint, 242
Water pollution: Bangladesh leather industry, 77; Flint, Michigan, crisis, 254; near Standing Rock, North Dakota, 254; nonpoint vs. point-source, 30, 93; regulations, 92–94; and scarcity from fracking, 70–71; wells contaminated with benzene, 43
Water resources management: considerations regarding floods, 154; improving quality and quantity, 287; key pillars, 145

Water scarcity: 40% of world's population already affected, 58; after tsunami, 133; key concept, 94–95; sea level rise and lack of drinking water, 118
Water stress, 95
Watersheds, definition, 29
Weather: extremes, 58, 112, 148, 155, 165; forecasting of tornadoes, 184; patterns, xi, 57, 95, 101, 138, 148; stations, 58, 144
Wetlands, 94–95, 148
White, Gilbert, "range of choice," xv
Wicked vs. tame environmental problems, xv–xvi
Wildfire, key concept, xv–xvi, 188–190
Wildland urban interface (WUI), 188
Wildlife: illegal trading/trafficking, 157, 160; population reductions, 121, 157, 158–161. *See also* Biodiversity loss; Endangered species; Habitat
Wildlife Conservation Society (WCS), snow leopard research, 124
Wilson, E. O., on protecting biodiversity, 5
Wind power: expansion, 65; jobs, 274; prices dropping, 213; pros and cons, 89; and wildfire, 188. *See also* Renewable energy
Windbreaks provided by trees, 233
World Bank: aid to improve solid waste management, 82; Kiribati climate mitigation effort, 117–118; linkages between environmental protection and sustainable development, 248
World Commission on Environment and Development, 192
World Conservation Monitoring Centre (UNEP), 181
World Health Organization (WHO): air pollution above recommended levels, 209; climate change health impacts, 58; deaths from toxic waste, 77; definition, 296; drinking water guidelines, 93
World Urban Parks (WUPO), 179
World Wildlife Fund (WWF), 64, 203
Worldview and environmental sustainability, 204–205
Worldwide dead zones, 28–29

Xeriscaping, 285

Yard waste banned from landfills, 245
Yellowstone National Park, 180, 182, 186
Yosemite National Park, 3

Zero emissions vehicles (ZEV) mandates, 249–250
Zero-waste: definitions, 38, 296; goal of circular economy, 202; Zero Waste Europe, 244; Zero-waste communities, 244; Zero-waste homes, 272

About the Author

Leslie A. Duram, PhD, is professor of geography and director of environmental studies at Southern Illinois University Carbondale. She is an award-winning author of over 50 peer-reviewed journal articles and three books: *Good Growing: Why Organic Farming Works*; *The Encyclopedia of Organic, Sustainable, and Local Food*; and *America Goes Green: An Encyclopedia of Eco-Friendly Culture in the U.S.* Dr. Duram was selected as a Fulbright Scholar, recognized by the American Association of Geographers Rural Geography Specialty Group, and elected to several governance positions in environmental organizations and policy councils.